Jane's

AIR FORCES OF THE WORLD

THE HISTORY AND COMPOSITION OF THE WORLD'S AIR FORCES

Jane's

AIR FORCES OF THE WORLD

THE HISTORY AND COMPOSITION OF THE WORLD'S AIR FORCES

Collins

DAVID WRAGG

In the UK for information please contact:
HarperCollinsPublishers
77-85 Fulham Palace Road
Hammersmith
London W6 8JB

everything clicks at www.collins.co.uk

In the USA for imformation please contact:
HarperCollinsPublishers
10 East 53rd Street
New York
NY 10022

www.harpercollins.com

Jane's Information Group
www.janes.com

First published in Great Britain by HarperCollinsPublishers 2003

1 3 5 7 9 10 8 6 4 2

ISBN 000711567-9

Printed and bound in Britain by Bath Press

Acknowledgements

In writing any book of reference, the author is always indebted to those who provide assistance, and especially those who provide good photographs. I am very grateful for their help to those in the many air forces and air arms, and of course those in the aircraft manufacturers and their archivists, as well as the service attachés and advisers in the embassies and high commissions who have put me in touch with their headquarters. It would be difficult to name all of them, but the following have been of particular assistance. In alphabetical order: Dick Atkins and Lynne Warne of Vought; Laura Barrett-Oliver and Bob Bishop of Northrop Grumman and Laurence A Feliu of the same company's Heritage Centre; Group Captain Normando Costantino of the Argentine Embassy; Patricia Contreras of the Chilean Naval Mission; Ricardas Degutis of the Lithuanian Embassy; Colonel P Fouilland of the French Embassy; David Gibbings of Westland Helicopter; Barry Guess of the BAe Heritage Centre; Andreas Karlsson for his help on the Royal Swedish Air Force; Henning Krisstensen of the Royal Danish Air Force; Sergeant MacCullagh of the Irish Air Corps; Marko Malec for his help with Slovenian aircraft; Brigadier M. Qudah of the Jordanian Embassy; Brigadier Derrick Page of the South African Air Force; Patrick Rietz of the German Bundesministerium de Verteidigung and Tim Travis at Raytheon. These are all due my thanks for entering into the spirit of the exercise.

David Wragg
Edinburgh, May, 2002

Contents

Introduction

LEFT: *An F-14 Tomcat (left) launches from the waist catapult as four F/A-18 Hornets wait to launch from the bow catapults of the aircraft carrier USS Enterprise (CVN 65) as the aircraft carrier operates in the Adriatic Sea on Feb. 23, 1999. Elements of the Enterprise Battle Group are operating in the Adriatic and land based aircraft and support personnel are deploying to their forward staging bases in Europe for potential NATO air operations. (DoD photo by Petty Officer 1st Class Benjamin D. Olvey, U.S. Navy)*

RIGHT: *Ukranian M-8 in service with NATO in Kosovo June 2000 (NATO Photo)*

In any book such as this the contents are a snapshot of the situation at a particular time, although it has also been the intention to provide a brief history whenever possible and relevant for the air forces and air arms covered. History is important, for the last century saw two global conflicts and many other more localized, but often serious, wars and campaigns. It is easy to forget those wars that don't directly affect ourselves such as the Falklands Campaign, the war between Iraq and Iran, the conflict between Pakistan and India or the Arab states and Israel, but the lessons of these conflicts is that war happens all too easily. It is as well to be prepared. Deterrence is so often associated with massive nuclear retaliation that people forget that conventional forces also have their part to play. Well balanced, well equipped and well trained armed forces are a deterrent against disagreements flaring up into wars.

Given this, what are we to say about the so-called 'peace dividend' that so many of the Western democracies have taken from the collapse of the Warsaw Pact and the break up of the former Soviet Union? A dividend is a return on an investment. Yet, there has been little investment in many of the armed forces covered in this survey. The dividend, if one is deserved at all, has been taken in advance before any benefits have accrued. The terrorist attacks on New York and Washington and the situation in the former Yugoslavia have shown that the world is, if anything, more unstable and uncertain today than during the second half of the twentieth century. The threats are more varied. It is no longer a case of stopping a massive Warsaw Pact thrust across the central front, if it ever was that simple. Democratic governments delude themselves and fail to provide leadership by burying their heads in the sand, claiming that there will be as much as ten years' notice of an emerging threat. No such period of warning existed before hi-jacked airliners, as devastating as any conventional bomb or missile, slammed into the twin towers of the World Trade Center and the Pentagon. Governments anxious to remain in office accord higher priorities to expenditure on social welfare than to defence, but without adequate defence, democracy itself is threatened and effectively taken hostage. There are other means of providing for education and health, but only governments can provide defence!

Twice during the twentieth century, democracy was saved by the timely intervention of the United States. The new century has started with what has to be regarded as a *Pax Americana.* The lesson learned should have been greater preparedness on the part of the democracies, but the United States had to take the lead in the Gulf War and in action over the former Yugoslavia. The question is not one of American leadership, since that would probably be a fact of life given the massive economic and military power of the United States. The problem is that the USA has had to provide more than the sum of its European allies in terms of effort, time and time again. As I write, the United States is increasing its annual defence expenditure by more than the total of that of the United Kingdom. Too few European countries have the mili-

tary airlift essential for operations outside of their own territory. In Europe, only the RAF has a heavylift capability, and that of just four aircraft. Aircraft and personnel are grudgingly deployed in penny packets. Instead of increasing defence expenditure, scarce funds are diverted into creating what amounts to a European army; and it is a European army if the comments of German and French leaders are to be believed. New command structures, duplicating those of NATO, divert scarce funds and skilled and experienced staff officers. NATO is the longest lasting and most successful alliance in the whole of history, but it is undermined at great risk by European politics. The fact that the word 'army' is used rather 'armed forces' tells us much about Continental attitudes. In the late 1930s, the Dutch, Belgian and French governments knew, as did the British, that in the event of a major war, British intervention would be essential. Yet, there were no joint exercises, no attempts at collaboration. NATO has provided all of this, and a command structure as well, and no less important, it formally tied the USA and Canada into protecting Europe. Given current rates of defence expenditure, in terms of a percentage of GNP often half what they were a decade or so ago, lessons have been forgotten.

The British like to believe that they always 'punch above their weight', but there are limits. Only two out of three aircraft carriers are operational. The size of the carriers in service is less than half that of the ships operated in the 1950s through to the early 1970s. In 2006 the Royal Navy will lose its Sea Harriers with their Blue Vixen radar leaving the Fleet without an air defence fighter. How is it that the United States, with less than five times the UK's population, can afford eleven large aircraft carriers, each more than four times the size of a British carrier, and keep a twelfth one in reserve? The French now maintain just one carrier, and would be embarrassed if the ship met with an unfortunate accident. Yet, the French, most of all amongst the Europeans, like to pretend that they can do without the United States. The malaise is global. New Zealand might not face an obvious threat today, or even tomorrow, but she has scrapped her combat squadrons, and has had to hire preserved aircraft to provide air defence training for her small navy. Meanwhile, her nearest neighbour, Australia, has suddenly started to worry about power projection, having scrapped her last aircraft carrier two decades ago.

History repeats itself, but never exactly. The British idea of ten years to prepare in the 1930s seems to have returned. That was a politicians' and civil servants' ten years, not that of an airman, sailor or soldier. Given the lengthy time taken to develop modern combat aircraft and get them into service, ten years is not enough. As events in the United States in September 2001 have shown, it does not even take three or four years for a major new threat to emerge. Given the situation in the states of the former USSR, where the populations are suffering a hangover from the old regime without feeling the benefits of a free society and market economy, a reversal of the new freedoms could have popular appeal. Where would that leave the democracies?

Afghanistan

POPULATION: 22.6 million

LAND AREA: 250,000 square miles, 647, 497 sq km

GDP: $2.0bn (£1.4bn), per capita $700 (£490)

DEFENCE EXPENDITURE: $250m (£175m)

SERVICE PERSONNEL: Officially None.

ABOVE: *MiG-21 wreck at Kabul (Jeremy Flack/Aviation Photographs International)*

AFGHAN AIR FORCE

Formed: 1937

Officially, the Afghan Air Force ceased to exist with the breakdown of central government during the closing years of the twentieth century, when some two-thirds of the country was controlled by the Taliban, Islamic fundamentalists. The nation's military air power was divided between the Taliban and the government supported National Islamic Movement. Some aircraft were also in the possession of the anti-Taliban Northern Alliance, including former Iraqi aircraft flown to safety in Iran during the Gulf War. The new regime installed with United Nations support following the US-led police action against the Taliban and the al-Qa'eda terrorists is preparing a new army and air force. These forces will be heavily dependent on outside aid given the country's parlous financial state, and will require outside assistance in training as well.

Internal dissent is not new in Afghanistan. An army air arm had been formed in 1924, using two Bristol F.2B fighters flown by German pilots. Afghan pilots were trained by the Soviet Union, which donated a squadron of R-2 reconnaissance aircraft. This fledgling air arm disappeared when the aircraft were destroyed during a civil war in 1928-29. A Royal Afghan Air Force was formed in 1937, receiving eight Hawker Hart bombers, sixteen Meridionali Ro 37 reconnaissance aircraft and eight Breda Ba 25 trainers in 1938. British and Italian instructors established a flying school, and Afghan pilots were trained by the Royal Air Force in India. Twenty Hawker Hinds supplemented the initial batch of aircraft in 1939, with some remaining in service until 1957. Out of 12 ex-RAF Avro Ansons received in 1948, five were still in service in 1968.

A return to Soviet equipment followed a 1955 agreement with the USSR. In 1957, MiG-17 fighters, Ilyushin Il-28 bombers and Il-14 transports, Antonov An-2 transports and SM-1 (Mi-1) were delivered with MiG-15UTI and Yakovlev Yak-18 trainers. Russian instructors arrived, and both the USSR and the USA built airfields. Afghan personnel were trained in the USSR and India.

MiG-21 interceptors replaced the MiG-17s during the early 1970s, while Mi-4 helicopters and Il-18 transports were also received. A republic was declared in 1973, but a further twist to the country's turbulent history came in December, 1979, when it was invaded by Soviet troops. The Afghan Air Force suffered considerably during the intense fighting following the invasion. Rebel activity forced the withdrawal of Soviet forces during the late 1980s, but afterwards Russian equipment included MiG-23s and Sukhoi Su-7s, Su-17s and Su-22s, while Mi-8, Mi-17 and Mi-25 helicopters were received, along with Aero L-39 Albatross trainers and ground-attack aircraft. Many of these aircraft have since been destroyed in the fighting. However, Iran is believed to have helped the Northern Alliance, part of the National Islamic Movement, with aircraft flown there from Iraq during the Gulf War. Although Pakistan has been a supporter of the Taliban regime, no military aircraft were ever supplied. US military action during late 2001 and early 2002 met little Afghan resistance, although two Chinook helicopters were lost.

Few aircraft are now operational, although there are believed to be some MiG-21, Su-17 and Su-22 aircraft, with L-39s, in the fighter and ground-attack role. Apart from a shortage of spares and fuel, many aircraft were destroyed on the ground during US air attacks. It is not known how many of the 80 armed helicopters, mainly Mi-8, -17 and -25, present in Aghanistan in late 2001 have survived the air attacks. There are believed to be several hundred SA-2, SA-3, Stinger, SAM-7 and SAM-14 missile launchers available.

ABOVE: *Su-22 wreck at Bagram (Jeremy Flack/Aviation Photographs International)*

Albania

POPULATION: 3 million

LAND AREA: 11,097 square miles, 28, 741 sq km

GDP: $3.8bn (£2.7bn), per capita $5,539 (£3,873)

DEFENCE EXPENDITURE: $139m (£97.2m)

SERVICE PERSONNEL: 27,000 active.

ALBANIAN PEOPLE'S ARMY AIR FORCE

Formed: 1947

An Albanian Air Corps was founded in 1914, but the Austrian Army seized the aircraft on the outbreak of World War I. Financial difficulties hindered attempts to form an air arm between the wars, and in April, 1939, Italian forces annexed the country. At the end of World War II Soviet forces occupied Albania, leading to the formation of the Albanian People's Army Air Force in 1947. This was achieved with Soviet instructors and key personnel accompanied by a gift of twelve obsolete Yakovlev Yak-3 fighters and some Polikarpov Po-3 biplane trainers. A founder member of the Warsaw Pact, in 1955, Albania received MiG-15 fighters and MiG-15UTI trainers. Relations with the Soviet Union cooled in favour of closer links with Communist China. By the early 1970s, Shenyang F-6 and F-4 fighters were introduced, although Soviet An-2 and Il-14 transports, and Mi-1 and Mi-4 helicopters remained until replaced by Chinese equivalents.

In recent years, the country has been affected by events across the border in the former Yugoslavia and by internal unrest. Some western equipment has been introduced in recent years, while NATO based some elements of its Operation Allied Force in Albania during 1999. As Europe's poorest country, Albania finds aircraft and fuel expensive, with flying hours at just 15 hours per annum, far too low for operational effectiveness. The APAAF remains part of the army, with 4,500 personnel, and may be reformed under the much-delayed Plan 2000. Three air regiments operate 22 Shenyang F-7A (MiG-21) and 23 F-6 (MiG-19) interceptors, and11 F-5 (MiG-17) fighter-bombers. There are eight FT-5 and 11 FT-2 (MiG-15) trainers, as well as six Nanchang CJ-6. Transport aircraft include 13 Shizaizhuang Y-5 (An-2), as well as helicopters, including 40 Harbin Z-5 (Mi-4) and a Bell 222UT. Helicopters on liaison duties include four SA316/319 Allouette III and three AS350b Ecureuil. There are a small number of SA-2 batteries.

Algeria

POPULATION: 32 million

LAND AREA: 919,590 square miles, 2,381,741 sq km.

GDP: $44.2bn (£30.9bn), per capita $7,300 (£5,104)

DEFENCE EXPENDITURE: $3bn (£2.1bn)

SERVICE PERSONNEL: 124,000 active (66 per cent conscript), plus more than 150,000 reserves.

ALGERIAN AIR FORCE

Formed: 1962

Algeria became independent of France in 1962, and almost immediately started to create an air force out of the National Liberation Army, with assistance from Egypt the USSR and Czechoslovakia. Egyptian pilots and technicians were seconded to fly the first aircraft, five ex-Egyptian MiG-15 fighters. Initially, it was known as the *Force Aerienne Algerienne*, but now has an Arabic title. More sophisticated equipment was introduced during its first decade. By the early 1970s, it had a total of 140 interceptors in ten squadrons, including MiG-17 and MiG-21s, as well as thirty Ilyushin Il-28 jet bombers in two squadrons, and a transport squadron of An-12s and Il-18s. Western equipment was also introduced, including 28 Potez Magister armed trainers and 20 SA330 Puma helicopters, the latter operating alongside 30 Mi-4s.

Russia has remained the main source of equipment for the AAF, but aircraft are also obtained second hand from other sources, including Belarus. South Africa has also become a major supplier, upgrading Mi-24 armed helicopters. It is possible that Roivalk armed helicopters will be obtained in the near future. Recent years have seen American, French and Dutch transport and communications aircraft introduced.

There are 10,000 personnel. Flying hours average 160 annually. There are five fighter, three fighter-bomber and two reconnaissance squadrons operating 33 MiG-29s. These replace over 50 MiG-21bisMF/UMs, 29 MiG-29BN/MS/U, 14 MiG-25/R/U, refurbished in the Ukraine during 2000-01, 13 Su-24 Fencer and 22 Su-22 Fitter, using AA-2, AA-6 and AA-7 AAM. The combat aircraft are supported by four Il-78 tankers. There are more than 30 Aero L-39ZA/C Albatros armed trainers. Two MR squadrons use 16 Beech Super King Air B-200T, while six 1900 provide SIGINT. Helicopters include 33 Mi-24 attack helicopters, upgraded in South Africa to MkIII standard, four Mi-6 and over 60 Mi-8/Mi-17, as well as nine AS355F Ecureuil and three Ka-27. Transport aircraft include 16 C-130H Hercules, three Il-76MD and six Il-76TD, and a VIP unit with two F27-400M Friendship, three Gulfstream III and two Falcon. Training aircraft include six T-34C Turbo Mentor, 30 Algerian-assembled Zlin 142, and 20 Mil-2 helicopters. There are three SAM regiments with SA-3, SA-6 and SA-8.

Future acquisitions may include maritime surveillance aircraft, from Russia.

Angola

POPULATION: 13.3 million

LAND AREA: 481,226 square miles, 1,246, 369 sq k

GDP: $6.6bn(£4.6bn), per capita $1,600 (£1,119)

DEFENCE EXPENDITURE: $1,100m (£769m)

SERVICE PERSONNEL: 130,500 active.

ANGOLAN PEOPLE'S AIR FORCE/ FORCE AEREA POPULAIRE DE ANGOLA

Formed: 1975

The Angolan People's Air Force was formed in 1975 following independence from Portugal. Strong support from Cuba included MiG-21 fighters and Mi-8 helicopters. The country was racked by civil war between different guerrilla groups, the strongest of which was UNITA, Union for the Total Independence of Angola, until a peace agreement was reached in 1995. The armed forces were reduced in strength following the peace deal, although the APAF received MiG-23s, Mi-17s and Mi-24 Hinds shortly before hostilities restarted in 1998. These were later joined by MiG-23s and Su-22s from Belarus. Eight ex-RAF Hercules C-130Ks are on order from Lockheed, while there have been reports of additional aircraft, Mi-17s, MiG-29s, and Su-27s being obtained from Eastern Europe, as well as a number of Mi-35s supplied by Russia in 2001.

The threat posed by UNITA has now diminished, which should affect future procurement programmes and the size of the APAF, currently 8,000 personnel. Fighter and strike aircraft include 26 MiG-23s, 36 Su-22M4/Us and 15 Su-25s, possibly some Su-27s, and 20 MiG-21MF/bis fighters. Training and counter-insurgency operations are handled by nine Pilatus PC-7 and six EMB-312 Tucanos bought from Peru in 2002. Attack helicopters include up to 30 Mi-25s and -35s, and six missile-armed SA-342s in addition to eight AS-565, more than 25 IAR-316 and up to 20 Mi-8/Mi-17 transport helicopters. MR is provided by seven C212MPA Aviocars, with another ten for transport operating alongside eight Hercules C-130Ks as well as two civil L-100-20s, eight An-26s, two VIP Boeing 707s and four communications BN-2T Islanders and four Pilatus PC-6Bs. In addition to the PC-7s, training uses three Cessna 172s, six Yak-11s and six L-29 Delfins. HOT anti-tank missiles and AT-2 ASM are used, with AA-2 AAM. A number of transport helicopters may be bought, possibly UH-60s or Super Pumas, although this order might be affected by the reduction in internal unrest.

Argentina

POPULATION: 37.6 million

LAND AREA: 1,084,120 square miles, 2,807,857 sq km

GDP: $282bn (£197bn), per capita $10,106 (£7,067)

DEFENCE EXPENDITURE: $4,8bn (£3.4bn)

SERVICE PERSONNEL: 70,100 active.

ARGENTINE AIR FORCE/FUERZA AEREA ARGENTINA

Formed: 1944

Argentine military aviation dates from September 1912, when a military aviation school was established using Farman, Bleriot and Morane aircraft. Aircraft appeared on manoeuvres during 1914. Additional aircraft were difficult to obtain during World War I because Argentina was a neutral country. It was not until 1919, that an Italian aviation mission provided six Ansaldo SVA Primo fighters; four Caproni Ca.33 tri-motor bombers; two Fiat R2 and two Ansaldo reconnaissance aircraft; and two Savoia trainers, to establish a Military Aviation Service. In 1920, a French military aviation mission visited. A Military Aviation Factory was established in 1927, building 100 Avro 504R trainers, followed by 40 Bristol F2Bs and 40 Dewoitine D21C-1 fighters. This substantial increase enabled the MAS to organize into two fighter, two bomber and two reconnaissance groups, reorganizing again into three regiments in 1938.

Nationally designed aircraft appeared during the mid-1930s, the Ae MB-1 light bomber, Ae T-1 liaison aircraft, and Ae MO-1 training and observation aircraft. Further licence-built aircraft followed in 1937, with 200 Curtiss Hawk 75-O fighters and 500 Focke-Wulf Fw44J Stieglitz trainers, while 35 Martin 139-W medium bombers were bought from the USA. External sources virtually dried up again during World War II, apart from some elderly DC-2 transports received in 1944. The Military Aviation Factory, renamed the Instituto Aerotecnico, produced 100 IAe DL-22 trainers.

ABOVE: *During the late 1950s, the Beech T-34 Mentor became the Argentine's standard trainer, with more than 90 built locally. (FAA)*

Argentina

ABOVE: *The FAE was one of the many operators of the Canberra jet bomber, in this case the B62 variant. (BAE Systems)*

The MAS became a separate service, the Fuerza Aerea Argentina in 1944. Re-equipment had to await the end of the war, when 100 Fiat G55 fighters, 20 Avro Lancaster and Lincoln bombers, Douglas C-47 and C-54 transports, and 30 Beech AT-11 Kansan trainers arrived. The FAA placed the first export order for the Gloster Meteor jet fighter, followed by 50 de Havilland Dove, 30 Vickers Valetta and 30 Bristol 170 transports, 200 Percival Prentice and 30 Fiat G46 trainers. Locally produced aircraft included 100 IAe Calquin light bombers and 150 DL-22 observation aircraft.

Argentina became a member of the Organisation of American States in 1948.

After an interval of some years, re-equipment restarted in 1957, with 90 Beech T-34 Mentor and 48 Morane-Saulnier MS760 Paris trainers built under licence. In 1959, North American F-86F Sabre fighter-bombers were introduced, followed later by more than 40 Douglas A-4 Skyhawk fighter-bombers, de Havilland Canada Beaver and Twin Otter transports, and the Argentine Dinfia Guarani and Huanquero transport and general-purpose aircraft. In the 1970s, these were joined by Dassault Mirage III fighter-bombers, 80 Dinfia Pucara COIN aircraft, Lockheed C-130E Hercules and Fokker F-27M Troopships, as well as Sikorsky S-55, Bell UH-1H Iroquois and 47G Sioux and Hughes 269-HM helicopters. Later, Mirage IIIs were joined by Mirage 5 fighter-bombers, and Israeli-built Daggers.

For many years, the Argentine had claimed the British colonies of the Falkland Islands, some 400 miles east of Argentina, and its dependency, South Georgia, 1,500 miles away. On 2 April, 1982, its forces invaded the Falklands and South Georgia, defended respectively by just 70 and 22 British marines. Initial landings were from the sea, but once the airfield at Port Stanley was secured, Hercules and F-27 transports, plus chartered civilian F-27s, air-landed troops. FAA and Commando de Aviacion Naval aircraft, including Skyhawks, Mirages and Etendards attacked a British naval task force, the leading ships of which set sail on 5 April. A number of Pucara, based at Stanley, were attacked by British naval aircraft. Operating from the mainland, Argentine air force and naval aircraft succeeded in sinking two British destroyers, two frigates and a container ship converted to carry most of the task force's troop-carrying helicopters. Argentine pilots showed considerable skill and courage, with the conflict lost by the poor performance of the army, despite frequently outnumbering the attacking British forces.

Afterwards, the Argentine dictatorship was overthrown leading to a gradual reduction in the size of the armed forces.

Argentina

ABOVE: *A UH-1 Huey captured during the battle for the Falkland Islands. (Jeremy Flack/Aviation Photographs International)*

Relatively little new equipment has been received, although Skyhawks and Mirages have been upgraded and it seems possible that ex-USAF F-16s or ex-Spanish Mirage F1EQ/BQs might be obtained, with a requirement of up to 40 aircraft, plus additional Skyhawks for modernisation to Fightinghawk standard. Nevertheless, the country's grave economic crisis must cast doubt on spending plans for the immediate future.

The FAA has 12,500 personnel, a reduction of almost a third over the past twenty-five years. Six squadrons operate in the fighter and strike role, with two having 24 up-rated Dagger/Neshers, and another having 8 Mirage 5Ps while a fourth squadron has 14 Mirage III/EA. Another two squadrons operate 36 upgraded A-4R/TA-4AR Fightinghawks. Two squadrons with 30 IA58 Pucara cover tactical operations. There are six air transport squadrons: one with four Boeing 707 tanker/transports; two have 11 C-130B/H/L-100-30 Hercules, including two KC-130H tankers; the remaining three squadrons include one with ten F-27s, one with four F-28s and one with six Twin Otters. Three IA50s are operated on communications duties. A Boeing 707, three Learjet 35As and two IA50s in a single squadron cover reconnaissance operations. Helicopters include four Bell 212s, nine UH-1Hs and 17 MD-500s, as well as two Chinooks. A Boeing 757 and a Sabreliner 75A provide VIP transport, with an S-70 Black Hawk. Training is based on 30 T-34C Turbo Mentors, 28 EMB-312 Tucanos, three MD500s and more than 20 IA63 Pampas, now manufactured locally by Lockheed Martina Argentina. Several MS760 Paris trainers remain in service. ASM-2 Martin Pescador are carried by FAA aircraft, while AAM include AIM-9B Sidewinder, R-530, R-550 and Shafir. As many as 30 Pucara may be in storage.

ARGENTINE NAVAL AVIATION/COMMANDO DE AVIACION NAVAL

Formed: 1919

An Argentine Naval Aviation Service was founded in 1919, encouraged by the arrival of an Italian mission. The first aircraft were two Macchi M7, two M9 and two Lohner L-3 flying boats. HS2L and F5L flying boats followed, while later Dornier Wal and Supermarine Southampton flying boats were operated. Training aircraft were obtained from Vickers, Avro and Savoia-Marchetti. Rationalization followed by major re-equipment in 1937 took the form of Douglas DB-8A-2 bombers, Vought V65F and V142 Corsair reconnaissance-bombers, Consolidated P2Y-3 flying-boats, and for operation from cruisers, Grumman J2F-2 and Douglas Dolphin amphibians.

Few changes took place until 1956, when ten F4U Corsair fighter-bombers arrived, followed in 1957 by Lockheed Neptunes for MR, Consolidated PBY-5A Catalina amphibians and Martin PBM-5 Mariner flying-boats. Grumman F6F-5 Hellcat fighter-bombers, JRF Goose and J2F-2 amphibians came later, with North American T-6 Texan, Vultee 13T-13 and Beech AT-11 trainers. The fighter-bombers were for the aircraft carrier *Independencia*, formerly HMS *Warrior*, which arrived in 1958. The Royal Netherlands Navy carrier *Karel Doorman*, another former British carrier, was bought in1969, renamed *Veinticinco de Mayo*. *Independencia* was withdrawn in 1970.

During the 1970s, *Veinticinco de Mayo* operated Grumman F-9B Panther fighter-bombers and six Grumman S-2A Tracker ASW aircraft. The Neptunes and Catalinas remained, while other aircraft included 24 Aermacchi MB.326K, Grumman TF-95 Cougar, Beech C-45 and North American T-

BELOW: *The IA58 Pucara was the only FAA combat aircraft able to be based in the Falklands after the invasion. (FAA)*

ABOVE: *Argentian Navy's Super Etendard (Michael J. Gething)*

28 Fennecs in the training role, plus Short Skyvan, DHC-6 Twin Otter, Douglas C-47 and C-54 transports. Helicopters included Sikorsky UH-19s and SH-34Gs, and Bell 47 Sioux, while the arrival of two new Type 42 guided missile destroyers from the UK also introduced Westland Lynx to the Argentine. Aboard the carrier, Douglas A-4 Skyhawk and Dassault Super Etendard aircraft replaced the strike aircraft at the end of the decade.

During the invasion of the Falklands, naval aircraft had to operate from shore bases because the *Veinticinco de Mayo* was having her catapult serviced. A naval Super Etendard fired the Exocet missile that sank the British Type 42 destroyer HMS *Sheffield*. Operations were limited by a shortage of Exocet missiles, in course of delivery at the start of the campaign. An Alouette III helicopter covering the invasion of South Georgia was forced down by fire from the defending marines.

The carrier was placed in reserve in 1997 and scrapped in 1999, but there have been improvements in capability elsewhere, with the introduction of 8 P-3B Orion long-range MR. These operate in one squadron with the remaining Trackers, modernized with turboprops. An Electra is operated for ELINT, with two Electra transports. Nine Sikorsky SH-3D/H Sea King helicopters are operated from Argentina-class (German *MEKO-360*) destroyers and an icebreaker. Naval aircrew are reported to have kept their carrier landing skills honed on the Brazilian carrier, *Minas Gerais*, using 11 Super Etendards and five Skyhawks. Nine Beech King Air/Super King Air provide additional MR as well as VIP transport. Helicopters include AS555 Fennec and Alouette IIIs aboard the other destroyers and the four *MEKO-140* frigates. Eight ex-US UH-1H utility helicopters have been acquired recently. Training uses MB326 and MB339, EMB-326GB Xavante, and T-34C Turbo Mentor aircraft as well as AS555 helicopters. Miscellaneous aircraft include a Queen Air and PC-6B Turbo Porter communications aircraft, and A109A utility helicopters. The Lynx have been scrapped due to an embargo on spares, and serviceability of the Type 42 destroyers is also likely to be poor.

The air arm accounts for a substantial proportion of the navy's 12,500 personnel. Supporting the navy is a coastguard service, the *Prefectura Naval*, with 14 aircraft, including C212-300 Aviocar transports, Puma, Panther and Dauphin helicopters for transport and SAR. It may be merged into the navy in the near future.

ARGENTINE ARMY AVIATION/COMMANDO DE AVIACION DEL EJERCITO

Formed: 1959

An air branch was formed in the Argentine Army in 1959 for AOP, liaison and light transport duties. At first, Cessna 182J, 310 and Skymaster aircraft were operated, joined by Piper Apache and Aztecs, Bell 206 JetRanger helicopters, three DHC-6 Twin Otter and three C-47s. It now operates small and medium helicopters, including six A-109, which may be joined by twelve ex-US AH-1F Cobras. Transport and utility helicopters include a Bell 212; 44 ex-US UH-1H and eight Hiller UH-12; three AS332B Super Pumas operated in the Antarctic; five SA-315B; two SA-330 Pumas and 20 AS-532 Cougar transports. Fixed-wing aircraft include a C212-300; three G-222; two DHC-6 Twin Otter; five Cessna 207; six Merlin IIIA and IVs; a photo-survey Queen Air and Citation I; a Sabreliner for VIP duties; 23 OV-1D Mohawks for surveillance and six Cessna T-41s for training.

BELOW: *A Commando de Aviacion Naval Argentina SH-3 Sea King hovers off the destroyer Heroina. (CANA)*

Armenia

POPULATION: 3.5 million

LAND AREA: 11,490 square miles, 29,758 sq km

GDP: $1.9bn (£1.3bn), per capita $3,703 (£2,589)

DEFENCE EXPENDITURE: $151.5m (£106m)

SERVICE PERSONNEL: 42,060 (inc 75 per cent conscripts) active.

ARMENIAN AIR FORCE

Formed: 1990

One of the many new states spun off in the break-up of the former Soviet Union, the Armenian Air Force has some 3,200 personnel and is part of this land-locked state's army. As elsewhere in the former Soviet Union, initial equipment largely consisted of Soviet material *in situ* at the time of the collapse of central control. Locked for much of its post-independence history in a struggle with neighbouring Azerbaijan, equipment has been relatively poor and in limited quantities, although this may now be changing following a new defence pact with Russia. A fighter squadron operates 13 MiG-29s, five Su-25s, a MiG-25 and two L-39s. Another squadron operates 12 Mi-24P/K/RKR attack helicopters, with seven Mi-8MTs transports, two Mi-9 command helicopters and two Mi-2s on utility duties. Fixed-wing transport is provided by an An-24 and an An-32. Training uses six An-2s, ten Yak-52s and six Yak-55/18Ts.

Operational effectiveness is likely to be poor.

Australia

POPULATION: 19 million

LAND AREA: 2,967,909 square miles, 7,682,300 sq km

GDP: $380bn (£265bn), per capita $24,500 (£17,132)

DEFENCE EXPENDITURE: $7.1bn (£5bn)

SERVICE PERSONNEL: 50,700 active, plus 21,340 reserve.

ABOVE: *For many years the RAAF operated Dassault Mirage IIIs, and after being withdrawn from frontline service, the Mirage IIIs they were used for defence research and weapons development. (RAAF)*

ROYAL AUSTRALIAN AIR FORCE

Formed: 1921

Australian military aviation originated with the formation of the Australian Flying Corps in 1913, with two Royal Aircraft Factory BE2a and two Deperdussin aircraft. One of the BE2as and a Farman seaplane accompanied an Australian contingent to German New Guinea shortly after the outbreak of World War I in 1914. A small force was attached to the Royal Flying Corps in Mesopotamia in 1915 fighting against Turkish forces. Further service with the RFC came in 1916 and 1917, with four squadrons serving in Egypt and the UK with FE2bs and SE5as. Post-war, the AFC was disbanded.

It was reformed in 1920, and early in 1921 became the Australian Air Force, an autonomous service that acquired the 'Royal' prefix in June. It used Australia's share of the 'Imperial Gift' of war surplus aircraft, more than 100 aircraft including DH9s, SE5as, ten Sopwith fighters, 26 Avro 504K trainers and six Fairey IIID seaplanes. Two squadrons were formed in 1925 for army co-operation duties, partly manned by regulars and partly reservists of the Citizen Air Force, while a fleet co-operation flight operated six Supermarine Seagull MkIII amphibians. Supermarine Southampton flying boats, Australian-built de Havilland Cirrus Moth trainers, Bristol Bulldog fighters and 28 Westland Wapiti general-purpose biplanes were added later. The Depression prevented further equipment purchases until 1934, when Hawker Demon fighters, Supermarine Walrus amphibians, Avro Cadet trainers and Anson reconnaissance aircraft were ordered. The last of these arrived in 1937, when a three-year expansion programme started. In addition to the de Havilland factory, Australian businessmen formed the Commonwealth Aircraft Corporation in 1936, building the North American NA-33 trainer, known in Australia as the Wirraway.

Only 12 squadrons out of a planned 32 were operational at the outbreak of World War II, with 164 aircraft and 3,500 personnel, including a new squadron of Lockheed Hudson bombers. A squadron of nine Short Sunderland flying boats was transferred to the RAF. Plans to contribute six squadrons to the RAF proved impractical, but many Australians served with the RAF, and eventually the RAF included 13 Australian squadrons, including No3, the top scoring Allied fighter unit in the Mediterranean. The RAAF also organised and operated the Empire Air Training Scheme, training 280 pilots, 184 observer-navigators and 320 wireless-operator/air gunners each month, plus basic training of others. Wirraways and Tiger Moths were joined by 200 Australian-designed Commonwealth CA-2 Wacketts. Several hundred Ansons and Fairey Battles were sent from the UK.

After the Japanese invasion of the Dutch East Indies, the RAAF received a number of ex-Royal Netherlands Air Force aircraft, including Curtiss Helldivers and Vultee Vengeances, Republic P-43s, Douglas C-47s and Dornier Do24 flying boats. Wartime aircraft included de Havilland Mosquito, Tiger Moth and Dragon Rapide; Boulton-Paul Defiant; Bristol Beaufort; Consolidated Liberator; Curtiss A-40 Kittyhawk; Douglas Boston; Handley Page Hampden and Halifax; Hawker Hurricane; Lockheed Lightning and Ventura; North American Mitchell and Supermarine Spitfire. The 1942 Japanese naval aircraft raid on Darwin, which lacked fighter cover, led to the development and production of the Commonwealth CA-12 Boomerang, an effective fighter which later did sterling work as a ground attack aircraft. RAAF units operated in every theatre of the war, while

ABOVE: *On exercises, a Boeing F/A-18 drops a 500-lbs bomb. (RAAF)*

personnel strength peaked at 200,000 men and women.

Post-war, a maximum personnel strength of 15,000 was set. The RAAF participated in the British Commonwealth Occupation Force in Japan. Re-equipment started, with Commonwealth Aircraft building 100 North American P-51 Mustang fighters, while the Government Aircraft Factory built 75 Avro Lincoln heavy bombers. This was followed by 80 de Havilland Vampire jet fighters to replace the Mustangs, which joined the Wirraways in the Citizen Air Force squadrons. Douglas C-47s joined the Berlin Airlift. A Gloster Meteor F8 squadron in Japan was attached to the US Fifth Air Force for operations over Korea, while in Malaya, a squadron of Lincolns and one of C-47s operated with British forces against Communist bandits.

During the early 1950s, North American F-86 Sabres re-equipped the fighter squadrons, while the bomber squadrons received English Electric Canberra jet bombers and Vampire trainers. All three types were built in Australia. The Sabres differed from the standard aircraft through using Rolls-Royce Avon engines. Lincolns were replaced on MR duties by Lockheed P2V-5 Neptunes. In 1955, Australia became

ABOVE: *The RAAF's Lockheed C-130E/H Hercules have been replaced by the C-130J Hercules II. (RAAF)*

Australia

A U S T R A L I A

ABOVE: *A formation of three RAAF BAe Hawks used for the lead-in fighter, LIFT, programme, bridging the gap between trainers and the frontline F/A-18 Hornet fighters and strike aircraft. (RAAF)*

a founder member of the South-East Asia Treaty Organisation.

The next decade saw continued upgrading of the RAAF's capabilities, with manufacture under licence of 100 Dassault Mirage IIIs and 75 Aermacchi MB.326H, while Lockheed C-130 Hercules and de Havilland Canada DHC-4 Caribou were also obtained. In the early 1970s, the Canberras were replaced by leased McDonnell Douglas F-4E Phantoms as an interim aircraft before deliveries of General Dynamics F-111s, while Orions started to replace the Neptunes. Two squadrons of Bell UH-1 Iroquois squadrons were also introduced and operated in support of Australian forces fighting alongside the US in Vietnam from 1966 until Australia decided to withdraw in 1970. A Mirage squadron was based at RAAF Butterworth in Malaysia for many years. During 2000, the RAAF assisted in the United Nations occupation of East Timor, overseeing the removal of Indonesian forces, and providing air transport support.

The RAAF has 14,050 personnel, plus 1,800 reservists. The main strike and reconnaissance capability lies with 35 extensively upgraded F-111C/RF-111C/F-111G in two squadrons. Three squadrons with 55 F/A-18A Hornets provide fighter and tactical air cover, with 16 F/A-18Bs in an OCU; while 33 BAe Hawk lead-in fighter trainers equip two tactical training squadrons where they replaced MB326Hs in 2001. FAC is provided by three PC-9As. Two squadrons operate 17 P-3Cs on maritime-reconnaissance duties, with two EP-3Cs for ELINT, and another three aircraft on training duties. Four Boeing 737 Wedgetail AEWC aircraft are on order for 2007, with another three on option. A transport and tanker group operates two squadrons with 12 C-130H and 12 C-130Js; one with five Boeing 707 tanker/transports; two with 14 DHC-4s; and a VIP squadron with five Falcon 900s. Eight BAe 748s provide navigation training. VIP transport includes two Beech 200s and a Beech 1900D. Training takes place on PC-9s. Missiles include AIM-7 Sparrow, AIM-9 Sidewinder and AMRAAM, with AGM-84A and AGM-142 ASM.

Future equipment is likely to include Lockheed Martin JASSM stand-off weapons, while the F-111s and F/A-18s are likely to be replaced by the Lockheed Martin F-35 JSF, and possibly a number of UCAVs. C-17s may be ordered to replace some of the remaining C-130Hs.

ROYAL AUSTRALIAN NAVY FLEET AIR ARM

Formed: 1948

While the Royal Navy made extensive use of Australian bases to support the British Pacific Fleet during the closing stages of World War II, the decision to form a Fleet Air Arm for the Royal Australian Navy was not taken until 1948. Prior to this, RAAF amphibious aircraft had operated from cruisers and the seaplane carrier HMAS *Albatross*. Initial equipment comprised two squadrons each of Hawker Sea Fury fighter-bombers and Fairey Firefly anti-submarine aircraft based ashore at Nowra, near Sydney, and aboard the light fleet carrier

ABOVE: *An RAAF General Dynamics F-111 accelerates away from the camera. (RAAF)*

HMAS *Sydney* (formerly HMS *Terrible*). A second carrier, HMAS *Vengeance* was leased in 1953 pending delivery of HMAS *Melbourne* (ex HMS *Majestic*) in 1956.

Delivered in 1948, HMAS *Sydney* played an important part in the Korean War, joining the British carriers in operations off the coast. Later, the Sea Furies and Fireflies were replaced by de Havilland Sea Venom jet fighters and Fairey Gannet ASW aircraft, while the RAN also acquired Vampire jet trainers and Bristol Sycamore helicopters. After the arrival of HMAS *Melbourne*, HMAS *Sydney* became a training carrier and then a fast troop transport during Australia's involvement in the Vietnam War. HMAS *Melbourne*'s aircraft were updated with the addition of McDonnell Douglas A-4G Skyhawk fighter-bombers and Grumman S-2E Tracker anti-submarine aircraft, as well as 27 Westland Wessex (S-58) ASW helicopters and nine Bell UH-1H Iroquois, while the Vampires were replaced by Aermacchi MB326H jet trainers in 1970-71. The RAN originally adopted the Royal Navy's 700/800 squadron numbering scheme before switching to the USN scheme.

Plans to replace HMAS *Melbourne* with the Royal Navy's HMS *Invincible* and acquire Sea Harriers were abandoned after the UK decided to keep the ship after the Falklands campaign. Wessex helicopters had been replaced by Sea Kings ashore and aboard the carrier. HMAS *Melbourne*'s withdrawal in June,1982, marked the end of Australian fixed-wing naval aviation, leaving only helicopters, operated from American-built frigates.

The RAN operates a single anti-submarine helicopter squadron equipped with 16 S-70B2 Sea Hawks, complemented by11 SH-2G(A) Super Seasprites. Another squadron operates seven Sea King Mk50 in the utility role with six AS350B Ecureuil on training duties. Three Kiowas and two HS748EW complete the force. Concerns over 'force projection' have led to consideration being given to operation of an aircraft carrier once more, while amphibious warfare units have space for up to four helicopters.

ABOVE: *An Australian Army S-70A Blackhawk lands aboard the amphibious transport HMAS* Manoora, *ready for a deployment to East Timor. (Australian Army)*

AUSTRALIAN ARMY AVIATION

Formed: 1968

Established as a separate corps in 1968, Army Aviation consisted of 50 Bell 47G Sioux and a small number of Alouette III helicopters for AOP and liaison duties. These were soon joined by 12 Boeing CH-47D Chinook heavy lift helicopters, and by Bell OH-58A Kiowa light helicopters assembled in Australia. Between 1966 and 1970 army helicopters joined the Australian contingent in Vietnam where, at its peak, a fifth of the country's small army was based. As part of the rationalization of Australian defences Iroquois helicopters were transferred from the RAAF to the Army to operate in the gunship and utility roles. Battlefield mobility has been enhanced by use of Sikorsky S-70A-9 Black Hawks, while Eurocopter AS350B Ecureuil were introduced for training.

The Army operates 35 Black Hawks; 40 Bell 206B-1 Kiowa, being replaced under Air 87 by 22 Tiger combat helicopters with the first due during winter 2004/5 and 25 UH-1H Iroquois. The Iroquois are due to be withdrawn in 2006-08, but no replacement has been ordered. There are six CH-47D Chinooks, including two delivered early in 2001. Training uses 17 Ecureuil. There are four Beech Super King Airs, leased for transport and surveillance work, and joined in the former role by two DHC-6 Twin Otters.

ABOVE: *The CH-47C Chinooks have been up-graded to CH-47D standard. (Australian Army)*

Austria

POPULATION: 8.3 million

LAND AREA: 32,366 square miles, 83,898 sq km

GDP: $194bn (£136bn), per capita $24,235 (£16,947)

DEFENCE EXPENDITURE: $1.6bn (£1.1bn)

SERVICE PERSONNEL: 34,600 active (50% conscript), plus 72,000 ready reserves.

AUSTRIAN AIR FORCE/OSTERREICHISCHE LUFTSTREITKRAFTE

Formed: 1955

A short-lived Deutschosterreichische Fliegertruppe, German-Austrian Flying Troop, was formed in 1918 on the break-up of the Austro-Hungarian Empire. After World War I it fought successfully against Yugoslav forces during the Corinthian War of 1919, but was disbanded by the Allied Control Commission that prohibited Austria from maintaining a military air service.

Austria regained full sovereignty in 1936 with plans for an air force with ten operational squadrons including fighter and bomber units plus training squadrons. Fiat CR20, CR30 and CR32 fighters, Junkers Ju86D bombers, Ju52/3M transports, Messerschmitt Bf108B Taifan and Focke-Wulf Fw58 Weihe communications aircraft and Fw44 Stieglitz trainers were quickly acquired and the combat squadrons were fully operational by 1938, when Austria was incorporated into Nazi Germany. The air force was incorporated into the Luftwaffe, with the fighter squadrons giving notable service in Europe and North Africa as the Jagdgruppen I/135 and I/138. Post-war, Austria was briefly divided into four occupied zones by the Allies, and again banned from maintaining a military air arm.

Austria was re-established as a sovereign state in 1955, compelled to maintain neutrality under post-war agreements. It immediately created the Osterreichishe Luftstreitkrafte, Olk, Austrian Air Force, within the Army. First aircraft were eight Yakovlev Yak-11 and Yak-18 trainers donated by the Soviet Union. Shortly afterwards Austria purchased some Zlin trainers. In 1957 three de Havilland Vampire jet trainers were obtained along with Fiat G46-5Bs, three Bell 47G Sioux helicopters, six Piper Super Cubs and two Cessna 182s for liaison and communications. By 1960 these aircraft had been joined by six Westland S-55 Whirlwind and six Alouette II helicopters. Fighter aircraft arrived in 1961, with SAAB J-29F Tunnan fighter-bombers, with a second squadron formed in 1962, when 30 Potez Magister trainers also arrived. Over the next decade, 40 SAAB 105OE strike aircraft replaced the J-29s. The Olk also established a transport squadron with six DHC Beavers and two Short Skyvans, two Sikorsky S-65, 14 Alouette III and nine Alouette II, 22 Agusta-Bell 204B and 12 206 helicopters, backed by SAAB Safir trainers and Magisters.

More potent aircraft followed, with 24 ex-Swedish SAAB J-35OE Draken fighters acquired in 1988, with AIM-9P3 AAM, relegating the 105OEs to the training role. Five additional Drakens were acquired in 1999 as spares. To enhance training in the confined Austrian air space and encourage experienced fast jet pilots to stay, much Draken training takes place in Sweden.

Today, the Olk has 6,500 personnel. The most recent deliveries are nine UH-60 helicopters, with three more on option. The Skyvans remain in service, likely to be replaced by either C-27 Spartans or CN235s, of which one is currently leased. Liaison is handled by 12 PC-6 Turbo Porters and 11 Kiowa helicopters, which also undertake reconnaissance. Mainstays of the transport fleet are 22 AB 212 helicopters, while 23 Alouette IIIs operate in the utility and SAR roles. In addition to the 105OEs, training is conducted on 16 PC-7s and 11 AB206A JetRangers. The replacement of the Draken by twenty four Eurofighter Typhoons has been postponed indefinitely, due to flooding in 2002.

Azerbaijan

POPULATION: 7.8 million

LAND AREA: 33,430 square miles, 86,661 sq km

GDP: $4.8bn (£3.4bn), per capita $2,181 (£1,525)

DEFENCE EXPENDITURE: $217m (£152m)

SERVICE PERSONNEL: 72,100 active, 575,700 reserves.

AZERBAIJAN AIR FORCE

Formed: 1990

Created on the collapse of the Soviet Union in 1990, the Azerbaijan Air Force used former Soviet aircraft and aircraft purchased from other sources, an interesting variety but a logistics and maintenance nightmare. The AAF was involved in the fighting with neighbouring Armenia over the disputed territory of Nagorno-Karabakh until Russia mediated a cease-fire in 1994.

There are some 8,000 personnel. Serviceability is believed to be extremely low. A squadron operates 28 MiG-25s of various variants on fighter and reconnaissance duties, while there is a ground attack squadron with four Su-17M, five Su-24 and two Su-25s. Transport aircraft include an An-12, an An-24, three Ilyushin Il-76s and a Tupolev Tu-134A. A helicopter regiment has seven Mi-2, 13 Mi-8 and 15 Mi-24. Training is provided on 18 Aero L-29s and 12 L-39s. There are more than 100 SA-2, SA-3 and SA-5 SAM batteries.

Bahamas

POPULATION: 313,000

LAND AREA: 4,404 square miles, 11,406 sq km

GDP: $4.7bn (£3.3bn), per capita $14,428 (£10,090)

DEFENCE EXPENDITURE: $26m (£18.2m)

SERVICE PERSONNEL: 860 active.

ROYAL BAHAMAS DEFENCE FORCE

Formed: 1973

Primarily a naval force, the Royal Bahamas Defence Force was created after independence from Britain in 1973. It operates two light aircraft, a Cessna 404 and 421, and in the past has had two Fairchild C-26s.

Bahrain

POPULATION: 626,000

LAND AREA: 231 square miles, 570 sq km

GDP: $6.9bn (£4.8bn), per capita $10,300 (£7,202)

DEFENCE EXPENDITURE: $444m (£310m)

SERVICE PERSONNEL: 11,000 active.

BAHRAIN AMIRI AIR FORCE

Formed: 1986

The small Gulf state of Bahrain maintains relatively powerful armed forces for its size, with modern equipment. The BAAF has 1,500 personnel, and has grown in strength since 1986, when it received its first Northrop F-5E fighter-bombers. Currently, it has one squadron with eight F-5Es and four F-5F; two fighter squadrons with 32 F-16C/Ds, with the older aircraft being upgraded to match the last ten aircraft delivered early in 2002; a helicopter squadron with 12 armed AB-212s and another with 12 AH-1E Cobras. A Boeing 747SP and a 727, two UH-60A/L Black Hawk, and two Grumman Gulfstreams (an II and an III) provide VIP transport. Other transports include two BO105s, an S-70A and an Avro RJ85. It is currently creating its own pilot academy with British help using BAe Hawks. Once the original F-16s complete their upgrade, the F-5Es might be retired.

Separate from the BAAF, the Bahrain Navy has two SAR BO105s.

Bangladesh

POPULATION: 130.8 million

LAND AREA: 55,126 square miles, 142,766 sq km

GDP: $37bn (£25.9bn), per capita $1,800 (£1,260)

DEFENCE EXPENDITURE: $684m (£478m)

SERVICE PERSONNEL: 137,000 active.

DEFENCE FORCE AIR WING

Formed: 1972

Until 1972, Bangladesh comprised East Pakistan, and participated in Pakistan's armed forces. Independence was seized after the war between India and Pakistan in 1971. Initially, the air force, part of a unified tri-service defence force, operated aircraft abandoned on independence, although most Pakistani aircraft had been destroyed by Indian aerial attack. Since that time, aircraft purchases have mainly been from China, adopting Pakistan's earlier procurement policy, no doubt because the country finds Western equipment too expensive. More recently, Russian equipment has been introduced, while the transport element has been boosted through purchase of ex-USAF Hercules.

Currently, the DFAW has 6,500 personnel. Flying hours are fairly low, at around 100 to 120 per annum, doubtless due to the strain fuel purchases place on the frail economy. Fighter and ground-attack operations are handled by 4 squadrons with 6 MiG-29 Fulcrum, plus another 2 for training, 21 Nanchang A-5C Fantan, 10 F-6s and 22 Chengdu F-7MG/M. AA-2 AAM are used. Training is conducted on 10 Shenyang FT-6, 36 Nanchang CJ-6s, 9 L-39ZA Albatros (which are steadily replacing FT-6s at the rate of one per year), and 12 Cessna T-37Bs. Transport is provided by 4 upgraded ex-USAF C-130B Hercules, 3 An-26s, 3 An-32s, 15 Mi-17s and 12 Bell 212s, while 3 206s are used for helicopter training.

Belarus

POPULATION: 10.2 million

LAND AREA: 80,134 square miles, 209,790 sq km

GDP: $9.4bn (£6.6bn), per capita $7,960 (£5,566)

DEFENCE EXPENDITURE: $373m (£260m)

SERVICE PERSONNEL: 82,900 active plus 289,500 reservists.

BELARUS MILITARY AIR FORCE

Formed: 1990

Formerly part of the USSR, Belarus became independent in 1990, equipped with aircraft left behind by departing Soviet forces or earmarked for locally-raised reserve forces. Reorganization has led to the Army's aviation units being absorbed into the Military Air Forces, VVS. Attempts are being made at standardization, while the VVS is perhaps unusual in having a surplus of Il-76 transport aircraft, many of which are in store and being sold. To raise revenue, some transports are operated in civil colours. Economic problems mean that flying hours are believed to be just 28 per year, suggesting low operational efficiency.

The VVS has 22,500 personnel. There seems to have been some inconsistency in aircraft procurement in recent years, with MiG-29s being sold to Algeria and Peru, but further aircraft of this type ordered! Fighters include 57 MiG-29s, 31 MiG-23s and 21 Su-27s; plus 87 Su-25 and 30 Su-24 strike aircraft, with 12 Su-24 for reconnaissance. Conversion training is provided on eight MiG-29UB, six MiG-23UB, four Su-27UB and ten Su-25UB. Many aircraft are for sale. Attack helicopters include 76 Mi-24/K/R, supported by more than 120 Mi-8, a dozen or so Mi-26 and up to 12 Mi-6. Transport is provided by 16 Il-76, with a similar number awaiting disposal, three An-12s, one An-24, nine An-26 and a VIP fleet of a Tu-134, a Tu-154 and a Yak-40. Training includes variants of the combat aircraft, plus Aero L-29s and L-39s.

Belgium

POPULATION: 10.2 million

LAND AREA: 11,778 sq m, 30,445 sq km

GDP: $238bn (£166.4bn), per capita $26,193 (£18,317)

DEFENCE EXPENDITURE: $3.4bn (£2.4bn)

SERVICE PERSONNEL: 39,420 active, plus 100,500 reserve.

BELGIAN AIR FORCE/FORCE AERIENNE BELGE/BELGISCHE LUCHTMACHT

Formed: 1946

Usually referred to by its French title rather than the Flemish, the Force Aerienne Belge originated in the creation of an army flying corps in 1911, using Farman HF3 and F20 biplanes manufactured under licence in Belgium. In 1912, the force used one of its aircraft for the first European trials of the new American Lewis gun. During its first two years, the flying corps was a section within the Army Balloon Company, but in 1913 it became the Compagnie des Aviateurs and its 16 aircraft were organized into 4 reconnaissance squadrons.

On the outbreak of World War I, the strength was hastily raised through pressing privately owned aircraft into service. A new title, Aviation Militaire, was adopted in 1915, but a shortage of pilots and the lack of a training organisation inhibited expansion as Belgium was almost overrun by German forces, with only a small area in the vicinity of Ostend left unoccupied. Pilots were often trained in Britain at their own expense, but enough became available to form a naval air squadron for service in Belgian Central Africa. Aircraft included Royal Aircraft Factory BE2as, RE8s, Sopwith 1/2-Strutters, Pups and Camels, DH9s,

ABOVE: *Colonial Central Africa saw the Belgian RE8s take over its skies (Jeremy Flack/Aviation Photographs International)*

Farmans, Spads, Hanriot HD1s, Caudron GIIIs and GIVs, and Breguet Br14s.

Post-war reorganization saw the Aviation Militaire with 26 squadrons in 8 groups: 1st Group operated balloons; 2nd (Observation) Group operated Ansaldo A300s and DH4s; 3rd (Army Co-operation) Group operated DH4s and F2Bs; 4th and 5th (Fighter) Groups operated Nieuport 27C and Spad 13 fighters; 6th (Reconnaissance) Group operated DH4s and DH9s, and Breguet Br14s; while 7th (Technical) and 8th (Flying School) groups operated an assortment of aircraft, including Avro 504Ks and Fokker DVIIs. Many of the aircraft were licence-built in Belgium. This was followed by a further reorganization in 1925, with the 26 squadrons formed into 3 wings, which became regiments in 1927. New aircraft types entered service, including Avia BH21 fighters and Breguet Br19 reconnaissance-bombers, Stampe et Vertongen RSV 32/100 and RS 26/180 and Morane-Saulnier MS230 trainers. The British Fairey concern opened a factory in 1931, producing Firefly and Fox fighters, while a Belgian manufacturer, Renard, produced the R31 reconnaissance aircraft.

During the mid-1930s, Gloster Gladiator fighters entered service, with Fokker FVII transports, Stampe SU-5 general-purpose biplanes and Koolhaven FK56 trainers. By the time World War II started in September 1939, Belgium was producing Hawker Hurricane and Fiat CR42 fighters and Fairey Battle light bombers, but most aircraft were obsolescent and destroyed on the ground when German forces invaded on 10 May 1940. Those that did get into the air fought against overwhelming odds. As Belgian resistance collapsed, the flying school squadrons escaped to French Morocco and other personnel managed to reach Britain from Belgium and Africa, to be absorbed into the RAF which eventually had 1,200 Belgian personnel. Personnel serving in the Belgian Congo joined the South African Air Force, many seeing service in North Africa. Later, two Belgian fighter squadrons were formed within the RAF, scoring 161 confirmed victories, 37 probables and 61 damaged enemy aircraft.

ABOVE: *The Fairchild C-119F Packets have been utilized for both long haul flights and dropping troops into the Congo region in resuce missions. (Jeremy Flack/Aviation Photographs International)*

The FAB was formed in 1946 on the disbanding of the RAF's Belgian units, as an autonomous air force. The initial structure included four 'day' fighter wings with Supermarine Spitfires and one 'night' fighter wing with de Havilland Mosquitoes. A transport wing operated Douglas C-47s, Avro Ansons, Airspeed Oxfords and de Havilland Dominies (Rapides), while the army co-operation wing operated Percival Proctors. The Spitfires were replaced by Gloster Meteors in 1949, and followed by Republic F-84 Thunderjets in 1951 when Belgium became a member of NATO, and eligible for US military aid. This released Meteors to replace Mosquitoes, before being replaced in turn by Avro Canada CF-100 all-weather fighters. Longer haul transports, such as Douglas C-54s and Fairchild C-119F Packets linked Belgium with the Congo. Meteor and Lockheed T-33A trainers were also introduced, and Bristol Sycamore helicopters. Prior to 1955, a number of pilots were trained in the United States, while training facilities were also established in the Congo to take advantage of the better flying conditions.

In the early 1960s, during a civil war in the Congo, the Packets dropped Belgian paratroops to rescue civilians held hostage.

By the end of the decade, the FAB was operating Republic F-84F Thunderstreak fighter-bombers and RF-84F Thunderflash reconnaissance aircraft as well as licence-built Hawker Hunters. During the late 1960s and early 1970s, these aircraft were replaced by 50 Lockheed F-104G interceptors, with additional F-104s for strike duties. More than 60 Dassault Mirage 5-BRs operated in two reconnaissance squadrons with 27 Mirage 5-BAs in a fighter-bomber squadron. Franco-German Transalls replaced the Packets, and 12 Lockheed C-130H Hercules entered service in 1973. Training used 36 SIAI-Marchetti SF260s, 40 Potez Magisters and 17 Mirage 5-BDs. Helicopters included ten Sikorsky S-58s, half of them produced under licence in France.

The Belgian armed forces were subjected to cutbacks and reductions in flying time even before the end of the Cold War, with FAB personnel falling from 20,500 in 1972 to 8,600 today. Belgium joined several other European NATO members in a joint programme to acquire Lockheed Martin F-16As, of which 72 F-16AM operate in six squadrons in the interceptor and attack role, equipped with AIM-9 Sidewinder and AIM-120 AMRAAM, AAM, and AGM-65G Maverick ASM, supported by an OCU with 18 F-16BMs. A transport squadron has the 11 remaining C-130H Hercules, while a sec-

Belgium

ABOVE: *Frontline strength of the FAB consists entirely of the Lockheed Martin F-16AM, currently receiving mid-life updates (Jane's).*

ond squadron has two Airbus A310-200s, three HS 748s, five Merlin IIIA, two Embraer ERJ-135 and two ERJ-145, two Falcon 20 and a Falcon 900. A squadron of five Sea King Mk48s provides SAR. Training takes two squadrons with a total of 29 Alphajets, a squadron of SF-260s and one with Magisters. The FAB is supposed to receive seven Airbus A400M transports around 2008, to improve the air mobility of Belgian forces and replace the C-130Hs. It plans to replace its F-16s with 48 aircraft in around 2015. Each successive modernization reflects a further cut in numbers. Many aircraft are in storage, including 39 F-16s, five Sea Kings and more than 60 trainers, but most of the F-16 force is currently receiving mid-life updates. Successive cutbacks in budgets mean that Belgium is no longer interested in the Lockheed-Martin F-35, and its F-16s will have to remain in service until 2015. There are 24 Mistral SAM launchers.

Belgium is in the process of merging its separate service staffs into a single command structure, but separate service structures are expected to remain at operational level.

NAVY/MARINE/ZEEMACHT

The Belgian Navy acquired its first helicopters during the early 1960s with two Sikorsky S-58Cs, soon joined by three Alouette IIIs for operation from warships. The Alouettes remain in service, but increasingly are assigned as a flight in the air force's SAR squadron. Expansion of this force is unlikely since Belgian Navy specializes in mine countermeasures.

ARMY AVIATION/FORCE TERRESTRE BELGE

On the re-establishment of the Belgian Army after the end of World War II, an aviation element was included, initially operating Auster AOP6s. These were later replaced by Alouette II and III helicopters, and eventually 38 IIs and 42 IIIs were operated alongside 12 Dornier Do27s in 4 squadrons. The Do27s were replaced by 10 BN-2A Islanders. Today, the FTB operates 28 armed Agusta A109HA helicopters with a further 18 in the observation role, while 26 elderly Alouette IIs remain on AOP duties with another 5 for training. The 10 Islanders are used for liaison and training duties, but are expected to be retired by 2005 when their role will pass to the A109 forces, which are expected to receive an MLU. Ambitious plans exist for acquisition of up to 15 helicopters, preferably heavy-lift Chinooks for inter-operability with neighbouring Netherlands, or possibly medium-lift Cougars, to enhance the army's air mobility.

Belize

POPULATION: 246,000

LAND AREA: 8,867 square miles, 22,965 sq km

GDP: $674m (£471m), per capita $2,978 (£2,412)

DEFENCE EXPENDITURE: $17m (£11.9m)

SERVICE PERSONNEL: 1,050 active, plus 700 reserves.

ABOVE: *Belize's Islander/Defender. (Jeremy Flack/ Aviation Photographs International)*

BELIZE DEFENCE FORCE AIR WING

Formed: 1983

The Belize Defence Force consists entirely of a small army, with two BN-2B Defender for maritime-reconnaissance and transport, also a T67-200 Firefly and a Cessna 182 for training. Support is provided by the presence of a British Army contingent, while US Drug Enforcement Agency aircraft within the country are nominally under BDF control.

Benin

POPULATION: 6.2 million

LAND AREA: 44,649 square miles, 115,640 sq km

GDP: $2.6bn (£1.8bn), per capita $2,200 (£1,538)

DEFENCE EXPENDITURE: $37m (£24.9m)

SERVICE PERSONNEL: 4,750 active.

PEOPLE'S ARMED FORCES OF BENIN/FORCE ARMÉES POPULAIRE BENIN

Formed: 1958

As Dahomey, Benin became an independent state within the French community in 1958, with an embryonic air arm within the Army, the Force Aerienne du Dahomey. This was equipped with the standard package of communications aircraft allotted to former French African colonies of a Douglas C-47, three Max Holste 1521M Broussard (Bushranger) and an Alouette II helicopter. The USA donated an Aero Commander 500B VIP transport. While the equipment has been updated since, economic problems have meant that two An-26 transports and a Twin Otter are shared with the national airline, Benin Inter Regional. The remaining aircraft include two C-47s, two Dornier Do128s and an Ecureuil helicopter as well as the Aero Commander and the Alouette II, a VIP flight of an F-28 and a Boeing 707-320. There are 150 personnel.

Bhutan

POPULATION: 1.5 million

LAND AREA: 19,000 square miles, 46, 620 sq km

ROYAL BHUTAN ARMY

This small Himalayan kingdom has a Dornier Do228 and two Mi-8 helicopters attached to its army for communications duties.

Bolivia

POPULATION: 8.4 million

LAND AREA: 424,160 square miles, 1,331,661 sq km

GDP: $9.4bn (£6.6bn), per capita $3,313 (£2,317)

DEFENCE EXPENDITURE: $130m (£90.9m)

SERVICE PERSONNEL: 35,000 (c.60 per cent conscripts) active.

ABOVE: *PC-7 Turbo Trainer on delivery. (Mason/ Aviation Photographs International)*

BOLIVIAN AIR FORCE/FUERZA AREA BOLIVIANA

Formed: 1941

Bolivian Army officers received flying training in 1917, but the Cuerpo de Aviacion was not formed until 1924. At first, a number of different aircraft were operated, including Breguet Br19A-2 and Fokker CV-C reconnaissance bombers, Caudron C97 and Morane-Saulnier MS139 trainers. Vickers 143 fighters and Vespa III AOP biplanes were added in 1929 which were later joined by Junkers W34 bombers, Curtiss Hawk 1A fighters and Falcon AOP aircraft and Curtiss-Wright Osprey general-purpose aircraft, eventually totalling some 60 aircraft. This force was engaged in a war with neighbouring Paraguay during the late 1920s and early 1930s; Paraguay won.

In 1937, an Italian air mission started reorganization but war in Europe intervened before Italian equipment could be delivered, although three Junkers Ju86 transports were expropriated from a German-owned airline. In 1940, Curtiss-Wright 19R and CW-22 trainers arrived. This was followed by an American air mission the following year that divided Bolivia into 4 air defence zones. The present title was adopted, although the FAB remained under army control. New equipment was provided by the USA, with Grumman OA-9 amphibians, Douglas C-47s, Stinson 105 Voyager and Interstate L-8 AOP aircraft and Beech AT-11 and North American NA-16-3 trainers, before American entry into World War II cut off this source.

In 1948, Bolivia joined the Organisation of American States and became eligible for American military aid, including Republic F-47D Thunderbolt fighter-bombers, North American B-25J Mitchell bombers and T-6 Texan trainers. The US provided eight Boeing B-17G Fortress bombers in 1958 for conversion into transports. Aircraft supplied over the next two decades, included 12 North American F-51D Mustang fighter-bombers and three AT-6 armed trainers, 20 North American T-6 and T-28 trainers, seven Cessna 185 liaison aircraft, 18 C-47s and a C-54. Also, more up-to-date, 12 Hughes 500M and 3 Hiller H-23 helicopters.

In 1973, the FAB received surplus Canadian Canadair T-33 and AT-33 armed-trainers, which were augmented by aircraft from other sources. Plans to replace these aircraft, possibly with ex-US A-4 Skyhawks, have been stalled by economic problems, but instead an upgrade of the AT-33s has been attempted, although a US/Canadian dispute over arms sales has created problems over engine overhauls. Additional AT/T-33s to maintain the force have been acquired through dealers. Nevertheless, the US has also supplied C-130 transports and UH-1H Iroquois helicopters for anti-drug operations, on which armed Pilatus PC-7 trainers are also used with Basler Turbo 67s. The two military-operated airlines, Transporte Aero Militar and Transporte Aero Boliviano use some of the transports.

Currently, the FAB has 3,000 personnel, although up to two-thirds are conscripts. Mainstays of the force are 18 Lockheed AT-33AN armed-trainers, supported by seven PC-7 armed-trainers, with others operating as advanced trainers. Another squadron has ten Hughes 500M armed-helicopters. Transport is provided by two squadrons operating 9 C-130A/B/H, five F-27-400, two Basler Turbo 67s, five Convair 580s, two King Airs and an IAI-201, a Sabreliner 60 and two Super King Airs while a VIP L-188A Electra and DC-8-54 are stored. Liaison and surveillance are covered by 31 Cessna 206s, seven 210s and four 402/421s, a Beech Bonanza and a Baron, and three Aero Commanders. Special operations, mainly drug enforcement, use two Bell 212s and 21 UH-1Hs. A SAR helicopter squadron has four HB315B and two SA315B Lamas and an UH-1H. Training aircraft include a considerable number of T-33s, a dozen or so Cessna 152s and some Piper Cherokees. There are no SAM defences.

Although land-locked Bolivia has a small navy maintaining river and lake patrols, Las Fuerzas Navales Bolivianos, which uses a single Cessna 402. The Bolivian Army, or Ejercito Boliviano, has a King Air and a CASA C212.

Bosnia-Herzegovina

POPULATION: 3,9 million

LAND AREA: 19,736 square miles, 51,199 sq km

GDP: $5.1bn (£3.6bn), per capita $8,557 (£5,983)

DEFENCE EXPENDITURE: $187m (£130.1m)

SERVICE PERSONNEL: 24,000 active.

ARMY OF THE FEDERATION OF BOSNIA-HERZEGOVINA/AbiH

Formed: c. 1995

The break up of the Yugoslav Federation following the death of the former President Tito led to civil war in Bosnia-Herzegovina between Orthodox Christian Serbs and the Muslim community. Eventually, Croatian Roman Catholics sided with the Muslims against Serbian ambitions to maintain the federation. The Bosnia-Herzegovina forces raised an army of some 30,000 personnel, with a small air arm initially operating ten Mi-8 helicopters, although after NATO-intervention in 1995, the United States provided 15 UH-1H helicopters. This force is now merging with the Croatian Defence Council force of around 10,000 personnel which also has a small number of Mi-8 helicopters. Some Mi-17s and a Mi-34 helicopter trainer are also operated, with a VIP Citation II and four UTVA-75 for communications.

REPUBLIKA SRPSKA AIR FORCE

Supported by neighbouring Serbia, the Army of the Serbian Republic of Bosnia and Herzegovina has some 14,000 personnel. It has a small air arm, the Republika Srpska Air Force, restricted by international peace treaty to 21 fixed-wing aircraft, and these are mainly remnants of the former Yugoslav Air Force's aircraft, including six Soko Oraos, 13 Jastreb, some Galebs and Super Galeb armed-trainers. There are 20 SA341 Gazelle light helicopters and ten Mi-8s. There are also many light aircraft. It can be regarded as an offshoot of the Serbian Air Force.

Botswana

POPULATION: 1.6 million

LAND AREA: 222,000 square miles, 574,980 sq km

GDP: $4.5bn (£3.1bn), per capita $7,200 (£5,034)

DEFENCE EXPENDITURE: $249.6m (£174.5m)

SERVICE PERSONNEL: 9,000 active.

BOTSWANA DEFENCE FORCE AIR ARM

After independence from Britain in 1966, Botswana established its own army, with an air transport and communications capability. In 1996, it acquired ten CF-5A (F-5A) fighters, and three CF-5B conversion trainers, augmented in the COIN role by seven Pilatus PC-7s (also used for training) and ten Britten-Norman BN-2A Defenders. Transport is provided by three ex-USAF C-130B Hercules, two CASA CN235 and two C212-300 Aviocar, and six Bell 412 helicopters, plus a VIP Gulfstream IV. There are two Cessna 152s and seven AS350 Ecureuil helicopter trainers, and nine anti-poaching Cessna O-2 Skymasters. Ground forces are increasing from 8,500 to 10,000 personnel in the near future, and the Air Arm may also increase from its 500 personnel.

Brazil

POPULATION: 171.9 million

LAND AREA: 3,287,195 square miles, 8,512,035 sq km

GDP: $643bn (£450bn), per capita $6,700 (£4,685)

DEFENCE EXPENDITURE: $17.9bn (£12.6bn)

SERVICE PERSONNEL: 287,600 active, 1,115,000 trained reserves.

BRAZILIAN AIR FORCE/FORCA AEREA BRASILEIRA

Formed: 1940

The Brazilian Alberto Santos-Dumont made the first flight in Europe in 1906, but it was not until 1913 that the Brazilian Navy established a seaplane school with an Italian Bossi seaplane. An army flying school formed shortly afterwards. A combined total of 17 aircraft was operated by both services in 1914, including Farmans and Bleriots. As a supporter of the Allies during World War I in 1917 Brazilian officers started training in the UK, and also received combat experience in Italy. The UK provided Brazil with 79 aircraft.

Post-war, French assistance was provided in the creation of a Brazilian Army Air Service. By 1922 this was operating a squadron of Breguet Br14A-2 and one of Spad S7 reconnaissance aircraft, as well as a variety of training aircraft. The Brazilian

Brazil

ABOVE: *An AMX strike aircraft, a joint venture with Italy, flies over the Amazon. (EMBRAER)*

Navy Air Service had three squadrons operating 14 Savoia reconnaissance seaplanes and 12 Felixstowe F5L flying boats, while there were also 12 Avro 504 trainers and the same number of Curtiss trainers. The inter-war years saw the Army operate Boeing 256 and 267 and Curtiss Hawk 75A fighters, North American NA-44 and Vultee V-119B light bombers, and Vought O2U-1 Corsair, de Havilland Fox Moth, Gipsy Moth, Beech D-17A, Morane-Saulnier and Avro 630 aircraft on training and communications duties. Meanwhile, the Navy operated Boeing F4B-4 fighters, Curtiss-Wright Osprey general-purpose aircraft, and Fairey Gordon reconnaissance-seaplanes. There were also Brazilian-designed Muniz M-7 and M-9 trainers. Other aircraft were built under licence, including Focke-Wulf Fw44J Stieglitz and Fw58B Weihe trainers. Given the vast size of the country and the natural barriers to surface communications of the Amazonian rain forests, the Army also found itself heavily involved in aerial survey work and communications duties. In 1940, Brazil followed the growing trend towards autonomous air services by merging the two air arms into the Forca Aerea Brasileira.

Brazil joined the Allies in 1942 after its shipping had been attacked by U-boats. Bases were put at the disposal of the Allies, and military assistance was received from the United States. The FAB started to receive aircraft in quantity, with 100 Curtiss P-40 Warhawk fighters, small numbers of Douglas B-18B and Lockheed A-28 bombers, 25 North American B-25 Mitchell bombers, 100 Vultee BT-15 Valiant, 200 licence-built Fairchild PT-19, 125 North American AT-6, ten Beech AT-7 and ten AT-11 trainers. Additional aircraft followed, and in 1944, a Brazilian squadron was sent to Italy with Republic F-47D Thunderbolts. Additional Thunderbolts entered service in 1945, along with 29 Vultee A-35B Vengeance dive-bombers, 25 Douglas A-20 bombers, 21 Piper L-4 and 40 Aeronca L-3 liaison aircraft, eight Beech C-45, 11 C-47, eight Lockheed C-60 and 33 Cessna UC-78 transports. At the end of the war, the FAB had around a thousand aircraft.

Peace meant the inevitable reduction in strength, but when Brazil joined the Organization of American States in 1948, American aid resumed. Additional Thunderbolt and Mitchell aircraft were delivered, plus Boeing B-17G Fortress and Lockheed P2V-7 Neptune MR aircraft, and North American T-6 Texan trainers. The first jets, 60 Gloster Meteor F8 fighters and T7 trainers, came in 1954. In 1956, 12 Fairchild C-82 transports were delivered, while Brazilian production of the Fokker S11 Instructor trainer started. These aircraft were soon followed by 50 Lockheed F-80C Shooting Stars and 50 T-33A jet trainers as well as Fairchild C-119 Packet transports. Additional deliveries included ten Lockheed C-130E Hercules, six HS748s, a Vickers Viscount and a number of Beech light transports.

Post-war, a separate naval air arm had been formed with an aircraft carrier, but in 1965 all aviation was once again merged into the FAB.

Steady modernization of the FAB saw 16 Dassault Mirage III fighters and 15 Douglas A-4F Skyhawk fighter-bombers enter service in the late 1960s. These joined Douglas B-26K bombers and 54 Lockheed TF-33 armed jet trainers, Neptunes and ex-naval Grumman S-2A Tracker MR. During the 1970s, 24 DHC-5 Buffalo as well as many surviving C-47s, C-54 and C-119 transports augmented the Hercules. Helicopters included six Bell SH-1D, six UH-1D and seven 206 JetRanger as well as 18 47G/J delivered earlier, and five Sikorsky UH-1D. A wide variety of training types included 150 Neiva IPD-6201 Universal and seven Potez Magister, as well as surviving examples of the earlier North American, Beech and Fokker trainers, which were replaced by 112 Aermacchi MB.326 trainers built in Brazil during the mid-1970s. Later, Brazil joined Italy in a collaborative ground-attack aircraft project, the AMX, also available as an operational trainer, and augmenting its earlier acquisition of Northrop F-5E Freedom fighters.

Despite having a growing indigenous aircraft industry with export sales of Tucano trainers, the cost of buying and maintaining aircraft has led to an ageing force and problems with serviceability in recent years. Brazil has borders with ten other states and a substantial offshore EEZ to police. In general, upgrading existing equipment has been adopted instead of replacement. Re-organization has also taken place, with army and naval air arms re-established, with the FAB operating Trackers from the aircraft carrier under naval control. Generally, purchases of foreign equipment are mainly

Brazil

ABOVE: *In addition to service with its home air force, where it is known as the T-25, the Brazilian Tucano is one of the most widely used intermediate trainers. (EMBRAER)*

second hand, but the domestic industry has shown considerable ingenuity in adapting its own designs to meet requirements, including adapting the EMB-145 regional airliner as an AEW platform. The domestically produced CLX light transport is planned as a Buffalo replacement.

Currently, the FAB has some 50,000 personnel, only about 10 per cent of whom are conscripts, an increase of some 20,000 in the past 30 years. Fighter defences are still provided by two squadrons with 18 Mirage F-103E, upgraded IIIE/Ds, with three squadrons operating 56 upgraded F-5E/F/B and 27 AMX in the strike role, supported in training by a further 14 AMX and some of the F-5s. The FAB has its own designation for many aircraft. Close support and COIN operations are covered by two squadrons with 53 AT-26s, the FAB designation for the EMB-326 (MB326), of which 33 are being upgraded. Plans exist to acquire 100 A-20 and AT-29 (EMB-319 Super Tucano) for COIN and as armed trainers to replace the ten RT-26 Xavantes which operate in two reconnaissance squadrons alongside four RC-95 Bandeirante, 12 Learjet 35s (which also undertake VIP duties) and three RC-130Es. Surveillance and radar calibration is provided by four Hawker 800XP, which also undertake Amazonian inspection flights. Seven liaison and observation squadrons, include one with 8 T-27 (EMB-312 Tucano), one with 29 Bell UH-1H armed helicopters, and five with 31 U-7 EMB-810 Senacas. MR is provided by three squadrons with ten P-16s, or EMB-110Bs (upgraded Grumman S-2 Trackers) and 21 P-95s (EMB-111 Bandeirantes). Most of the Trackers, other than those needed for carrier duties, are being replaced by ex-USN Lockheed P-3A/B Orions, although the planned 12 aircraft have been slashed to eight.

Given the size of the country and the difficulties in surface communication, there is a substantial air transport capability.

RIGHT: *The Embraer 145 regional jet is providing the basis for a family of aircraft including the RJ145SA for the Amazon surveillance programme. (EMBRAER)*

Brazil

ABOVE: *In common with many navies, Brazil's Marinha started small ship helicopter operation with Westland Wasp helicopters. (GKN Westland)*

Strategic transport is provided by a squadron with 17 C-130H Hercules and two KC-130H tanker/transports, with a squadron of four KC-137 (Boeing 707-320C) tanker/transports. Another squadron operates 12 C-91Cs (HS748s), a squadron has 17 C-95A/B/C Bandeirante, another has 17 C-115 (Buffalo), a VIP squadron has a VC-91 (HS748), 12 VC/VU-93 (BAe-125), five VC-96 (Boeing 737-200), five VC-97 (Brasilia) and five VU-9 (Xingu). There are also seven regional transport squadrons with seven C-115s, 86 C-95A/B/C and six EC-9s, a variant of the VU-9. The FAB has six Eurocopter AS332, eight AS355 and 27 HB350 helicopters, as well as four Bell 206s. Liaison is provided by 50 C-42 and 30 U-42 Neiva Regente, and three Cessna 208. Training uses 38 AT-26s, 97 C-95A/B/C, 25 T-23, 98 T-25 and 61 T-27 Tucano, in addition to the F-5Es and AMX already mentioned. Missiles include AIM-9B Sidewinder, R-530, Magic 2 and MAA-1 Piranha.

BRAZILIAN NAVAL AIR ARM/FORCA AERONAVAL DA MARINHA DO BRASIL

Originally formed in the early 1950s with US assistance, and three Bell 47J helicopters that were soon followed by two Westland Widgeon. The Royal Navy's light carrier HMS *Vengeance* was acquired in 1957 and renamed *Minas Gerais*, for which 13 Grumman S-2A Tracker ASW aircraft were acquired with a few North American T-28C armed trainers. Other aircraft were also introduced, including 15 Westland Whirlwind and five Wasp helicopters, while three Hughes 269A and nine 200 were acquired for training. The fixed-wing aircraft were transferred to the FAB in 1965, although carrier operations continued. Subsequently six Hughes 500 and four Sikorsky SH-3D Sea Kings were also bought.

Re-establishment of naval fixed-wing flying has taken place, with 20 ex-Kuwaiti A-4 (designated AF-1) and three TA-4 Skyhawks for operations from the carrier *Sao Paulo*, formerly the French Navy's *Foch*, which will enter service during 2002-2003 once the *Minas Gerais* is retired. Acquisition of four ex-Royal Navy Type 22 frigates has introduced the Westland Super Lynx to the Marinha, with earlier Lynx being upgraded, equipped with the Sea Skua ASM.

Currently, 1,150 of the 48,600 naval personnel are connected with naval aviation. In addition to the Skyhawks and Super Lynx, there are 13 Sea Kings for ASW and SAR, five Cougar for SAR and transport, and training uses 16 JetRanger III and 11 Esquilo, with another nine in the utility role. The Trackers continue to operate from the carrier flown by FAB personnel under naval control while Embraer is modifying ex-USN S-2/E-1 Trackers/Tracers for shipboard AEW.

BRAZILIAN ARMY AVIATION/AVIACO DE EXERCITO BRASILEIRO

Given its size, with a total of 189,000 personnel, the Brazilian Army has a relatively small air corps, and is still heavily dependent upon the FAB for support. It is structured into two units, both with a light transport and armed scout element. There are four Sikorsky S-70A Black Hawk helicopters, as well as 19 Eurocopter AS550 Fennec and 36 HB350-1 Esquilo, for attack and reconnaissance duties, and eight AS532 Cougars for transport.

BELOW: *Brazilian frigates operate Westland Mk21A Super Lynx helicopters, with earlier Lynxes up-graded to this standard. (GKN Westland)*

Brunei

POPULATION: 334,000

LAND AREA: 2,226 square miles, 5,765 sq km

GDP: $6.1bn (£4.27bn), per capita $8,100 (£5,664)

DEFENCE EXPENDITURE: $354m (£248m)

SERVICE PERSONNEL: 5,900 active, plus 700 reserves.

ROYAL BRUNEI ARMED FORCES

Part of an integrated defence force, the air arm supports the ground forces using 1,100 personnel. Two squadrons of helicopters include one with five BO105 armed helicopters and one with ten Bell 212 and an SAR 214, four Sikorsky S-70A and a VIP S-70C. Training uses two SIAI SF-260W, six Pilatus PC-7, two Bell 206 and an IPTN-CASA CN235. One MR IPTN-CASA CN235M operates alongside three transports. Strike capability has been enhanced by six BAe Hawk 100 armed trainers and four Hawk 200 trainers. Six additional Hawk 100s may be acquired in the near future.

Bulgaria

POPULATION: 8.2 million

LAND AREA: 42,818 square miles, 110,911 sq km

GDP: $12.8bn (£8.9bn), per capita $4,832 (£3,379)

DEFENCE EXPENDITURE: $354m (£248m)

SERVICE PERSONNEL: 77,260 active, plus 303,000 reserves.

ABOVE: *One of the two Bulgarian Tu-134s, the Tu-134a 'Crusty' (Jeremy Flack/ Aviation Photographs International)*

BULGARIAN AIR DEFENCE FORCE

Formed: c.1948

Bulgarian military aviation dates from the formation of an Army Air Corps to fight in the Balkan War of 1912-13, using 12 Bleriot and Bristol monoplanes flown by foreign pilots against Turkish forces. This improvised force was disbanded once hostilities ended, to be revived in 1915 with German and Austro-Hungarian assistance once Bulgaria allied herself with them during World War I. The Central Powers provided both aircraft and pilots. Allied victory in 1918 meant that the air corps had to be disbanded a second time and the Treaty of Neuilly banned Bulgarian military aviation.

In 1937, Bulgaria denounced the Treaty and formed the Bulgarian Air Force within the Bulgarian Army. Polish aircraft were bought, including 24 PZL P-24-G fighters, 43 PZL P-43 reconnaissance-bombers, an Avia B534 fighter and Letov S328 reconnaissance aircraft, while Focke-Wulf Fw44 Stieglitz trainers were built under licence in Bulgaria. Bulgaria allied herself with the Axis powers in 1941, allowing German forces to use Bulgarian bases for the invasion of Greece. Luftwaffe instructors and advisers were seconded, and military aid included Messerschmitt Bf109E fighters, Junkers Ju86D and Ju87B and Dornier Do17 bombers, as well as Fw58 Weihe communications aircraft and Arado Ar96 trainers. In 1943, 100 Dewoitine D520 and another 150 Bf.109Es were supplied so that the Bulgarians could replace Luftwaffe units transferred to the Russian front. Bulgaria never declared war on the USSR, but despite pleas of neutrality, the country was invaded at the end of 1944.

Post-war, Bulgaria found herself within the Soviet sphere of influence. Initially limited to an air force, BVVS, of no more than 5,000 men and 90 aircraft, Soviet assistance later allowed this to rise. Initially, obsolete Yakovlev Yak-9 fighters and Ilyushin Il-2 close support aircraft were provided with a few bombers, transports and trainers, including Yakovlev Yak-18s. In 1953, jets arrived, 24 Yakovlev Yak-23 and 60 MiG-15 fighters, as well as Il-10 ground-attack aircraft.

A member of the Warsaw Pact, Bulgaria continued to receive Soviet assistance and the air force strength grew to 12,000 by 1970. More modern equipment arrived, initially MiG-17s, then MiG-19 and eventually MiG-21 fighters, with Ilyushin Il-28 recon-

Bulgaria

naissance-bombers, and Il-12 and Il-14 transports. Mil Mi-4 helicopters were also supplied, while training used Yak-18s, Aero L-29 Delfin and MiG-15UTIs. Growth continued throughout the Cold War years, with personnel reaching some 20,000. Older aircraft were steadily replaced, with deliveries of MiG-23, -25 and -29 interceptors, and Sukhoi Su-22 and Su-25 attack aircraft. The bomber disappeared in favour of a tactical fighter and ground attack force able to operate in close support of ground forces aided by Mi-8 transport and Mi-24 attack helicopters.

The collapse of the USSR and of the Warsaw Pact have left Bulgaria struggling to maintain an increasingly elderly air force, cut off from cheap Soviet equipment and fuel, so that flying hours average between 30 and 40 per year and serviceability of aircraft is poor. The BVVS is being steadily slimmed down, with as many as a hundred aircraft a year being withdrawn and the number of operational bases reduced to just three from a dozen or so. Personnel numbers are 18,300, but likely to reduce substantially. Fighters are organized into three regiments with 17 MiG-29A and 4 -29UB, and a few remaining MiG-23s while more than 60 MiG-21s are still operated. Three strike regiments have 39 Sukhoi Su-25A/UBs. The MiG-29 and Su-25s are receiving upgrades. Reconnaissance aircraft are in a single regiment with 21 Su-22M/U and 10 MiG-21MF/UM. The frontline squadrons operate 28 L-29s as target tugs and for basic training, with advanced training on 35 L-39s. Helicopters include a regiment of 43 Mi-24s in the attack role, and another with eight Mi-8, 31 Mi-17, six Bell 206 and a 430. A transport regiment has two Tu-134s and a Yak-40 for VIPs, two An-24s, five An-26s and six Let L-410s.

As is usual with former Warsaw Pact countries, there is no separate army air corps, but the small Bulgarian Navy operates nine Mi-14 ASW helicopters.

Burkina Faso

POPULATION: 12.2 million

LAND AREA: 105,839 square miles, 274,123 sq km

GDP: $3.8bn (£2.6bn), per capita $1,000 (£699)

DEFENCE EXPENDITURE: $69m (£48m)

SERVICE PERSONNEL: 10,000 active.

BURKINA FASO AIR FORCE/FORCE AERIENNE DE BURKINA FASO

Formerly Upper Volta, Burkina Faso formed an air transport and communications force within the Army. Burkina Faso became a Soviet client state during the 1980s, and at one time the FABF received MiG-17s and MiG-21s, which have long since ceased to be operational. The current air arm has just 200 personnel, and its only armed equipment is five SIAI SF-260W/WL armed trainers, while there are also five Mi-8/-17 transport helicopters, a Eurocopter SA316B and an AS350, as well as two Cessna 152s for training and liaison. Transport is provided by a VIP Boeing 727, two Nord 262s, two HS748s, a Commander 500B and a Beech Super King Air.

Burundi

POPULATION: 6.8 million

LAND AREA: 10,747 square miles, 27,834 sq km

GDP: $1.2bn (£839m), per capita $600 (£419)

DEFENCE EXPENDITURE: $67m (£47m)

SERVICE PERSONNEL: 45,500 active.

BURUNDI NATIONAL ARMY/ARMEE NATIONALE DU BURUNDI

A small air wing with 200 personnel is operated within the army. Serviceability of the aircraft is believed to be extremely poor given the country's parlous economic state and the civil war that has raged since 1993. Combat aircraft include five SIAI SF260W Warrior armed trainers and two SA342L Gazelle helicopters. A Falcon 50 is kept for VIP use, with three Alouette III helicopters for light transport and liaison and two Cessna 150L trainers. Believed to be unserviceable are two C-47s and two Mi-8 helicopters.

Cambodia

POPULATION: 11.5 million

LAND AREA: 71,000 square miles, 181,300 sq km

GDP: $3.2bn (£2.2bn), per capita $730 (£510)

DEFENCE EXPENDITURE: $195m (£136m)

SERVICE PERSONNEL: 140,000 active.

LEFT: *This MS. 733 Alcyon, formerly a trainer, has been civilianised. (Jeremy Flack/Aviation Photographs International)*

ROYAL CAMBODIAN AIR FORCE

With its neighbours, Vietnam and Laos, Cambodia had a chequered history throughout much of the twentieth century. It was part of French-Indo China until independence in 1949, becoming an independent kingdom within the French Union until complete independence in 1955. An air force, Royal Khmer Aviation, was planned in 1953, for police and communications operations. Its first aircraft were seven Fletcher FD-25A/13 Defender light attack aircraft, followed in 1955 by seven Morane-Saulnier MS733 Alcyon trainers. American military aid started in 1956, with eight Cessna L-19 Bird Dog AOP aircraft and Douglas C-47 and DHC-2 Beaver transports. As attention centred on South East Asia, aircraft started to arrive from both sides of the Iron Curtain. France provided 30 surplus Douglas A-1D Skyraiders, Potez Magister trainers and Alouette II light helicopters; the United States, North American T-28D armed-trainers, Curtiss C-46 Commando transports, Sikorsky S-58 helicopters, and Cessna T-37B and North American T-6G trainers; from the Eastern Bloc MiG-15 and MiG-17 fighter-bombers and An-2 transports.

Cambodia claimed to be neutral, but allowed North Vietnamese troops to use its territory for operations against South Vietnam. In 1970, a pro-Western government took over, creating a republic and RKA became the Cambodian Air Force. The following year, Viet Cong guerrillas destroyed all the MiGs on the ground. The country suffered a major civil war from 1970 to 1975, during which most of the air force's remaining aircraft were destroyed and in 1975 the country became isolated as international opinion turned against the Khmer Rouge regime. This was overthrown in 1979 by invading Vietnamese troops, and a new constitution was established in 1981, with the country known as Kampuchea until the restoration of the monarchy.

The 2,000 strong Royal Cambodian Air Force operates a small number of combat aircraft, with a squadron of 14 MiG-21bis and five MiG-21UM, some of which have been upgraded by Israel's IAI. There are also five L-39 armed-trainers and at least three Mi-24 attack helicopters. Two transport squadrons operate one An-26, two HAMC Y-12s, a BN-2, a Cessna 401 and a 402, and a Socata Tobago light aircraft. A helicopter squadron has six Mi-8s, seven Mi-17s and a VIP Mi-8P, and a Eurocopter AS-355. Training uses five Tecnam P-92s and the L-39s. Overall, the size of the armed forces is set to reduce, but it is not clear how this will affect the RCAF, and it is possible that the main cuts will fall upon the army.

Cameroon

POPULATION: 15.4 million

LAND AREA: 183,000 square miles, 475,500 sq km

GDP: $11bn (£7.7bn), per capita $2,500 (£1,748)

DEFENCE EXPENDITURE: $155m (£108m)

SERVICE PERSONNEL: 22,100 active.

CAMEROON AIR FORCE/L'ARMEE DE L'AIR DU CAMEROUN

Originally a French African colony, Cameroon has a small air force with just 300 personnel, although it is active since the country's coastal borders are disputed with its neighbours Nigeria and Equatorial Guinea. A composite squadron operates armed trainers, five Alphajets, six Potez CM-170 Magisters and six ex-South African MB-326 Impala, although two are used for training, four armed SA-342L Gazelle helicopters and two maritime patrol Dornier Do128 Skyservants. Transport aircraft - some of which use civil registrations - include three Lockheed C-130H, a DHC-4 and four DHC-5D Buffalo, another Do128, a VIP Boeing 707 and Gulfstream III, an IAI-201 Arava and two Piper Aztecs, three Bell 206, three Eurocopter SA-318C Alouette II and three SA-319 Alouette III, two AS332L Super Puma and an SA-365 Dauphin.

Canada

POPULATION: 21.75 million

LAND AREA: 3,851,809 square miles, 9,976,185 sq km

GDP: $705bn (£493bn), per capita $24,381 (£17,050)

DEFENCE EXPENDITURE: $8.1bn (£5.7bn)

SERVICE PERSONNEL: 56,800 active, plus 35,400 reserve.

CANADIAN ARMED FORCES/FORCES ARMEES CANADIENNES

Formed: 1967

Officially Canada has a unified defence force, but in recent years many of the changes made when the Canadian Army, Royal Canadian Navy and Royal Canadian Air Force were amalgamated in 1967 have been reversed, so that Air Command is often referred to as the 'Canadian Air Force'.

Canadian military aviation dates from 1914 and the formation of a Canadian Aviation Corps with a single aircraft that was scrapped the following year after accompanying Canadian troops to France on the outbreak of World War I. The British Royal Flying Corps established a flying school in Canada and many Canadians flew with the RFC in Europe. Licence-production of Avro 504 and Curtiss JN-4 aircraft started in 1916 and by 1918 two Canadian squadrons were flying SE5A fighters and DH9A bombers. The Royal Canadian Naval Air Service was formed in 1918, with 12 Curtiss HS2L flying boats and 8 Sopwith seaplanes, but disbanded in 1919.

In 1920, the Canadian Air Force was formed within the Canadian Army with the British 'Imperial Gift' of war-surplus aircraft. Canada received 80 aircraft of Airco, Avro, Bristol, Curtiss and Sopwith manufacture. King George V bestowed the 'Royal' prefix in 1923. Initially, the new service was organized on a non-permanent basis to provide refresher training for personnel with wartime experience, but this proved unsatisfactory and the service was reorganized and re-established on a permanent basis in 1924. Many of the older flying boats were replaced in 1924 by 8 Vickers Viking Vs, and in 1925 the first Canadian-built Vickers Vedettes joined the force. In the early years, as much as two-thirds of the budget was devoted to civil operations, including fire-watching, surveying and communications, but this started to change as the depression took effect and a 20 per cent cut in strength occurred in 1932. New aircraft had been entering service in small numbers, although relatively few were combat types. Some Armstrong-Whitworth Siskin III fighters were delivered, but other aircraft deliveries included Ford Trimotor transports, Armstrong-Whitworth Atlas liaison and AOP aircraft, de Havilland Gipsy Moth, Curtiss-Reid Rambler, Avro Avian, Tutor and Fleet trainers. Just before the cuts, some Fairchild 71-C seaplanes were introduced.

In 1935, an expansion programme began, and in 1936, most civil duties were transferred to the Department of Transport. New squadrons were formed, some with obsolete ex-RAF Westland Wapitis. New aircraft included Vickers Vancouver flying boats, followed by 18 Supermarine Stranraer MR flying boats, and 20 Northrop Delta transports. As the international situation deteriorated, re-equipment and expansion accelerated, with many aircraft being built in Canada, often by subsidiaries of British companies such as Avro and de Havilland. New aircraft included Grumman GE-23 fighters, Bristol Blenheim bombers and Blackburn Shark torpedo-bombers, Noorduyn Norseman transports and North American NA-16-3 trainers. As it expanded, the RCAF was organised into 3 commands, Eastern, Western and Training, and became a separate service in 1938. A British air mission visited Canada, to arrange Canadian production of aircraft for the RAF, and prepare for Canadian participation in the Empire Air Training Scheme, producing pilots for the RAF and other air forces.

The outbreak of war found the RCAF with 270 aircraft in eight regular and 12 reserve squadrons, but only a relatively

ABOVE: *Canadian military aviation started with Canadians in the Royal Flying Corps during World War I, but post-war, aircraft such as this Avro seaplane provided early Canadian military aviation. (CAF)*

small number were modern, including 20 Hawker Hurricane fighters and ten Fairey Battle light bombers. Airspeed Oxford and Avro Anson trainers were other new acquisitions. One RCAF Hurricane squadron served with distinction in the Battle of Britain. By 1941, six RCAF squadrons were based in the UK. During the war, the RCAF served mainly in Europe and in protecting shipping in the North Atlantic, but a number of squadrons operated from its other coastline, over the Pacific, and in 1942, a squadron was posted to Ceylon (Sri Lanka). Allied governments decorated some 8,000 RCAF personnel, and more than 17,000 died in action. Wartime aircraft included Hawker Hurricane and Typhoon, Supermarine Spitfire, Curtiss Kittyhawk and Tomahawk, Bell Airacobra, de Havilland Mosquito, Bristol Beaufighter and North American Mustang fighters, night-fighters and fighter-bombers, Fairey Battle light bombers and Albacore torpedo-bombers, Vickers Wellington, Handley Page Hampden and Halifax, Avro Manchester, Lancaster and Lincoln, Bristol Blenheim and Consolidated Liberator bombers. Northrop Nomad, Lockheed Hudson and Ventura, and Douglas Digby MR aircraft, were joined in this role by Consolidated PBY-5A Catalina amphibians and Saunders-Roe Lerwick flying boats. Transport aircraft included Lockheed Lodestars and Douglas C-47s, while trainers included de Havilland Tiger Moths, Fleet Forts and Finches, Fairchild Cornells, Boeing-Stearman PT-27s and Cessna T-50s.

At the end of the war, RCAF strength was

cut sharply from 90 squadrons to 8 regular and 15 reserve squadrons, although limited expansion was also planned. The RCAF also gained its first peacetime overseas commitment, joining the occupation forces in Germany, then remaining as part of NATO's defences throughout the Cold War with a peak of 12 fighter squadrons based in the former West Germany during the 1950s.

First post-war aircraft were North American B-25 Mitchell light bombers and Beech C-45F Expeditor light transports, plus extra Mustangs and Lodestars. Long-range transport came with Canadair DC-4M North Stars - Canadian-built C-54s. The first jets, de Havilland Vampires, arrived in 1948. These were followed by Avro Canada CF-100 Canuk all-weather fighters and Canadair built F-86E Sabres. Lockheed T-33A trainers were also built in Canada. Canada became a founder member of NATO in 1949.

Post-war, Lancaster and Lincoln heavy-bombers were used for MR, but these were replaced first by Lockheed P2V-7 Neptunes and then by the Canadair Argus, a variant of the Yukon transport developed from the Bristol Britannia. Canadair also built the CC-109 Cosmopolitan, licence-built turbo-prop Convair Metropolitans to replace the C-47s and some 50 Fairchild C-119 Packet transports also entered RCAF service. Two de Havilland Comet 1A jet transports were

ABOVE: *Two Canadian-built McDonnell CF-101 Voodoo interceptors during the mid-1960s when they provided the frontline strength during the last days of the Royal Canadian Air Force. (CAF)*

bought, and later adapted for radar-calibration. Other aircraft included DHC-1 Chipmunk trainers, Vertol H-21A, Sikorsky S-51, S-55 and S-58, and Bell 47G helicopters. De Havilland Canada became the world's largest builder of short-take off aircraft, providing the RCAF and other air forces with DHC-2 Beavers, DHC-3 Otters, DHC-4 Caribou and DHC-5 Buffalo transports during the 1950s and 1960s.

On many occasions, Canada provided transport and communications support for UN forces, starting initially in the Suez Canal zone after the Anglo-French withdrawal after the Suez fiasco of 1956.

Canadian-built Lockheed CF-104 Starfighters replaced the Sabres in Germany during the 1960s, initially with six interceptor and two reconnaissance squadrons, but by the end of the decade, these had been reduced to two squadrons as part of a long series of defence cuts. The CF-100s were replaced by licence-built McDonnell CF-101 Voodoo interceptors during the early 1960s, but were replaced later by up-rated US-built versions. Canadair-built Northrop CF-5As were introduced during 1969 and 1970, although not all of these saw operational service.

Meanwhile, the Royal Canadian Navy had created an air arm, building on its wartime experience when its personnel had manned two escort carriers and four Royal Navy air squadrons. Post-war, the squadrons transferred to the RCN, and a light fleet carrier, HMCS *Warrior* was obtained from the UK in 1946, operating two squadrons, one with Supermarine Seafire fighters and one with Fairey Firefly ASW aircraft. In 1948, HMCS *Warrior* went into reserve and was replaced by HMCS *Magnificent*, on loan from the RN, while Hawker Sea Furies replaced the Seafires and a hundred Grumman TBM-3E Avengers, mostly shore-based, replaced the Fireflies. In 1957, HMCS *Magnificent* was returned to the RN and the former HMS *Powerful* was commissioned as HMCS *Bonaventure*. Until her withdrawal in 1968, this ship operated McDonnell F2H-3 Banshee fighters, Grumman S2F-1 Tracker anti-submarine aircraft, and Sikorsky S-55 helicopters. The S-55s and some Bell HTL-

ABOVE: *A Hawker Sea Fury is towed into position aboard HMCS* Magnificent *in 1953. (CAF)*

RIGHT: *Later, the Royal Canadian Navy showed a mix of British and US influence in its carier-based operations, with British light carriers and US aircraft offering the best combination. This is Canada's last carrier, HMCS Bonaventure, only recently commissioned into the Royal Canadian Navy as she sailed into Malta's Grand Harbour in 1958. Grumman Trackers are ranged on deck. She was withdrawn in 1968. (Harry Wragg)*

Canada

ABOVE: *Despite being part of a unified defence force, in recent years the Canadian Air Force has re-established its identity. A CF-188A Hornet armed with Sidewinder missiles patrols over Bosnia Herzegovinia.*

4s were also operated from some other Canadian naval vessels.

Canadian Army aviation restarted in 1946 with Auster AOP aircraft, replaced in 1954 by Cessna L-19A Bird Dogs, Bell 47 Sioux and Sikorsky S-51 helicopters.

In 1967, Canada's centenary year, the three services were formed into a unified defence force, with ranks modelled on army lines rather than those of the British armed forces. The new CAF was organised into commands, Headquarters, Training, Material, Mobile, Maritime, Air Defence, Air Transport and CAF Germany. Initially, the CAF had a personnel strength of 90,000, but this has since reduced to 59,100. At first, Air Defence Command had 66 McDonnell Douglas F-101 Voodoo interceptors in three squadrons, as well as Bomarc B SAM squadrons. Mobile Command operated 26 Canadair CF-5A fighter-bombers in two squadrons, 50 Bell CUH-1N Iroquois and 74 OH-58A Kiowa helicopters, and 15 DHC-5 Buffalo transports. Air Transport Command operated five Boeing 707-320C tanker/transports, 23 Lockheed C-130Es in two squadrons, a Canadair CC-106 Yukon squadron, and a Cosmopolitan squadron, with a handful of Caribou and C-47 transports. Six squadrons of DHC-3 Otters were in reserve. CAF Germany operated many of these aircraft on rotation, plus 40 CF-104 Starfighters in two squadrons. Maritime Command brought together the RCAF's four MR squadrons, and the RCN squadrons, with a squadron of Trackers and one of Sikorsky CHSS-3 Sea King ASW helicopters. The last mentioned were usually based aboard the nine destroyers then able to operate helicopters. Training Command replaced its Chipmunks with Beech Musketeers, Canadair CL-4 Tutors, CF-5Bs, Beech C-45 Expeditor and Douglas C-47 aircraft and Hiller UH-12 helicopters. Eight DHC-6 Twin Otters were delivered during the early 1970s for rescue duties.

The Argus was replaced in the early 1970s with 30 Lockheed P-3B Orions, known in Canada as the CP-140 Aurora. A purchase of CF-188A Hornets (F/A-18), standardized interceptor and strike force aircraft.

Given the size of the country and its many sparsely populated areas, and the need to patrol two oceans (a third coastline, the Arctic, is frozen for most of the year), transport and MR loom large in Canadian defence planning. The end of the Cold War has meant that its Arctic regions are no longer a 'front line' between East and West, but in common with many other countries, it has meant that defence planning has become more difficult, and threats more varied. Canadian involvement with UN operations has also meant that a long reach has become essential. Continued reductions in defence expenditure have seen decisions deferred, including new medium-lift and SAR helicopters, although these have now been resolved. Upgrades of the Aurora and Hornet have been accompa-

Canada

nied by a steady reduction in force sizes, with the official line being that 'less is better', on the basis that new or upgraded equipment is better than a greater quantity of older equipment.

Once again, Canada is taking a leading role in training for other air forces. In 1997, the Canadian government approved the NATO Flying Training in Canada Programme, NFTC, operated by Bombardier, which now comprises most of the Canadian airframe industry, using at least 26 CT-155 (BAe Hawk 115) jet trainers and 24 CT-156 Beech Harvard II turboprop trainers.

Today, the CAF operates 60 out of its 80 CF-188AB Hornets in five squadrons, using AIM-7 Sparrow and AIM-9 missiles, while the remaining 20 CF-188B are used for training and conversion. Four squadrons operate 18 CP-140 Auroras on maritime-reconnaissance, with this force padded out by three CP-140A Arcturus, essentially the same airframe but with a lower level of avionics for EEZ patrols. Four squadrons operate 32 CC-130E/H and 5 KCC-130H Hercules as transports and tankers, while long-range transport is provided by a squadron of five CC-150 Polaris (A310) airliners capable of conversion into freight or mixed configuration, while one can also be converted for VIP use. At the other end of the scale, the remaining seven Buffalo have been retired as an economy measure, six CC-144 Canadair Challenger jets provide transport and VIP cover. Four CC-138 Twin Otters remain for additional SAR and transport, especially in remote inland regions. The CH-113 Labrador SAR helicopters have now been replaced by 15 CH-149 Cormorant, the Westland Agusta Merlin, which is a contender for replacement of the 30 CH-124 Sea King helicopters in three squadrons providing ASW cover, usually afloat from frigates and destroyers. Ninety-nine CH-146 Griffon (Iroquois) helicopters, of which 30 are armed, provide light transport for ground forces and additional SAR cover. Training is provided on a wide variety of types, including 29 CT-133 (T-33) Silver Stars, for EW training, while at least nine more of these aircraft are in store. Mainstay of the training programme, at least until the NFTC operation is in full swing, are 130 CT-114 Tutors, of which 24 are in the aerobatic team. Training also uses two CT-145 Beech Super King Airs and 13 CH-139 Bell JetRangers, while 12 Slingsby T-67M Firefly basic trainers are operated by a contractor.

BELOW: *The new Canadian Armed Forces SAR helicopter is the Merlin, known in Canada as the Cormorant. (Westland Agusta)*

Cape Verde

POPULATION: 430,000

LAND AREA: 1,580 square miles, 4,040 sq km

GDP: $280m (£196m), per capita $2,800 (£1,958)

DEFENCE EXPENDITURE: $7.6m (£5.3m)

SERVICE PERSONNEL: 1,200 active.

CAPE VERDE AIR FORCE / FORCA AEREA CABOVERDAINE

A former Portuguese colony in the Atlantic off the West Coast of Africa, Cape Verde maintains a small air force, with less than 100 personnel. Three An-26 transports are owned, although serviceability might be low, but the main role is the provision of a Dornier Do228-201 and a Bandeirante on MR in support of the coastguard's three patrol craft.

Central African Republic

POPULATION: 3.7 million

LAND AREA: 234,000 square miles, 616, 429 sq km

GDP: $1.2bn (£839m), per capita $1,400 (£979)

DEFENCE EXPENDITURE: $44m (£30.8m)

SERVICE PERSONNEL: 4,150 active.

CENTRAL AFRICAN SQUADRON/ESCADRILLE CENTRAFRICAINE

Formerly part of French Equatorial Africa, the Central African Republic became independent in 1960, after which it received military aid from France. Originally created as the Central African Air Force, the small air arm initially had the more or less standard French departing gift of a single C-47, three Max Holste 1521M Broussards (Bushrangers) and an Alouette II helicopter. Severe financial difficulties have resulted in reductions since and the reduction in status to an air arm of the army with around 150 personnel. It has a single AS-350 Ecureuil helicopter and a Rallye Guerrier trainer for liaison duties. A VIP Dassault Falcon 20 and a Caravelle are not believed to be operational.

Chad

POPULATION: 7.9 million

LAND AREA: 488,000 square miles, 1,282,050 sq km

GDP: $1.7bn (£1.2bn), per capita $800 (£559)

DEFENCE EXPENDITURE: $48m (£33.6m)

SERVICE PERSONNEL: 30,350 active.

CHAD AIR SQUADRON/ESCADRILLE TCHADIENNE

Formerly the most northerly part of French Equatorial Africa, Chad became independent in 1960, after which it received military aid from France, including the standard French package of a single C-47, three Max Holste 1521M Broussards (Bushrangers) and an Alouette II helicopter. Since 1970, the country has been involved in numerous border disputes with its neighbour Libya, acquiring a number of Pilatus PC-7 Turbo Trainers via France in 1985, of which two are still operational. It managed to capture four SIAI SF260W armed trainers from Libyan forces and put these into service, although probably only two are still operational. Other aircraft operated by the ET, which has 350 personnel, include two Lockheed C-130H/H-30 Hercules and an An-26 transport, a Reims-Cessna FTB337, two Alouette III helicopters and a Pilatus PC-6B Turbo Porter on communications duties.

Chile

POPULATION: 15.4 million

LAND AREA: 286,397 square miles, 738,494 sq km

GDP: $87bn (£60.8bn), per capita $12,800 (£8,951)

DEFENCE EXPENDITURE: $2.9bn (£2bn)

SERVICE PERSONNEL: 87,500 active (30,600 conscripts), plus 50,000 reserves.

CHILEAN AIR FORCE/FUERZA AEREA DE CHILE

Formed: 1930

Chilean military aviation started with the foundation of a flying school in 1913, followed shortly afterwards by the establishment of the Chilean Military Aviation Service. Initially three Bleriot aircraft were operated, followed by three more, as well as three Sanchez-Besa, a Deperdussin and a Voisin. By 1915, there were ten aircraft in two squadrons. In 1917, six Bristol M1Cs were introduced, one of which made the first flight across the Andes in 1918. In 1919, the Naval Aviation Service was established to operate seaplanes, and an aircraft factory was also founded. The Military Aviation Service continued to grow steadily but slowly, buying war-surplus DH4 bombers in 1921.

The two air arms were merged to form the Fuerza Aerea de Chile in 1930. Early FAC aircraft included Vickers Wibault and Curtiss Hawk III fighters; Junkers R-34 and Dornier bombers; Vickers Vixen and Curtiss Fox general-purpose aircraft.

Dornier Wal flying boats and Fairey IIIF seaplanes, well as Loening C2 and Sikorsky S.38 amphibians. There were also de Havilland Gipsy Moth and Avro 504 trainers. Focke-Wulf Fw44, Avro 626 and Nardi FN305 trainers, and Arado Ar95 general-purpose aircraft later joined these. An American aviation mission reorganized the FAC in 1941, providing Curtiss P-40 and Republic F-47 fighters; Douglas B-24 and North American B-25 bombers; Sikorsky OS2U-3 seaplanes and Consolidated PBY-5A amphibians; as well as Fairchild PT-19, North American T-6 and Vultee BT-13 trainers. There were few new aircraft during World War II, but afterwards additional F-47s were delivered. During the 1950s, Beech T-45 Mentor and D18S trainers, and Twin Bonanza communications aircraft, entered service. Many of the aircraft bought during this period were transport aircraft, including DHC-2 Beaver and DHC-6 Twin Otters. The first jet aircraft, de Havilland Vampire T55 trainers, did not arrive until the end of the decade, and were joined by Douglas B-26 Invader bombers and 50 Chilean-built Chincol trainers.

Chile had become a founder member of the Organisation of American States in 1948, but never became entirely dependent on US aircraft.

Jet fighter-bombers were delivered during the 1960s. These included 20 Lockheed F-80C Shooting Stars, replaced in 1969 and 1970 by Hawker Hunter FGA9s, with some Hunter T7 trainers. A number of second-line aircraft were also obtained, including 14 Grumman HU-16B Albatross maritime-reconnaissance amphibians, 20 Beech C-45 and 25 Douglas C-47 and four DC-6 transports, as well as Bell 47 and UH-1D Iroquois, Hiller UH-12E and Sikorsky H-19 helicopters. A change of government in 1970 made Chile's relationships with Western countries difficult, which continued after a military dictatorship took over before an eventual return to democracy. During this period, the FAC managed to update itself with Dassault Mirage 5 and Northrop F-5 fighter-bombers, as well as a few Lockheed Hercules. The FAC joined the Coalition Forces fighting for the liberation of Kuwait during the Gulf War of 1990-91.

Today, the FAC has 12,500 personnel, its strength having increased by some 50 per cent over the past 30 years. Economic difficulties have delayed modernization, although the main aircraft types have been upgraded. Mirage 5s have been upgraded locally as the Pantera, or Mirage 50 interceptor/reconnaissance aircraft, of which there are 15. The Hunters were replaced in 1995-96 by ex-Belgian Air Force Mirage 5s upgraded before delivery and operated as the Elkan, with 20 5BA and five 5BD. These two types each equip a squadron, while a third uses 13 F-5E and three F-5F, upgraded in Israel. A fighter competion was eventually won by the F-16C/D in late 2000, but the planned order was cut from 24 aircraft to ten, further delaying negotiations, although a second batch of aircraft may be ordered around 2006. A single Israeli-modified Phalcon AEW aircraft based on the Boeing 707 supports these aircraft. Additional combat support comes from two squadrons with 24 A-37B and 21 A-36 armed trainers, while reconnaissance support depends on a King Air A-100 and two Learjet 35As. Seven Beech 99As share maritime-reconnaissance, transport and ELINT duties. Tanker and transport duties are shared between four Boeing 707-320s, while there are three C-130B and two C-130H Hercules. Smaller transports

LEFT & BELOW: *Early Chilean naval aviation used a variety of aircraft, including the Dornier Wal flying boat. (CANAC)*

BELOW: *The Embraer EMB-111 augments the maritime-reconnaissance work of the longer-range Lockheed P-3A Orions.*

Chile

ABOVE: *The Chilean Navy deploys Eurocopter AS532C Cougar helicopters aboard its frigates and destroyers for both anti-submarine and anti-surface vessel duties.*

ABOVE: *Chilean naval markings differ from those of the FAC through having an anchor on the tailplane.*

include 14 DHC-6 Twin Otters, used mainly in the Antarctic, as well as four C212 Aviocar. A Gulfstream III and a Beech 200 provide VIP transport.

The helicopter force is in a state of change, with a single S-70 delivered in 1998 as a replacement for UH-1Hs, of which seven remain in service after five were left behind during the Gulf War, having been found unsuitable. Up to 12 Bell 412s are planned instead. There are also eight Bo-105s and five SA315B Lamas, the latter used for SAR. Training uses 12 PA-28s, 19 T-35A/B Pillans, some remaining T-37Bs, as well as six Extra 300s, also used for an aerobatic team, and four Bell 206 helicopters. The main missiles are AS-11 and AS-12 for ASM duties, AIM-9B Sidewinder, Shafrir and Python III AAM, and some MATRA Mistral and Mygalle SAM launchers.

CHILEAN NAVAL AVIATION/COMMANDANCIA DE AVIACION NAVAL DE LA ARMADA DE CHILE

Chilean naval aviation was absorbed into the new air force in 1930, but reappeared post-war, initially with Bell 47G Sioux helicopters and Beech T-34 Mentors used on communications duties. Added impetus was given to naval aviation with the arrival of two new Leander-class frigates in 1973, with Westland Wasp helicopters. In recent years, the significance of naval aviation has grown with two additional Leanders acquired from the Royal Navy and the arrival of four ex-British County-class guided missile destroyers. A long-range MR and SAR capability has emerged with the addition of P-3A Orions, now being upgraded.

Today, the CAN has some 600 personnel out of the naval total of 24,000. It has an MR squadron operating six P-3As, six EMB-111 and another three used on communications duties. Armed helicopters included six AS-532 Cougars deployed at sea, two equipped with torpedoes and four with Exocet anti-shipping missiles. Additional SAR cover comes from seven BO105S and six Bell 206 JetRanger helicopters which are also used for Antarctic operations. Training uses a squadron with ten Pilatus PC-7s and another with ten Cessna 337 Skymasters.

BELOW: *Cessna O-2As are normally found in the COIN role, but the Chilean Navy also uses them for training.*

CHILEAN ARMY AVIATION/COMMANDO DE AVIACION DEL EJERCITO DE CHILE

The Chilean Army uses aircraft primarily for communications and support duties, with just five MD530F armed helicopters. Transport is provided by six C212 Aviocar, five CN235 and four DHC-6 Twin Otter aircraft, as well as twelve SA330F/L Puma and three AS332 Super Puma helicopters. Eleven SA315B Lama (Alouette II) and two Bell 206B helicopters provide communications, with two Cessna 337 Skymaster, eight Piper PA-28 Dakota and three Cessna 208 Caravan. Training is provided on 15 Enstrom 280FX helicopters and 16 Cessna R172 Hawk.

China

POPULATION: 1,293 million

LAND AREA: 3,768,000 square miles, 9,595,961 sq km

GDP: $794bn (£555bn), per capita $4,300 (£3,395)

DEFENCE EXPENDITURE: $160bn (£112bn)

SERVICE PERSONNEL: 2,310,000 active, plus up to 600,000 active reserves.

PEOPLE'S LIBERATION ARMY AIR FORCE

Formed: 1949

For most of the twentieth century China was split by warring factions, with a formal division occurring in 1949, when Communist forces gained control of the mainland and established the People's Republic of China, leaving the Nationalist Chinese to the island of Formosa.

Before the overthrow of the Emperor in 1911, and the declaration of a republic, one or two aircraft had already made an appearance in China. Nevertheless, it was left to a number of localized air arms, sponsored by opposing warlords, to introduce military aviation to the country. Yuan Shih-k'ai at Shanghai introduced 12 Caudron GIII and GIV aircraft in 1914, while in northern China, Tuan Chi-Jui formed a small air arm and joined the Allies in 1917. Official Chinese Army and Navy air arms also came into existence around this time.

British and American assistance brought the Chinese Aviation Service into existence in 1919, merging the Army and Navy air arms. In 1920, the CAS received 60 Avro 504K trainers, followed by 40 Handley-Page O/400 bombers and Vickers Vimy transports. Morane-Saulnier trainers followed, with Breguet Br14B-2 and Ansaldo A300 bombers. These aircraft were soon operating against insurgents throughout China. A Manchurian Air Arm also operated O/400s and Br14B-2 bombers, as well as Potez XXV bombers, and Caudron C59 and Schreck FBA flying boats.

Britain refused a request for assistance by the Central Government after Japan invaded Manchuria in 1931. An American aviation mission supplied Boeing 218 and Curtiss Hawk fighters and a six squadron Central Government Air Force was operational by 1934. The USA also supplied Northrop Gamma, Vought V65 Corsair and Douglas O-38 bomber and attack aircraft, as well as Fleet trainers. In 1935, an Italian mission brought Fiat CR30 and CR32, and Breda Ba27 fighters; Fiat BR3, Caproni Ca111, Savoia-Marchetti SM79 and Heinkel He111A bombers; and Breda Ba25 trainers. Eventually, there were also some Blackburn Lincock fighters and Vickers Vespa VI reconnaissance-bombers from the UK. Meanwhile, the provincial air arms survived, including the Kwantung Air Force, mainly operating Russian aircraft, and the Kwangai Air Force, mainly operating British aircraft, and which allied itself with the Central Government in 1936.

Renewed attacks by the Japanese started in 1937, with the invaders having the advantage in numbers, equipment and training. USSR support for the CGAF included Polikarpov I-15 and I-16 fighters and Tupolev SB-2 bombers, with Russian pilots manning six squadrons. The Chinese gained air supremacy until the Japanese moved their latest aircraft into the area, occupying all of central China as well as Manchuria, and some other areas as well. As Japanese authority was consolidated, some of the occupied provinces were allowed to operate air arms under Japanese control using Japanese equipment. In Manchuria, colonized by Japan as Manchouko, the Manchoukuoan Air Force operated Nakajima fighters and Kawasaki

BELOW: *Avro 504 trainer. (Jane's)*

China

ABOVE: *The Chinese built Yak 18 (Jane's)*

bombers, and many of its personnel were absorbed into the Japanese forces. After the fall of Nanking, the Cochin Chinese Air Force operated Nakajima Ki34 transports and Tachikawa Ki9 trainers.

Despite these setbacks, the CGAF remained.in the air, and was supported by an American Volunteer Corps, formed in 1941 with 90 Curtiss P-40B fighters and known as the Flying Tigers (post-war, veterans formed the airline of the same name). After the United States entered World War II, this force was absorbed into the USAAF and the Central Government received renewed US aid. US aircraft supplied to the CGAF included 129 Vultee 48C, 377 Curtiss P-40B/E, 108 Republic P-43, 15 Lockheed P-38 Lightning and 50 North American F-51 Mustang fighters; 130 North American B-25 Mitchell bombers; 28 Curtiss C-46 Commando, 80 Douglas C-47 and some C-54 transports; 20 North American AT-6, eight Beech AT-7, 15 Cessna AT-17, 150 Boeing-Stearman PT-17 Kaydet, 135 Fairchild PT-19, 70 Ryan PT-27 and 30 Vultee BT-13 trainers.

Post-war, the CGAF was reorganized and restyled as the Chinese Air Force. Surviving wartime aircraft were augmented by additional Mustangs, Lightnings and Mitchells, plus Republic F-47 Thunderbolt fighter-bombers and Consolidated B-24 Liberator heavy bombers, and 250 ex-RCAF de Havilland Mosquito fighter-bombers were purchased. Nevertheless, the rapid advance of Communist forces forced the nationalists to withdraw to the offshore island of Formosa in 1949.

BELOW: *One of the People's Liberation Army Air Forces F-7M Airguards comes in to land. (Jane's)*

Soviet aid for the Chinese Communists had not started until 1945 as the USSR had maintained a policy of neutrality against Japan until the last days of the war. For much of the time, the Air Force of the People's Liberation Army used captured CGAF and then CAF aircraft, including Mustangs, Mitchells, C-46s and C-47s. Soviet aid provided Yakovlev Yak-9 and Lavochin La-11 fighters, while the USSR established a flying school for the Air Force of The People's Liberation Army in Manchuria in 1948.

Pre-occupied with consolidating its victory, the new regime did not play a part in the Korea War at the outset in 1950, but China's first jets, MiG-15 fighters supplied in 1951, soon saw action over Korea.

The immediate post-Korean War period saw the two major Communist powers as allies, with the USSR supplying technology and equipment. Soviet aid included additional MiG-15s, and then its successors, the MiG-17, MiG-19 and MiG-21; Tupolev Tu-2 and Tu-4 (a B-29 copy) bombers, and Ilyushin Il-28 jet bombers; Lisunov Li-2 (C-47 copy), Antonov An-2, and Ilyushin Il-12 and Il-14 transports; Mil Mi-1 and Mi-4 helicopters; and Yakovlev Yak-12 and Yak-18, and MiG-15UTI and Il-28U trainers. China started to build these aircraft in addition to Soviet supplies, a policy that accelerated as relations between the two powers cooled rapidly over ideological differences and competition to take the lead in the Communist world. It could also be said that China's ability to produce replacements made the rift possible. Under Chinese manufacture, the MiG-15 series became the Shenyang F-2, the -17 the F-4, the -19 the F-6 and the -21 the F-7. Copying Soviet practice, aircraft operated by the Civil Air Bureau, or CAAC, were also available for military service. Later, China also started to produce transport aircraft and helicopters, and more recently has developed its own designs across the whole range of military aircraft.

Relatively little activity was undertaken by the AFPLA during the Vietnam War, with North Vietnamese forces relying largely on Soviet assistance, due as much to Vietnamese suspicions of China as rivalry between China and the USSR. Air power could contribute little, other than reconnaissance, to the Chinese annexation of Tibet due to the shortage of landing fields in the inhospitable terrain. Nor did air power feature greatly in the border clashes between Russia and China during the 1960s and 1970s. The break-up of the Soviet Union and Russia's abandonment of Communism has, paradoxically, led to an improvement in relations, with China once again looking to Russia, and also the Ukraine, for equipment. A new source of equipment and expertise, especially in upgrading equipment, is now Israel. China's own economy has also shown considerable growth in recent years, making the cost of maintaining the world's largest military air arm affordable. Nevertheless, despite its 420,000 personnel and 4,000 aircraft, the PLAAF, as it is now known, has severe shortcomings, with much of its equipment obsolete and serviceability limited. Flying hours are supposed to average around 100 per annum for fast jet pilots, but this is likely to be so heavily biased in favour of those in 'frontline' units, those facing Taiwan, that the cost of fuel may well mean that elsewhere hours are far short of those required for combat readiness. In 2002, it was announced that defence expenditure is to be increased, which may see accelerated modernization, although personnel numbers are being reduced.

ABOVE: *The close air support fighter - NAMC Q-5 'Fantan'. (Jane's)*

The PLAAF is divided into seven military air regions, administered from four air army headquarters. There are 33 air divisions, with 27 operating fighters and four bombers, with two transport divisions. Each air division usually has three air regiments, with up to four squadrons, usually of 10-15 aircraft, in each. The main concentration of strength lies in those air regions in the south-east.

The PLAAF fighter and interceptor strength is being boosted by deliveries from Russia of at least 50 Sukhoi Su-27P, while 200 licence-built examples are in course of delivery from Shenyang. There are also 400 J-8/J-811 Finback Chinese-designed interceptors; up to 600 J-7s (MiG-21) of different marks. In the FGA role, 40 Su-30MKKs have been delivered, while there are 500 Nanchang Q-5 (MiG-19) Fantan, and more than 1,500 Shenyang J-6A/B/C developments of the MiG-19 fighters in 60 air regiments. There are still up to 500 J-5/5A (MiG-17). The three bomber air divisions have 120 Xian H-6E/F (Tu-16), some at least of which are nuclear capable while 30 have been modified to carry the YJ-6 ASuV missile. Reconnaissance and ELINT is provided by a force of 40 Harbin HZ-5/HJ-5 (developed from the Il-28), and 100 Chinese-designed JZ-6 (MiG-19) and JZ-7, as well as two ELINT Tu-154. A Beriev A-50 development of the Il-76 operates as an AEW test bed, while a further 18 Il-76 operate in the transport role. Other transport aircraft include 15 Tu-154M, ten Il-18, 300 Harbin Y-5 (An-2), 30 Y-7 (An-24), 15 Y-11, eight Y-12, 12 Y-14 (An-26), six VIP Boeing 737-200 and two VIP CL-601 Challenger. Eight An-30 are used for survey work with two Y-12. The helicopter force includes 350 Harbin Z-5 and Z-6 (both Mi-4). Training uses more than 200 K-8 and 500 Chengdu JJ-5, as well as 50 JJ-7. There are 1,500 Nanchang CJ-6 trainers, 350 Shenyang J-4 (MiG-17F) and 150 Shenyang JJ-6.

China

AVIATION OF THE PEOPLE'S NAVY

The People's Navy aviation activities are primarily land-based, apart from a small number of helicopters aboard its more modern destroyers and landing ships. Nevertheless, this may change in the future. In 1998, a Chinese business concern acquired the ex-Soviet aircraft carrier *Varyag* from the Ukraine, ostensibly as an entertainment venue, but the vessel must have provided valuable insights to the APN, given Chinese ambitions for superpower status. As with army aviation, equipment is a mixture of Russian and French, including licence-built French helicopters, although at present, helicopters are a small part of this force. Roughly 25,000 personnel are engaged in naval aviation out of a total 220,000.

Substantial shore-based forces include 300 Shenyang J-6/JJ-6 (MiG-19), 100 Chengdu J-7I/II/III (MiG-21), 100 Shenyang J-5A (MiG-17PF), also used for training, and 100 Nanchang Q-5 Fantan (MiG-19) fighters and strike aircraft. Torpedo bombers include more than 150 Harbin H-5 (Il-28) and 20 Harbin H-6 (Tu-16), with another ten H-6-III tankers. A dozen Beriev Be-6 flying boats provide long-range MR, with Y-8s (An-12). ASW helicopters include 20 Kamov Ka-28 Helix, 25 Z-9 (SA-365 Dauphin), 20 Z-8 (SA321 Super Frelon) and four Harbin SH-5, with the last two types also having a transport role, with at least six Harbin Z-5 (Mi-4). Fixed-wing transport aircraft include ten Y-7 (An-24) and 40 Y-5 (An-2).

PEOPLE'S LIBERATION ARMY AVIATION CORPS

Formed: 1988

The PLAAC was formed in 1988 by the transfer of utility and transport helicopters from the PLAAF. There is as yet only a limited combat capability available with the small number of Gazelles. The PLAAC seems to be looking to the west for its requirements, even building the Eurocopter SA-365 Dauphin under licence.

The PLAAC has 300 Harbin Z-5 (Mi-4), 35 Mi-8 and 20 Mi-17. Eight SA342L-1 Gazelle helicopters are used for training, but can also use HOT anti-tank missiles. There are 28 Sikorsky S-70C2 Black Hawk transport helicopters, plus 25 Harbin Z-9 (SA-365 Dauphin) and 20 Changhe Z-11 (AS352) utility helicopters, while six AS332 Super Puma are used for VIP transport.

Colombia

POPULATION: 43.8 million

LAND AREA: 462,000 square miles, 1,139,592 sq km

GDP: $81bn (£57bn), per capita $5,400 (£3,776)

DEFENCE EXPENDITURE: $2.0bn (£1.4bn)

SERVICE PERSONNEL: 158,000 active (45 per cent conscripts), plus 60,700 reserves.

COLOMBIAN AIR FORCE/FUERZA AEREA COLOMBIANA

Formed: 1943

Military aviation did not start in Colombia until after World War I, with a flying school founded in 1922 with a single Caudron GIIIA biplane, while a naval flying boat flight was also formed. At first, both air arms remained small. The Navy operated a handful of Seversky SEV-3MWW amphibians, while the Army operated Curtiss Hawk fighters and Falcon reconnaissance aircraft; Bellanca 77-140 bombers; Curtiss-Wright CT-32 Condor, Ford 4-AT-E and Junkers W33 and W34 transports. North American NA-16-3, Consolidated PT-11 and Curtiss Fledgling were used for training.

In 1943, an American mission proposed a merger to form an autonomous air force, the Fuerza Aerea Colombiana. Aircraft were in short supply during the war years, but in 1948, Colombia joined the Organisation of American States and American military aid followed. Initially, this included a squadron each of Republic F-47 Thunderbolt fighter-bombers; Boeing B-17G Fortress and North American B-25 Mitchell bombers; Convair PBY-5A Catalina amphibians and Douglas C-47 transports; Boeing-Stearman PT-17 Kaydet and North American T-6 Texan trainers. These were followed by DHC-2 Beaver transports in 1951. In 1954, the first jets, six Lockheed T-33A trainers arrived, plus Beech T-34 Mentors to replace the other earlier trainers. Operational jets, Canadair CL-13 B Sabre Mk6 fighters, arrived in 1956. Many of these aircraft remained in service for some considerable time, with deliveries of additional North American Sabres and Douglas B-26 Invader bombers. Growing emphasis on COIN operations saw Cessna T-37C and T-41D armed-trainers enter service. Douglas C-54 and Lockheed Hercules were obtained, with Otters for the utility role, and Aero Commander 680 for communications. Helicopters included 20 Bell 47G/D/J Sioux, and smaller numbers of UH-1D Iroquois, and Kaman HH-43B Huskie, as well as 12 Hughes OH-6A and six 269s, with four Hiller UH-23s for training.

During the 1980s and 1990s, the FAC gained stronger frontline combat capability with the arrival of Mirage 5 and then IAI Kfir fighter-bombers, recently upgraded to Mirage 50 and Kfir C7 standard. COIN remained important, often supported by US aid, to assist in operations against drug producers and traffickers. FAC personnel strength has grown over the past 30 years

Colombia

from 6,000 to around 7,000. Many of its aircraft seem to overlap in capabilities and functions with those under army control.

The FAC is structured in four commands, for air combat, tactical air support, military air transport and training. Air Combat Command has two squadrons, one with 13 Mirage 50 and one with 13 Kfir. Tactical Air Support has five AC-47/AC-47T, three IA-58A, 22 A-37B and six AT-27, plus a Fairchild C-26 whose conversion was funded by the US government. Reconnaissance aircraft include eight Schweizer SA2-37A and five SA2-37B, 15 OV-10A and three Fairchild C-26 for anti-drug operations. Helicopters are vitally important in this role, with 12 Bell 205, five 212, two 412, and two UH-1B helicopters, plus 13 Sikorsky UH-60A and seven S-70, ten Mil Mi-17, 11 MD-500ME, two MD-500D and three MD-530F. Military Air Transport Command has a Boeing 707-320 and two 727, one Fokker F-28, 14 C-130B and two C-130H Hercules, a C-117, two CASA C-212, three CN-235, two Douglas C-47 and two Embraer Bandeirante, plus a further 17 UH-1H helicopters. It has a number of Piper, Beech and Cessna light aircraft - some seized from drug traffickers - for communications. Training Command has 14 T-27 Tucano, three T-34M, 13 T-37 and eight T-41, with two UH-1B, four UH-1H and 12 Enstrom F28F helicopters. The FAC will share in a further 25 UH-1Hs being provided by the USA for the air force, police and army for anti-drug operations. AIM-9 Sidewinder and R-530 AAM are deployed.

COLOMBIAN NAVY AIR ARM/AVIACION ARMADA DE COLOMBIA

The Colombian Navy uses aircraft for communications and liaison, including five Piper Navajo and four Cherokee, a Beech King Air, and three Aero Commander 500, while helicopters include two AS555SN Fennec and four BO105CB, all based ashore.

ABOVE: *Both the Colombian Air Force, FAC, whose aircraft is seen here, and the Colombian Army use Sikorsky UH-60 Black Hawks, many of which have been provided as US aid to help combat drug trafficking. (Sikorsky)*

COLOMBIAN ARMY/EJERCITO DE COLOMBIANA

The Colombian Army has a relatively small aviation element that is receiving rapid expansion as the United States has provided substantial numbers of UH-60 and UH-1H helicopters to the army and police for the campaign against drugs. The Colombian government earlier provided five UH-60s and six OH-6A, giving a total of around 40 UH-60L Black Hawks, 33 UH-1Ns, 10 Mi-17s, and an assortment of light aircraft, mainly seized from drug traffickers, including four Piper Seneca and four Cherokee. Training uses five UTVA-75s.

Comoros Islands

POPULATION: 400,000

LAND AREA: 838 square miles, 2,170 sq km

COMOROS DEFENCE FORCE

Since independence from France, a small defence force has been established in this small group of islands off the coast of East Africa. An AS350B Ecureil helicopter is operated.

Congo

POPULATION: 3 million

LAND AREA: 132,000 square miles, 341,880 sq km

GDP: $2.9bn (£2bn), per capita $1,900 (£1,329)

DEFENCE EXPENDITURE: $73m (£51m)

SERVICE PERSONNEL: 10,000 active.

CONGO AIR FORCE/FORCE AERIENNE CONGOLAISE

Formed: 1958

The former French Congo, the southernmost part of French Equatorial Africa, reached independence in 1958, and was provided with the standard military aid package of a Douglas C-47, three Max Holste 1521M Broussards (Bushrangers) and a Sud Alouette II helicopter. It soon became the People's Democratic Republic of the Congo, and fell within the Soviet sphere of influence, receiving MiG-17s and MiG-15UTI trainers. These were later joined by 16 MiG-21s in 1986, plus An-24 transports.

Today, it has 1,200 personnel and serviceability of its aircraft is believed to be low due to a shortage of spares and the difficulty of keeping aircraft which are little used in the open in a hot and very humid climate. Twelve of the MiG-21s are believed to have survived, with eight MiG-17Fs, but are unlikely to be operational. VIP transport is provided by a Boeing 727, while there are also five An-24s, one An-26 and an N-2501. Helicopters include two Mi-8s, an SA365 Dauphin, two Alouette II and two Alouette III.

Democratic Republic of Congo

POPULATION: 53.3 million

LAND AREA: 895,000 square miles, 2,345,457 sq km

GDP: $4.7bn (£3.3bn), per capita $400 (£279)

DEFENCE EXPENDITURE: $400m (£279m)

SERVICE PERSONNEL: 81,400 active.

CONGO AIR FORCE

Formed: 1961

The former Belgian Congo became independent in 1961 and almost immediately descended into civil war as the province of Katanga, rich in mineral resources, attempted to break away. Known initially as the Force Aerienne Congolaise, the air force was formed with light aircraft to counter aircraft operated by the rebel forces. After the civil war ended, the government and rebel air arms were merged, and the latter's foreign pilots provided an initial pool of experience. An Italian air mission soon replaced Belgian advisers and instructors. In 1969, 17 MB326GB armed-trainers were delivered, augmenting up to 20 North American T-6 Harvard and T-28 Trojan armed-trainers. Douglas C-47, DC-4, and DHC-4 Caribou transports were also acquired, as well as light aircraft including de Havilland Dove and Beech 18. Six Alouette III helicopters were joined in 1971 by the first of nine SA330 Pumas. Training took place on Piaggio 148 trainers.

The country renamed itself Zaire and acquired more potent aircraft in the form of Dassault Mirage Vs, as well as Lockheed C-130Hs. Equipment purchases were relatively easy given the country's wealth. Nevertheless, civil war once again struck, with the overthrow of the central government in 1997. The country now has foreign forces fighting both on the side of the new government, with troops from Angola, Namibia and up to 12,000 from Zimbabwe, and on the side of the rebels, including Angolan UNITA rebels, troops from Burundi, Rwanda and Uganda. Little use seems to have been made of the air force in this internal conflict. There are reported to be 1,500 personnel, but it is suspected that the spares and support organisation has broken down, leaving many aircraft unserviceable. Ten Su-25 are believed to have been ordered, as well as 22 Mi-24 attack helicopters, but the mainstay remains six surviving Mirage 5s and perhaps as many as 15 of the MB326GB/K, officially serving both the attack and training roles. There are two Hercules, a VIP 727, three DHC-5D Buffalo, eight C-47s and a CASA C212-200 Aviocar, as well as nine Eurocopter SA330 Puma, an AS332 Super Puma and seven Alouette III helicopters. Any training takes place on five SIAI SF260MZ and twelve Reims Cessna FRA150.

Costa Rica

POPULATION: 4.1 million

LAND AREA: 19,653 square miles, 50,909 sq km

GDP: $11.1bn (£7.8bn), per capita $7,000 (£4,895)

DEFENCE EXPENDITURE: $86m (£60m)

SERVICE PERSONNEL: 8,400 paramilitary active.

CIVIL GUARD AIR SURVEILLANCE UNIT/GUARDIA CIVIL SECCION AEREA

Although Costa Rica maintained a small air force until the end of a civil war in 1948, it has since depended on paramilitary security forces, with a small air surveillance unit with 300 personnel and unarmed observation and transport aircraft. The largest aircraft are two DHC-4 Caribou transports and a Mil Mi-17 helicopter. Light aircraft include a Piper Navajo and a Seneca, four Cessna U206G Stationairs and an O-2 Bird Dog AOP aircraft. There are two MD500E helicopters.

Cote d'Ivoire

POPULATION: 15 million

LAND AREA: 124,510 square miles, 322,481 sq km

GDP: $14.3bn (£1bn), per capita $2,000 (£1,398)

DEFENCE EXPENDITURE: $134m (£100m)

SERVICE PERSONNEL: 13,900 active, plus 12,000 reserves.

IVORY COAST AIR FORCE/FORCE AERIENNE DE COTE D'IVOIRE

Formed: 1960

Part of the former French West Africa, the Ivory Coast became independent in 1960, receiving the standard independence military aid package of a C-47, three Max Holste 1521M Broussards and an Alouette II helicopter. It soon embarked on a limited expansion programme, acquiring another two C-47s and two more Broussards, an additional Alouette II and three Alouette III. It received a gift of an Aero Commander 500 from the US. Two SA330 Puma helicopters were bought later, followed by a Fokker F-27M Troopship and three Cessna 337s in 1971.

In recent years, the FACI has acquired a limited combat ability with five Alphajet armed-trainers. Biggest expansion of the FACI, which has 700 personnel, has been in VIP transport, for which it has a Fokker 100, Gulfstream III and IV, a Puma and a Dauphin. In the communications role is a Cessna 421 and a 401, and a Beech Super King Air, while four F33C Bonanza are used for training, with two Reims Cessna 150H.

Croatia

POPULATION: 4.4 million

LAND AREA: 21,824 square miles, 56,524 sq km

GDP: $19.4bn (£13.6bn), per capita $7,192 (£5,029)

DEFENCE EXPENDITURE: $520m (£363.4m)

SERVICE PERSONNEL: 58,300 active, plus 140,000 reserves.

CROATIAN AIR FORCE

Formed: 1991

Formerly a Yugoslav constituent republic, Croatia declared its independence in 1991, initially using aircraft that had been part of the Yugoslav Air Force. It has 4,600 personnel and flying hours are low, at an average of 50 annually. Financial problems have made re-equipment difficult, with an upgrade of the MIG-21s in Israel rejected as too expensive, and an offer of 14 ex-USAF F-16A/Bs rejected for the same reason. Mainstays of the force remain two squadrons with 24 MiG-21bis/UM, augmented by Soko J1 Jastreb and J-20 Kragul ground attack aircraft. There are at least ten Mi-24 Hind attack helicopters. Transport uses four An-2s, two An-26s, 2 An-32, five CN235, five UTVA and two Do28. There are 16 Mi-8/-17 transport helicopters, as well as one UH-1H, nine Bell 206B and two MD-500s. There are two CL215 and one CL415 fire bombing amphibians. Many light aircraft are used in communications and training roles, as well as a Sabreliner 75A and a CL601 Challenger for VIP use. Training uses some MiG-21U, but is mainly based on 17 PC-9s and a few surviving G-2 Galeb.

Cuba

POPULATION: 11.3 million

LAND AREA: 44,206 square miles, 114,494 sq km

GDP: $16.8bn (£11.75bn), per capita $2,600 (£1,818)

DEFENCE BUDGET: $750m (£524m)

SERVICE PERSONNEL: 46,000 active, plus 39,000 ready reserves.

ABOVE: *Early Cuban military aviation included this Vought QO-1 scout and observation aircraft for the 'Air Corps-Cuban Army' in 1924. (Vought)*

ANTI-AIRCRAFT DEFENCE & REVOLUTIONARY AIR FORCE/DEFENSE ANTI-AEREA Y FUERZA AEREA REVOLUCIONARIA

Formed: 1959

A Cuban army air arm was planned in 1915, but it was not until 1917 that the first aircraft, six Curtiss JN-4D trainers, arrived. There do not seem to have been any further acquisitions until 1923, when six DH4B bombers and a number of Vought VO-2 AOP aircraft entered service with the Aviation Corps. Most of these aircraft were lost in a hurricane in 1926, leaving the Aviation Corps to rebuild. In 1934, it was reorganized into Army Aviation and Naval Aviation. By this time it was operating Waco D-7 general-purpose biplanes, Bellanca Aircruiser and Howard transports, with Stearman A73 and Curtiss-Wright 19R-2 trainers.

Cuba offered bases to the Allies during World War II, and in return received Grumman G21 Goose amphibians, Aeronca L3 AOP aircraft and trainers, including Boeing-Stearman PT-13 and PT-17 Kaydet, and North American T-6 trainers. When Cuba became a member of the Organisation of American States in 1948, further US military aid followed, including North American F-51D Mustang fighters and B-25J Mitchell bombers, and Douglas C-47 transports. In 1955, a reorganisation resulted in the entire force once again passing into army control, becoming the Cuban Army Air Force, Fuerza Aerea Ejercito de Cuba. That same year, the first jets, Lockheed T-33A trainers, entered service, followed later by ex-Royal Navy Hawker Sea Furies. Three DHC-2 Beaver transports entered service in 1957, followed by two Westland Whirlwind (S-55) helicopters in 1958.

From 1956 to 1959, Cuba endured a revolution involving fierce fighting with Communist forces under Fidel Castro, whose victory early in 1959 saw both a change of government and a political reorientation towards the USSR. The title of Fuerza Aerea Revolucionaria was adopted and Soviet military advisers flowed into Cuba, although the US prevented the island from becoming a base for ballistic missiles. Less successful was a US-backed attempt at counter-revolution by Cuban exiles, with the invaders being attacked by FAR fighters.

Soviet equipment and cheap fuel meant a rapid increase in the size, although initially the equipment provided was obsolescent, such as MiG-15 fighter-bombers. Later MiG-17 and MiG-19s arrived followed by MiG-21 interceptors during the early 1970s. By this time, the FAR had 12,000 personnel, operating 50 MiG-21s, 40 MiG-19s, 75 MiG-17s and 20 MiG-15s. Transport aircraft were also provided, the total of 50 including Antonov An-2 and An-24s, as well as Ilyushin Il-14s. Helicopters included 25 Mil Mi-4 and 30 Mi-1, with 30 MiG-15UTI and Zlin 226 trainers.

New aircraft during the 1980s included MiG-23 and, later MiG-29 attack and fighter aircraft, as well as Mi-24 Hind attack helicopters. The collapse of the USSR led to the withdrawal of advisers and an end to the supply of cheap aircraft and, more important, cheap fuel. It created a shortage of spares, aggravated by Cuba's weak economy.

Today, the FAR has 8,000 personnel, with an average of just 50 flying hours annually. Defence analysts differ over the number of aircraft that are no longer airworthy, but it is at least 25 per cent and rising. Four fighter/ground attack squadrons operate a total of 30 MiG-21F and 50 MiG-21bis, as well as 20 MiG-29/UB and 69 MiG-23M/BN/U. ASW is provided by 14 Mi-14, although only five are believed to be operational. There are also 36 Mi-8 and 14 Mi-17 transport helicopters and 12 Mi-24D attack helicopters. Transport aircraft include 20 An-2s, four An-24s and 20 An-26, as well as two An-32s and two VIP Yak-40. One An-30 is used for survey work. Training aircraft include 18 Mi-17F, 25 L-39 Albatros, 20 Zlin 326 and ten Zlin 142. Missiles include AS-7 ASM, AA-2, AA-7, AA-8, AA-10 and AA-11, while there are 13 sites with SA-2 and SA-3 SAM.

Cyprus

POPULATION: 800,000

LAND AREA: 3,572 square miles, 9,251 sq km

GDP: $9.6bn (£6.7bn), per capita $15,409 (£10,775)

DEFENCE EXPENDITURE: $462m (£323m)

SERVICE PERSONNEL: 10,000 (87% conscript) active, plus 60,000 reserves.

CYPRUS AIR FORCE

Formed: 1973

An air wing was created in the Cyprus National Guard following the Turkish invasion and subsequent partitioning of the island in 1973. Officially, all units are classified as non-active under an agreement reached between the two Cypriot authorities at Vienna. The National Guard Air Wing was renamed the Cyprus Air Force in 1996, when it adoped the national military markings of Greece. It has approximately 140 personnel, operating four SA342L-1 Gazelle helicopters capable of firing HOT anti-tank missiles, plus three Bell 206L3 LongRanger and two MD500 helicopters; two PZL-140, known locally as the Karnia, a BN-2B Maritime Defender and two PC-9s for patrol and pilot training. The Greek Cypriot police also have two Bell 412 helicopters and a BN2T Turbine Islander.

In Turkish Northern Cyprus, an army of 5,000 personnel has no aircraft of its own, but Turkish Army Cessna U-17s and Bell UH-1D Iroquois are at their disposal, stationed permanently.

Czech Republic

POPULATION: 10.2 million

LAND AREA: 30,461 sq m, 78,864 sq km

GDP: $52bn (£36.4bn), per capita $14,163 (£9,904)

DEFENCE EXPENDITURE: $1,155m (£808m)

SERVICE PERSONNEL: 53,600 active.

CZECH AIR FORCE & AIR DEFENCE/CESK LETECTVO A PROTIVZDUSNA OBRANA

Formed: 1995

Formerly the larger and more prosperous part of the former republic of Czechoslovakia, the Czech Republic split with the other main element, Slovakia, in 1995. Czechoslovakia itself had only dated from 1918 and the federation of Bohemia, Moravia, Slovakia, Ruthenia and part of Silesia on the break-up of the Austro-Hungarian Empire. On its creation, a Czechoslovak Army Air Force was formed almost immediately, based on air units that had operated with the Czech Legions that had existed in France and Russia during World War I. War surplus aircraft were pressed into service with assistance from France. The Czechs were given licences to produce French aircraft, but were soon producing their own designs, including Aero A-18 fighters, Smolik Sm-1 and Sm-2 bombers and Letov S-10 trainers. Within a decade, the CAAF had 400 aircraft in 25 squadrons. The main types included Avia BH21 and Letov S-20 fighters; Aero A-24 and Letov S-16 bombers; Aero A-11 and A-12 AOP aircraft; Avia BH10 and BH11, and Letov S-18 trainers.

During a period when aircraft became obsolescent far more quickly than today, this force changed significantly over the next decade. Avia BH33 and BH34 fighters entered service, joined by Potez 63 fighter-bombers, and Fokker FVII and FIX tri-motor bombers. Other newcomers were Aero A-30 and A-100 reconnaissance aircraft and A-32 AOP aircraft. Towards the end of the 1930s, the Tupolev SB-2 bomber was built under licence as the Aero B-17, while the CAAF also received Avia 135 fighters and A-300 bombers. The Munich Agreement of 1938 bought a year's breathing space for Britain and France before the start of World War II, but at the cost of Czechoslovakia ceding a third of her territory to Germany. In 1939, Germany occupied the rest of the country, dismantled the Czechoslovak state, and made Slovakia a separate state. No resistance was offered to the invader's overwhelming force, but Czech personnel escaped to join the French and Polish air forces, and after the fall of France, many flew with the Royal Air Force, which had three Czech fighter squadrons and a bomber squadron. Other Czechs fought with the Soviet forces. The Germans eventually formed a Slovak Air Force to fight alongside the Luftwaffe, flying Messerschmitt Bf109G fighters.

The country was reunited at the end of World War II, with a new Czechoslovak Air Force using abandoned Luftwaffe and Slovak Air Force equipment, as well as the aircraft flown by the former RAF squadrons, including Liberator bombers. Small numbers of de Havilland Mosquito fighter-bombers, Lavochin La-7 fighters and Petlyakov Pe-2 bombers soon joined these aircraft.

A successful Communist *coup d' etat* in 1948 saw Czechoslovakia forced into the Soviet sphere of influence, and the CAF purged of ex-RAF elements as Russian advisers moved in and Soviet forces were stationed in the country. The country became a member of the new Warsaw Pact when this was established in 1955, but even before this, the USSR provided aircraft. Initially, these consisted of Ilyushin Il-10 ground-attack aircraft and Li-2 (C-47) and Ilyushin Il-12 transports. The first jets,

Czech Republic

ABOVE: *The Aero L-159 provides a light strike capability for the Czech Air Force. (Jane's)*

almost 200 MiG-15 fighters, entered service in 1951, at about the same time as Antonov An-2 transports and Zlin 226 trainers. Eventually, all of the MiG series of fighters and interceptors entered service up to and including the MiG-21 accompanied by Sukhoi Su-7B ground-attack aircraft and Ilyushin Il-28 bombers. Constant up-dating of the CAF came to an abrupt end in 1968, when the Soviet armed forces intervened in Czechoslovak internal politics after support for a less authoritarian regime was perceived as undermining the cohesion of the Warsaw Pact. There was a gap of some years before Czechoslovak forces received further equipment in the form of Sukhoi Su-22M4s and then Su-25BKs, as well as Mil Mi-24 attack helicopters.

Post-war strength peaked in the early 1970s at around 18,000 personnel, by which time it had 150 MiG-21s and 100 MiG-19s, as well as 150 Su-7Bs, alongside older aircraft, including 80 each of MiG-17s and MiG-15s. The break-up of the USSR and the collapse of the Warsaw Pact saw Czechoslovakia return to its western links. The CAF underwent further change following Slovakia's vote for independence, ending the federation, in 1995. In 1999, the Czech Republic became a member of NATO, and is receiving NATO assistance in making its aircraft compatible with NATO standards. Nevertheless, more than this is needed given the country's economic problems and the size of the task. New aircraft, the ideal solution, are being considered. But the largest programme, the purchase of 72 Aero L-159 fighter/ground-attack aircraft, the first of which were received during 2000/2001, is being extended over a longer period since it was accounting for 80 per cent of the procurement budget. As many as half of the L-159s may be sold, possibly to Slovakia, to ease the burden on the defence budget.

The CAF has 13,400 personnel and is organised into a tactical air force and an air defence command, the latter with SA-2, -3 and -6 SAMs. Flying hours are low at an average of 60 per annum. Squadron strength is far larger than in most air forces, with a single squadron operating 36 MiG-21s, originally to be replaced starting in 2004 by 24 Saab/BAe Gripen. The Su-25s and Su-22s have been replaced by the L-159s in two squadrons, although these might be cut to a single squadron if disposal of half the fleet goes ahead. Transport aircraft are in two squadrons with 15 L-410M Turbolets, with another three on survey work, two An-24, four An-26 and an An-30, while there is one Challenger CL-600 and three VIP Tu-154s. Helicopters include 36 Mi-24 attack helicopters, and 40 Mi-8/Mi-17 for transport, AEWand ELINT. Eleven PZL W-3 helicopters are used on transport and SAR duties, with 33 Mi-2 on liaison and training duties. A number of the transport aircraft and helicopters are likely to be sold, overall reducing numbers by up to a quarter, and halving the numbers the obsolete Mi-2. Training uses 37 L-39 Albatros and 24 L-29 Delfin, as well as eight Zlin 142. Combat aircraft are equipped with AA-2, AA-7 and AA-9 AAM, although this may change with the Gripen.

New transport aircraft are also being considered, including C-130, C-27J and CN235. Procurement funding remains a problem and will remain so for some time after the floods of 2002 affected the economy.

BELOW: *Mil Mi-8 helicopters provide the backbone of many East European helicopter fleets, including the Czech Republic.*

Denmark

POPULATION: 5.3 million

LAND AREA: 16,611 square miles, 49,932 sq km

GDP: $162bn (£113bn), per capita $25,900 (£18,112)

DEFENCE EXPENDITURE: $2.4bn (£1.7bn)

SERVICE PERSONNEL: 21,810 active, plus 64,900 reserves.

ABOVE: *Training and liaison duties for the RDAF are handled by a small force of Saab T-17 Supporters. (RDAF)*

ROYAL DANISH AIR FORCE/KONGELIGE DANSKE FLYVEVBEN

Formed: 1950

Although the RDAF dates from a merger of army and naval aviation in 1950, Danish military aviation dates from 1912, when the two services received their first aircraft. The Royal Danish Navy received a Henri Farman and the Army received a Danish-designed B&S monoplane. In 1913, the RDN received two Donnet-Leveque flying boats, while the Army gained a Henri Farman and a Maurice Farman in 1914. Danish neutrality through World War I meant that the services were dependent on aircraft built at the Royal Army Arsenal. Post-war, the RDN obtained ex-German aircraft and some Avro 504 trainers purchased in 1920. The Army bought LVG BIII, Fokker C1 and Potez XV aircraft.

Reorganization in 1923 saw the creation of a Naval Flying Corps and an Army Flying Corps. The Naval Flying Corps was further reorganized in 1926 with the formation of two squadrons, No1 Luftflotille, with Hansa-Brandenburg reconnaissance seaplanes replaced in 1928 by Heinkel He8s; and No2 Luftflotille with Hawker Danecocks (licence-built Woodcocks), replaced by Hawker Nimrods in 1935. The inter-war aircraft of the NFC also included Hawker Dantorp torpedo-bombers and de Havilland Gipsy Moth and Avro 621 trainers. An ex-Lufthansa Dornier Wal flying boat joined three He8s on survey work in Greenland, later proving of considerable use to the wartime Allies when they established a chain of bases to ferry aircraft across the Atlantic. In 1932, further reorganization created an Army Aviation Corps, Havaens Flyvertropper, with five squadrons or eskadrille. Bristol Bulldog fighters equipped No1 Eskadrille until replaced by Gloster Gladiators in 1935, Fokker CV reconnaissance aircraft equipped No2 Eskadrille, and No3 when its Fokker CI were replaced in 1934, as well as No5 when it formed in 1935. Defence cuts meant that No4 was never formed.

Denmark hoped to remain neutral during World War II, but the country was overrun by German forces in spring, 1940. Having lost most of their aircraft on the ground, a number of pilots managed to reach the UK to join the RAF.

Post-war, the two air arms were re-established, initially sharing six Percival Proctor and some liaison aircraft. A joint flying school was formed in 1946, and a joint air staff in 1947, in preparation for the creation of an autonomous air force. In the meantime, many new aircraft were delivered, including Supermarine Spitfire IX fighters and Sea Otter amphibians, and North American AT-6 Harvard trainers. The first jets, Gloster Meteor fighters, arrived in 1949. When formed in October, 1950, the new RDAF had five Squadrons: Eskadrille No721 operated Convair PBY-5A Catalinas, Sea Otters and Airspeed Oxfords; No722 with Spitfires and Oxfords; No723 had the Meteor F4s until these were replaced by Meteor NF11 night-fighters in 1952; No724 had Meteor F8s; while No725 had Spitfires. New aircraft in 1950 included 27 DHC-1 Chipmunk basic trainers. This force was soon increased to eight squadrons, as Denmark became a founder member of the North Atlantic Treaty Organisation, NATO. In 1951-53, The USA supplied 200 Republic F-84E/G Thunderjets and a quantity of Lockheed T-33A jet trainers, initially replacing 725's Spitfires, and then equipping new squadrons 726-730. Other deliveries at this time included Douglas C-47 transports and Bell 47D helicopters.

This much enhanced force was updated during the late 1950s with 30 Hawker Hunter Mk51 fighters and ten Republic RF-84F Thunderflash reconnaissance-fighters, followed later by North American F-100D Super Sabres and, during the early 1960s, Lockheed F-104G Starfighters. These acquisitions were accompanied by ex-Canadian Fairey Fireflies for target towing, Hunting Pembroke C52 communications aircraft and Sikorsky S-19 helicopters. The Catalinas soldiered on until replaced by Sikorsky S-61A helicopters in 1970. By this time, the RDAF had 11,000 personnel. Squadrons included two with F-104G interceptors; two with F-100D Super Sabre and one with SAAB F35 Draken fighter-bombers; a Draken RF35 reconnaissance-fighter squadron and a Hunter ground-attack squadron. Most squadrons had 16 aircraft, but the Draken squadrons had 23 each. Transport aircraft included eight Douglas C-

Denmark

ABOVE: *The Danish Army is developing a helicopter force, which currently includes 12 MD500M (OH-6 Cayuse) for liaison. (RDAF)*

47s and five C-54s, while an SAR squadron had eight S-61As. Training was conducted on Chipmunks and T-33s, with conversion trainers for the fast jet combat aircraft. During the 1970s, naval and army aviation reappeared, largely due to the growing role for the helicopter, but at this stage the RDAF still controlled 12 Hughes 500M helicopters and KZ VII Larks used by the Army, and eight Alouette IIIs for the RDN.

Even before the end of the Cold War, Denmark cut defence expenditure, despite occupying an important strategic position at the exit from the Baltic to the North Atlantic, and also having the remote Faroes Islands and the vast expanses of Greenland to police. Collaboration on procurement with other NATO European countries enabled it to replace its Starfighters and Drakens with Lockheed Martin F-16 Fighting Falcons, but combat squadron strength was cut to just three. These aircraft have received an MLU. Despite being a member of the Nordic Standard Helicopter Programme, Denmark broke ranks with the other members and opted for the EH101 to replace the S-61s. In common with a number of other NATO members, Denmark announced in 1999 that its defence posture would move from that of the Cold War to the provision and support of expeditionary troops within international operations, and to this end has ordered three C-130J Hercules II., with an option on a fourth aircraft.

Currently, the RDAF has just 4,900 personnel; although flying hours are reasonable at 180 per annum. Three fighter squadrons operate 60 F-16s, with nine in reserve, with AIM-9 Sidewinder and AIM-120A AMRAAM AAM, and AGM-12 Bullpup ASM. A transport squadron operates Hercules, three Canadair CL604 Challenger (two for MR and one VIP), that have replaced Gulfstream IIIs in this role. The SAR squadron operates helicopters, with eight S-61As being replaced by EH101s, although at least five of the 14 aircraft on order will be used for troop transport. Flying training uses the SAAB T-17. Denmark is involved with the American-led JSF project, although likely aircraft numbers have not been announced. SAM defences include the Hawk missile.

ROYAL DANISH NAVAL FLYING SERVICE/SOVAERNETS FLYVETJENESTE

Having replaced the eight Alouette IIIs, with which Danish naval aviation was re-established, with eight Lynx helicopters, the RDNFS is responsible for fisheries protection both off the Danish coast and its off-shore territories of the Faeroes and Greenland. Up to four helicopters are embarked at any one time, usually on support vessels, as the RDN does not have frigates capable of carrying helicopters. The Lynx were updated during the early 1990s and are expected to remain in service until 2015.

ARMY FLYING SERVICE/HAERENS FLYVETJENESTE

The Danish Army returned to aviation with 12 MD500Ms originally supported by the RDAF, later augmented by 12 Eurocopter AS550 Fennecs. It will also use around five of the EH101s on order.

BELOW: *The Royal Danish Air Force has been one of the participants in the European Lockheed-Martin F-16A Fighting Falcon programme. (RDAF)*

Djibouti

POPULATION: 783,000

LAND AREA: square miles, 23,051 sq km

GDP: $460m (£322m), per capita $924 (£646)

DEFENCE EXPENDITURE: $23m (£16.1m)

SERVICE PERSONNEL: 9,600 active.

DJIBOUTI AIR FORCE/FORCE AERIENNE DJIBOUTIENNE

Formed: 1977

Djibouti gained independence from France in 1977. The country received the standard package of three Max Holste Broussards (Bushranger) communications aircraft, a C-47 and an Alouette II helicopter while a significant French presence has remained. The region is important strategically, at the narrow southern end of the Red Sea, and unstable politically. The small air arm has just 200 personnel. At least one of the two Mi-2 helicopters has been armed with rocket pods. Other aircraft include one An-28 received from Poland, a VIP Falcon 50, a Cessna 402 communications aircraft and U206G liaison aircraft, five Mi-8s and two AS355F Ecureuil.

Dominican Republic

POPULATION: 8.7 million

LAND AREA: 18,699 square miles, 48,430 sq km

GDP: $13.6bn (£9.5bn), per capita $5,800 (£4,055)

DEFENCE BUDGET: $114m (£80m)

SERVICE PERSONNEL: 24,500 active.

ABOVE: *One of the three CASA C212-400 transports. (Mader-Acaes Collection)*

DOMINICAN AIR FORCE/FUERZA AEREA DOMINICANA

The Aviacion Militar Dominicana was formed by the Army during the early 1940s, with some Aeronca L-3 AOP and American training aircraft, including Boeing-Stearman PT-17 Kaydets, Vultee BT-13s and North American AT-6s. It was not until 1948 that an expansion plan was authorised, buying obsolete Bristol Beaufighters as well as slightly more modern de Havilland Mosquito and Republic F-47D Thunderbolt fighters-bombers and a small number of Boeing B-17G Fortress bombers. By the early 1950s, these aircraft were replaced by a squadron of 20 North American F-51D Mustang fighter-bombers, later augmented by the AMD's first jets, 20 de Havilland Vampire fighter-bombers, as well as seven Douglas B-26 Invader bombers. This period also saw the arrival of two Convair PBY-5A Catalina amphibians; six Douglas C-47 and six Curtiss C-46 Commando, three DHC-2 Beaver and three Cessna 170s for transport and communications aided by Dominica's membership of the Organisation of American States. Helicopters arrived, initially two Bell 47s and two Sikorsky H-19, followed by an Alouette III and two Alouette II. Training used North American T-6 Harvard and Beech T-11 aircraft. In the early 1970s, seven Hughes OH-6A helicopters were obtained.

Army control ended with the creation of the Dominican Air Force, but COIN and support for ground forces have become more important, plus SAR. Currently, the FAD has 5,500 personnel, a 50 per cent increase over 30 years. Combat strength includes six Cessna A-37 Dragonfly armed jet trainers, two O-2 Super Skymaster, likely to be replaced by ten Super Tucano armed-trainers currently on order. Five Ipanema crop-dusting aircraft are used on anti-drug operations. There are three CASA C212-400 transports and a VIP Commander 680. A handful of Beech Queen Air and King Air, Piper Navajo and Cessna 207/210 are used on communications duties. There are two AS365 Dauphin VIP helicopters, and ten AS350 Ecureuil and six OH-6A, and nine UH-1H Iroquois, which also handle SAR. Training uses eight T-35B Pillan, acquired from Chile, five T-41D Mescalero, and a few T-6s.

Ecuador

POPULATION: 12.8 million

LAND AREA: 104,505 square miles, 454,752 sq km

GDP: $20bn (£14bn), per capita $4,500 (£3,146)

DEFENCE EXPENDITURE: $320m (£224m)

SERVICE PERSONNEL: 59,500 active, plus 100,000 reserves.

ABOVE: *The Ecuadorian Air Force, FAE, operates 19 Beech T-34C Turbo Mentor trainers. (Raytheon)*

ECUADORIAN AIR FORCE/FUERZA AEREA ECUATORIANA

Formed: 1935

In 1920 an Italian aviation mission visited Ecuador and the Cuerpo de Aviadores Militares was formed within the Army with Ansaldo, Aviatik and Savoia aircraft. It remained primarily a training and liaison operation even after 1935, when the present title was adopted and a Curtiss-Wright Osprey AOP aircraft purchased with some of the same manufacturer's 16E trainers. Meridionali reconnaissance aircraft were introduced in 1938. Fighter aircraft followed a visit by an American aviation mission in 1941, which supplied Seversky P-35s as well as Fairchild, Ryan PT-20 and North American NA-16 trainers. The first transport aircraft were a German-owned airline's Junkers Ju52/3 transports. There were no further aircraft deliveries while World War II continued.

Ecuador became eligible for American military aid on joining the Organisation of American States in 1948. This included Republic F-47D Thunderbolt fighter-bombers, Convair PBY-5A amphibians, Douglas C-47 and Beech C-45 transports, and North American T-6 trainers. The first jets arrived in 1954, 12 Gloster Meteor FR9 fighters and six English Electric Canberra B6 bombers. Helicopters were not introduced until the US provided Bell 47 helicopters in 1960, as well as 12 Lockheed F-80C Shooting Star interceptors and two Douglas DC-6B transports, Lockheed T-33A, North American T-28 and Cessna T-41A trainers. By the early 1970s, personnel strength had grown to some 3,500. During the early 1960s, Cessna A-37B Dragonfly armed jet trainers, Hiller FH-1100 helicopters, a Short Skyvan light transport, were introduced, with Cessna 180 AOP aircraft operated for the Army.

Border disputes with Ecuador's neighbours led to further expansion, despite a reduction in US military support in an attempt to reduce tension. Sepecat Jaguar strike aircraft and BAe Strikemaster armed jet trainers entered service during the 1970s, followed by Mirage F1 interceptors and Israeli Kfir attack aircraft. The FAE's operation of two national airlines, TAME, Transportes Aereos Militares Ecuatorianos, and Ecuatoriana introduced Boeing and Fokker airliners. Separate naval and army air arms came into existence during the 1970s.

Mirage, Kfir and Jaguar aircraft have been extensively upgraded in recent years, with attrition deliveries of Kfirs to maintain operational strength. Currently, the FAE, despite the relatively small personnel strength of 4,000, has five squadrons of combat aircraft; one with eight Jaguar EB and ES, one with 17 Kfir CE/TC-2, and a third with 23 A-37B. A fourth squadron operates 14 Mirage F-1JE/1JB, while a fifth has nine remaining Strikemaster Mk89A armed jet trainers. The Kfirs and Jaguars have been upgraded as attempts to buy used A-4M Skyhawks from the US and MiG-29s from Russia have failed. Transport aircraft are mainly operated through the airlines, but the FAE does have six Lockheed C-130B/H/L100, a single DHC-5 Buffalo, and three DHC-6 Twin Otters. TAME has six Boeing 727s, a DHC-6, an F-28 and two HS748s, while Ecuatoriana has three Boeing 707-320s, a DC-10-30 and two Airbus A-310 airliners. There are four North American Sabreliners for communications and VIP work. Helicopters include 26 Bell UH-1B/H, a VIP 212, nine 206B, and five Alouette III. Training and COIN share 23 AT-33 armed trainers, and there are 19 Beech T-34C Turbo Mentor and 20 Cessna 150. Missiles include R-550 Magic, Super 530, Shafrir, Python 3 and Python 4, all AAM.

BELOW: *A Jaguar International 'S' of the F.A.E flying over the south coast of Ecuador. (Michael J. Gething)*

ECUADORIAN NAVAL AVIATION/AVIACION NAVAL ECUATORIANA

The ANE developed for transport and communications, and funding difficulties pre-

Ecuador

vented it acquiring naval helicopters in 1999 for operation from two ex-British Leander-class frigates. It operates five Bell 206N and two 222 communications helicopters. Fixed-wing aircraft include a VIP Citation I and Beech King Air 300, with a King Air 200 and a Cessna 320E for communications. Two Casa CN235M transports provide offshore surveillance. There are three T-34C Turbo Mentor trainers.

ECUADORIAN ARMY AIR SERVICE/SERVICIO AEREO DEL EJERCITO ECUATORIANA

The Ecuadorian Army operates a light aircraft on survey and liaison duties, and attack and transport helicopters. Armed helicopters are 13 SA342K/L Gazelle, with four Super Puma, two Puma and a Bell 214B for transport, four Eurocopter AS350B Ecureuil on communications with four SA315B Lama on SAR and surveillance, which also uses a Cessna Citation II, a Learjet 24D and two Beech King Air 100/200. Transport aircraft include a DHC-5D Buffalo, six IAI-201 Arava, three PC-6B Turbo Porter and a CN235M. There is a VIP Sabreliner 40R. Training uses three Cessna 172. The Army is considering helicopters in the S-70 category.

Egypt

POPULATION: 70.6 million

LAND AREA: 386,198 square miles, 999,740 sq km

GDP: $90bn (£62.9bn), per capita $5,000 (£3,496)

DEFENCE EXPENDITURE: $2.9bn (£2bn)

SERVICE PERSONNEL: 443,000 active (70 per cent conscripts) plus 254,000 reserves.

ABOVE: *One of the newer additions to the Egyptian strength is the Aero L-59E, which can be used as an advanced trainer or light attack aircraft. (Aero Vodochody)*

ARAB REPUBLIC OF EGYPT AIR FORCE

Formed: 1954

Egyptian military aviation dates from the formation of the Egyptian Army Air Force in 1932, equipped with five de Havilland Gipsy Moths for training, surveying and army co-operation. During its first five years the EAAF was largely operated by RAF personnel, but it received new equipment during this period, including ten Avro 626s in both 1933 and 1934, with a Westland Wessex in the latter year. Hawker Audaxes, Avro Ansons and Miles Magisters further expanded this force. The Anglo-Egyptian Treaty of 1936 made Egypt a sovereign state. In 1939, the link with the army was broken, and the title of Royal Egyptian Air Force adopted. This coincided with another round of equipment purchases, with ex-RAF Gloster Gladiator and Hawker Hart fighters, some Bristol Blenheim bombers, and Westland Lysander and Percival Q-6 AOP aircraft, plus additional Ansons. These aircraft formed two fighter and three AOP squadrons. The outbreak of World War II found the REAF assisting the RAF in protecting the Suez Canal for the first two years, until Egypt declared itself neutral in 1941, so that the REAF saw no action throughout the war years. Despite neutrality, it received two squadrons of Hawker Hurricanes and one of Curtiss Tomahawk fighters.

Post-war, RAF assistance resumed allowing considerable expansion as war-surplus aircraft were supplied. These included Supermarine Spitfire VB and IX fighters; Handley Page Halifax, Short Stirling and Avro Lancaster heavy bombers (seldom used due to a shortage of suitable aircrew); Curtiss C-46 Commando, Douglas C-47 and de Havilland Dove transports; and North American T-6 Harvard and Miles Magister trainers. This assistance ended abruptly with an arms embargo after Egypt invaded Israel in May 1948, although the REAF had little success and the fighting ended in January 1949. Egypt next received de Havilland Vampire FB5 and Hawker Fury fighter-bombers; Gloster Meteor F4 and Macchi C205 fighters and Fiat G55B fighter-trainers and Meteor T7 trainers. A follow-on order for a further 20 Meteor F4s was banned by the British, but the REAF obtained 30 Vampire FB52s from Italy via Syria in 1953, and obtained additional Meteors and Vampires once the embargo was lifted.

Egypt

ABOVE: *Frontline strength of the Egyptian Air Force includes Mirage 2000EM/EB interceptors seen here. (Jane's)*

An abrupt change in procurement followed the ending of the monarchy in 1954, with the 'Royal' prefix dropped from the title. Soviet equipment was ordered, initially from Czechoslovakia, in 1955, including MiG-15 fighters, Ilyushin Il-28 jet bombers, Il-14 transports, and MiG-15UTI and Yakovlev Yak-11 trainers. Most of these aircaft were destroyed on the ground by British and French air attacks during the Suez crisis of November, 1956. The aircraft were soon replaced, with MiG-15s, MiG-17s and Il-28s supplied during 1957. Training was provided by Russian and Indian advisers in Egypt, and in Czechoslovakia. Through the end of the decade and into the early 1960s, additional Soviet equipment was provided, with additional MiG-17s followed by MiG-19s and a small number of Tupolev Tu-16 heavy bombers and Antonov An-12B transports, some of which were shared with the then national airline, United Arab Airlines. Mil Mi-4 and Mi-6 helicopters were also supplied. Even though it was still receiving assistance itself, the EAF helped the new Indonesian Air Force, the AURI, and the Yemen Air Force, which at one time consisted mainly of Egyptian personnel. For a period, Egypt and Syria united in the short-lived United Arab Republic, and the Syrians also received Egyptian help. As the scale of support grew, the USSR provided a complex air defence system to cover both the Suez Canal and the Aswan Dam as well as Cairo and Alexandria. Egypt was the first country outside the Soviet Bloc and China to receive the MiG-21 interceptor. Having grown to 31 squadrons, most of the EAF's aircraft were destroyed on the ground during the Arab-Israeli War of June, 1967, and the Suez Canal was closed.

Once again, the USSR replaced the EAF's heavy losses, providing SA-2 and SA-3 SAM missiles, and many of the aircrew for the aircraft. By the early 1970s, the EAF had grown again to 20,000 personnel. Its aircraft by this time included 150 MiG-21 interceptors; 105 Sukhoi Su-7B and 180 MiG-15 and MiG-17 fighter-bombers; 28 Il-28 and 15 Tu-16 bombers; 40 Il-14 and 20 An-12 transports; 40 Mi-4 and smaller numbers of Mi-6 and Mi-8 helicopters; and a total of 150 MiG-15UTI, L-29 and Yak-18 trainers. Later, MiG-23 interceptors were supplied, flown by Russian personnel, while the air defences were strengthened further with the addition of SA-6, -7 and -9 missiles.

Despite a change of leadership in Egypt in 1970, a further Egyptian and Syrian attack on Israel followed in October 1973. On this occasion, Israeli losses were far heavier, losing 200 aircraft as against just 30 in 1967. Nevertheless, as SAM missile stocks ran low and the Arab ground forces overstretched their supply lines, the conflict once again moved against the Arabs.

An uneasy peace prevailed until diplomatic relations were established between Egypt and Israel, including the re-opening of the frontier in 1980. After purchasing Shenyang F-6s and Chengdu F-7As, this period also saw a move away from Soviet influence and renewed links with the West. Gradually, Soviet equipment was replaced by Western aircraft, including 15 Boeing CH-47C Chinooks. Army and naval aviation also emerged, with the Army having almost 100 Gazelle light helicopters, some of which were loaned to the Navy, and 30 Westland

Egypt

Commando (S-61) troop-carrying helicopters, while the Navy acquired ASW Westland Sea Kings (S-61). The EAF acquired substantial numbers of Mirage 5s, but most of its aircraft were from the United States, with F-4 Phantoms being followed by more than a hundred F-16s. Naval and army aviation returned to the EAF, acquiring suitable helicopters, including AH-64A Apache attack helicopters, for the ground support role and ex-USN Super Seasprites for the Egyptian Navy. More recently, the Chinese-Pakistan K-8 Karakorum basic trainer has been introduced.

The EAF has 29,000 personnel, 580 combat aircraft and more than a hundred armed helicopters, a very high ratio of aircraft to personnel, with another 75,000 in a separate Air Defence Command based on AAA and SAM. There are 30 F-16A in two squadrons and 136 F-16C in seven squadrons supported by six F-16B and 42 F-16D twin-seat aircraft, mainly used for conversion training. There are also 17 Mirage 2000EM/EB interceptors and conversion trainers. Older aircraft include some 80 Mirage 5DE/E2/SDR/SDD in three interceptor and reconnaissance squadrons, while 30 F-4E Phantoms equip two squadrons. Two squadrons operate 60 MiG-21PF/PM/R/U, three have 60 Chengdu F-7A interceptors and two 45 Shenyang F-6/FT-6 (MiG-19) attack aircraft. While most of the 38 Alphajets are used for advanced jet training, 13 of them can also be used in the attack role. Currently, the 35 AH-64A Apache attack helicopters are being upgraded to Longbow standard, although the US has delayed a decision on the sale of the Longbow radar to Egypt. The interceptors are supported by five Grumman E-2C Hawkeye AEW aircraft. In addition to the Apache, helicopters include 79 SA342L/M Gazelle which can be used in the attack role, 19 CH-47D Chinook heavy-lift helicopters, while the medium-lift role is filled by 28 Westland Commando, although some are converted to the VIP and ELINT roles, 40 Mi-8 and 20 Mi-17. Two Agusta AS-61 are used in the communications role, while there are four Sikorsky UH-60A/L Black Hawk VIP helicopters. Ten remanufactured ex-USN Super Seasprites provide ASW cover for the Egyptian Navy, and are the only examples so far to be fitted with dipping sonar, complementing five remaining Westland Sea King Mk47 and twelve Gazelle. Transport aircraft include 26 C-130H/H-30 Hercules, with two converted for ELINT. A single Airbus A340 and two of the Hercules, three Falcon 20s and seven Gulfstream III/IV/IV-SP provide VIP transport, as do two Beech 1900C, with another four used on ELINT duties. Five DHC-5 Buffalo are in the transport role with another four for navigational training. Apart from the Alphajets and conversion versions of the combat aircraft, training uses 46 L-59E, which are also suitable for ground-attack, 46 EMB-312 Tucano, 17 Hiller UH-12E, 80 K-8 Karakorum and 74 Grob G115EG. Missiles include a combination of Western and Russian AAM and SAM, but ASM are predominatly Western.

El Salvador

POPULATION: 6.4 million

LAND AREA: 8,236 square miles, 21, 331 sq km

GDP: $10.6bn (7.4bn), per capita $3,000 (£2,098)

DEFENCE EXPENDITURE: $171m (£120m)

SERVICE PERSONNEL: 16,800 active.

SALVADORIAN AIR FORCE/FUERZA AEREA SALVADURENA

Latin America's smallest and most densely populated country, El Salvador established a Military Aviation Service in 1923 with five Aviatik trainers. Small numbers of Waco and Curtiss-Wright Osprey general-purpose aircraft and Fleet trainers followed these. The first combat aircraft, four Caproni ground-attack aircraft, arrived in 1939. Post-war Beech AT-11, Fairchild PT-19, Boeing-Stearman and Vultee trainers were obtained. After joining the Organisation of American States in 1948, C-47 transports and North American T-6 trainers were provided, followed by six Vought F4U Corsair and six North American F-51D Mustang fighter-bombers for two squadrons, and Beech T-34 trainers.

Jet equipment arrived much later, second hand Dassault Ouragan fighter-bombers and Magister armed jet trainers, followed by Cessna A-37B Dragonfly armed trainers and O-2A Super Skymasters. COIN operations became increasing important during a civil war, during which the government forces relied heavily on US support, including MD500E and Bell UH-1 Iroquois helicopters.

Today, the FAS has 1,100 personnel. Ouragans and Magisters are grounded, although six of the latter are believed to be still airworthy. There are eight surviving A-37B Dragonfly armed jet trainers and 13 Super Skymasters in a single squadron. A squadron operates seven MD500D/E and three UH-1M armed helicopters, with another operating 18 UH-1H in the transport and SAR roles; many more UH-1Hs are in store. Eight C-47s have been converted as turboprops for transport and COIN. Other transport aircraft include a VIP DC-6B, an Arava 201, and a Fairchild Merlin IIIB. Training uses ten ex-Chilean Air Force T-35Bs, as well as five Socata Rallye 235S, six Hughes 269 and a T-41D. Further equipment purchases have been affected by the impact on the economy of a major earthquake in 2000.

Equatorial Guinea

POPULATION: 535,000

LAND AREA: 10,830 square miles, 28,051 sq km

GDP: $800m (£559m), per capita $3,346 (£2,340)

DEFENCE EXPENDITURE: $12.7m (£8.9m)

SERVICE PERSONNEL: 1,320 active.

NATIONAL GUARD/GUARDIA NACIONAL

The National Guard of this former Spanish colony operates a small air arm with 100 personnel on transport and communications duties. Aircraft include a VIP Falcon 900, which replaced a Yak-40, an An-32, three C-212s and a Cessna 337, as well as two SA-316 Alouette III helicopters.

Eritrea

POPULATION: 3.9 million

LAND AREA: 45,754 square miles, 118,503 sq km

GDP: $710m (£496m), per capita $441 (£308)

DEFENCE EXPENDITURE: $360m (£252m)

SERVICE PERSONNEL: c.171,900 active, plus 120,000 reserves.

ERITREAN AIR FORCE

Formed: 1994

After many years of internal unrest, Eritrea became independent from Ethiopia in 1993. Internal unrest continues, and border disputes with Ethiopia have resulted in conflict. The small air force was formed in 1994 with L-90 Redigo trainers, later joined in 1996 by Aermacchi MB339FD armed jet trainers. Late in 1998, ten MiG-29s were acquired, possibly from Moldova, and flown by Ukrainian mercenary aircrew, but half of these have been lost since. A Mi-35 helicopter was captured from the Ethiopian Air Force and pressed into service.

Currently, the EAF has around 800 personnel. It has the five surviving MiG-29s and five MB339 in the combat role, while some sources suggest that there could also be a handful of MiG-23 and MiG-21 as well. Transport is provided by four Harbin Y-12, plus a VIP Dornier Do228. There are two Mi-8 and four Mi-17 in the SAR and transport roles, and an Mi-35 attack helicopter. There are seven Redigo trainers still in use.

Estonia

POPULATION: 1.4 million

LAND AREA: 17,410 square miles, 45,610 sq km

GDP: $5.6bn (£3.9bn), per capita $9,753 (£6,820)

DEFENCE EXPENDITURE: $81m (£56m)

SERVICE PERSONNEL: 4,450 active, plus 14,000 reserves.

ESTONIAN ARMY AVIATION

Formed: 1991

In common with the other small Baltic states Estonia was occupied by the Soviet Union in 1940, having enjoyed independence only since 1920. After being occupied by German forces from 1941 to 1945, the country became a constituent republic of the USSR and did not achieve independence until 1991. The army has a small air arm, with 110 personnel and is equipped with three An-2 transports and a PZL-140 Wilga for training, while the three Mi-2 helicopters were joined in 2000 by four Robinson R44s, a gift from the USA.

ABOVE: *Mi-2 (Mader-Acaes Collection)*

Ethiopia

POPULATION: 63.7 million

LAND AREA: 432,403 square miles, 1,096,900 sq km

GDP: $6.7bn (£4.7bn), per capita $571 (£399)

DEFENCE EXPENDITURE: $457m (£319m)

SERVICE PERSONNEL: 252,500 active.

RIGHT: *MiG-23 (Jane's)*

ETHIOPIAN AIR FORCE

Formed: 1975

Ethiopian military aviation dates from 1930 and the delivery of six Potez 25 bombers and three de Havilland Gipsy Moth trainers to the then Imperial Ethiopian Aviation. In 1935, Italy invaded, with the far stronger and better-equipped Regia Aeronautica, destroying the IEA. British forces liberated Ethiopia in 1941, but no attempt was made to re-start military aviation until 1946, when the Imperial Ethiopian Air Force was established with help from a Swedish aristocrat. Its first aircraft were ten de Havilland Tiger Moth trainers, soon replaced by 30 more modern SAAB 91-A Safirs, and a handful of Cessna AT-17 Bobcat trainers. In 1948, 30 SAAB-17A bombers were delivered, with a similar number delivered during the early 1950s, and eight Fairey Firefly fighters, Stinson L-5 AOP aircraft and Douglas C-47 transports.

American military aid commenced in 1960, providing the first jets, 12 North American F-86F Sabre fighters and a number of Lockheed T-33A armed-trainers. Re-equipment during the late 1960s included later versions of the Safir, Northrop F-5A fighters and de Havilland Dove and Ilyushin Il-14 transports. Canberra jet bombers were also acquired. By the early 1970s, the IEAF had 3,000 personnel, and was operating the F-5As in one squadron of eight aircraft, with another squadron operating the 12 Sabres, while a bomber squadron had six Canberras. COIN operations were becoming increasingly important and involved two squadrons, one with eight SAAB-17A and another with six North American T-28 and three Lockheed T-33A. There were six C-47 and two C-54 transports, plus a solitary Il-14 and three Doves, in a single squadron. Relatively few helicopters included three Alouette III and an Agusta-Bell 204B. Training used SAAB 91C/D Safir, North American T-28 and Lockheed T-33A. After the monarchy was overthrown in 1975, the present title was adopted. The USSR became the main supplier of equipment, with MiG-17 fighter-bombers, followed by MiG-21s and An-12s, and Mi-8 helicopters.

Increasingly, the EAF found itself involved in operations against Eritrean secessionists, resulting in independence for Eritrea in 1993. This was followed by war between the two states over disputed borders. After the break-up of the USSR, Russia continued to be the regime's main arms supplier, with MiG-23s and Su-27s, as well as Mi-24 and Mi-35 attack helicopters and Mi-17 transport helicopters. The EAF suffered relatively heavy losses during the fighting in 1998, with Eritrea claiming to have shot down as many as seven MiG-23s. During the conflict, the EAF relied on Russian pilots to fly most, if not all, of its Su-27s.

Today, the EAF has some 2,500 personnel. It operates ten Su-27A/U in one squadron, while there are also 18 MiG-23BN, supported by four MiG-23UB trainers, and 30 MiG-21MF, with five MiG-21U trainers. A number of MiG-17s are still believed to be operational. There are 11 Mi-24 attack helicopters. Transport is provided by four C-130B Hercules acquired in 1998, 11 An-12 and an An-32, as well as 21 Mi-8/-17 and two Mi-14 helicopters, with 20 Alouette III on communications duties. A VIP Yak-40 is also operated. Training uses 12 SF260TP and 15 L-39Z Albatros. It is not known whether there are any additional Mi-35 attack helicopters. Rumours of Kamov Ka-50 being delivered remain unsubstantiated. The Mi-14s were acquired for ASW duties, but now that the country is land-locked are used as transports.

The Ethiopian Army operates two DHC-6 Twin Otter and a Cessna 401 on light transport and communications duties.

Fiji

POPULATION: 825,000

LAND AREA: 7,055 square miles, 18,272 sq km

GDP: $1.5bn (£1bn), per capita $6,400 (£4,475)

DEFENCE EXPENDITURE: $32m (£22.4m)

SERVICE PERSONNEL: 3,500 active, plus 6,000 reserves.

FIJI MILITARY FORCES AIR WING

Formed: 1989

The extensive group of islands that is Fiji gained its independence from Britain in 1970, and established a small army and navy. An air element was created within the army when France donated an AS355 Ecureuil II helicopter in 1989, although this machine and an SA365 Dauphin were both sold in 2000 and the Air Wing disbanded.

Finland

ABOVE: *Frontline strength of the Finnish Air Force is the Boeing F/A-18 Hornet, which equip three squadrons. (Jane's)*

POPULATION: 5.2 million

LAND AREA: 130,120 square miles, 360,318 sq km

GDP: $120bn (£84bn), per capita $23,772 (£16,623)

DEFENCE EXPENDITURE: $1.5bn (£1.05bn)

SERVICE PERSONNEL: 32,250 (some 50% conscripts) active, plus reserves of 430,000.

FINNISH AIR FORCE/ILMAVOIMAT

Formed: 1920

Finland seized her independence from Russia during the Russian Revolution, and in the war that followed the Finnish forces flew a Swedish-built Albatros reconnaissance-bomber which was sooned joined by additional aircraft of the same make. Other aircraft included a Friedrichshafen seaplane, a flying boat of unknown origin, Nieuport 12s and 17C-1s and Morane Parasols. Most of the pilots were Swedish volunteers. The war ended in 1920, and the new air arm, that had become known as the Finnish Flying Corps, became a separate service, the Finnish Air Force, Ilmavoimat. France provided support, including 20 Breguet Br14B-2 reconnaissance-bombers, 12 Georges-Levy flying boats, Caudron GIII, C59 and C60 trainers, while more than a hundred Hansa-Brandenburg A22 seaplanes were built under licence. A British aviation mission visited 1924, when a small number of Gourdou-Leseurre C1 and Martinsyde F4 Buzzard fighters entered service.

The USSR did not accept Finnish independence and demanded the use of Finnish bases. The continuing tension between the two countries ensured that a steady stream of new aircraft was introduced, including Blackburn Ripon II reconnaissance and torpedo seaplanes, Gloster Gamecock II and Bristol Bulldog fighters, de Havilland and Letov trainers, and licence-built Fokker DXXI fighters and CX reconnaissance aircraft. There were also the Finnish-designed Viima and Tuisku trainers. While Germany detained 35 Fiat G50 fighters en route to Finland before World War II broke out in 1939, 18 Bristol Blenheim bombers were delivered from the UK. The Soviet Union invaded Finland on 30 October, 1939, amassing vastly superior forces both on land and in the air. At first, the Ilmavoimat held its own against some 900 obsolescent Soviet aircraft, but the USSR transferred additional aircraft to the Finnish front and the numbers reached 2,000 by 1940. Despite deliveries of additional Blenheims, Gloster Gladiator, Hawker Hurricane, Brewster 239 and Curtiss Hawk 75A and A-4 fighters and Westland Lysander AOP aircraft, Finland was forced to cede territory to the USSR, in 1940. The following year, when the Russo-German alliance collapsed, Finland allied herself with Germany, and received Morane-Saulnier MS206 and Messerschmitt Bf109G fighters, Junkers Ju88 and Dornier Do17 bombers, and Do22W seaplanes. As the war turned against Germany, Finland surrendered to Russia a second time in 1944, and was forced into war against Germany.

Post-war, the Ilmavoimat received sufficient surplus Bf109G fighters to standardize on the type with four squadrons. The 1947 Treaty of Paris restricted Finland to 60 combat aircraft and 3,000 personnel, amongst other measures enforcing Finnish neutrality during the Cold War. The first jets were not received until 1955, when de Havilland Vampire FB52s were delivered, as well as two Hunting Pembroke C53 communications aircraft. In 1958, Folland Gnat light fighters were introduced, and licence production of Potez Magister jet trainers started. The following decade saw high performance aircraft with 20 Mikoyan MiG-21 interceptors equipping two squadrons, while a third squadron operated nine Gnats and another operated Magister armed-trainers. Transport aircraft included ten Douglas C-47 Dakota as well as the Pembrokes, DHC-2 Beaver, as well as two Mi-8, four Mi-4 and two SM-1 (Polish-built Mi-1), a Bell JetRanger and two Alouette II helicopters. Training was provided on 30 SAAB-91C Safir, 55 Potez Magister, and a small number of MiG-15UTI and MiG-

21UTI trainers.

SAAB Drakens eventually replaced the MiG-21s, while BAe Hawks were introduced to replace the Magisters, with basic training using Valmet L-70 Vinka aircraft. During the late 1990s F/A-18 Hornets were introduced, with the last deliveries in August 2000, when the Drakens were retired. In 1997, the helicopter force was transferred to the army. The Ilmavoimat's strength remains within the Paris Treaty limits.

Today, the Ilmavoimat has 2,700 personnel, although its wartime strength would be 35,000, mainly employed on AA duties. It has 57 F/A-18C Hornets in three squadrons, with another seven F/A-18D for conversion training. There are 52 Hawk 51/51A, used mainly for training but with 20 available in an emergency for reconnaissance and ground-attack duties. A Fokker F-27-100 provides electronic support and two F27-400M transport. Three Learjet 35s are used for VIP transport, ECM training and target-towing. Nine L-90 Redigo, seven Piper Arrows and six Chieftains are used on liaison and communications duties, mainly in support of the army. In addition to the Hawks and Hornets, 28 Vinka are used for training. SAMs include SA-11, SA-16 and SA-18, as well as Crotale.

FINNISH ARMY AVIATION/MAAVOIMAT

Formed: 1997

The Ilmavoimat's helicopter force, seven elderly Mi-8 transport helicopters plus a couple of MD500D for training, transferred to the Finnish Army in 1997. Ambitious long-term expansion plans include creating a force of up to 15 attack helicopters and as many as 40 transport helicopters, with the latter role being filled by an initial 20 NH90s selected under the Nordic Standard Helicopter Programme.

ABOVE: *The Finnish Air Force was an early customer for the BAe Hawk advanced jet trainer. (BAE SYSTEMS)*

France

POPULATION: 59.3 million

LAND AREA: 212,919 square miles, 550,634 sq km

GDP: $1.3tr (£909bnr), per capita $25,300 (£17,692)

DEFENCE EXPENDITURE: $35bn (£24.5bn)

SERVICE PERSONNEL: 273,740 (7% conscript) active, plus 419,000 reserves.

FRENCH AIR FORCE/ARMEE DE L'AIR

Re-formed: 1943

French forces used a balloon for spotting at the Battle of Fleurus in 1794, but the origins of the present Armee de l'Air date from 1910. In that year, the French Army established the Aviation Militaire, with a Bleriot, two Farman and two Wright aircraft, some flown by naval officers preparing for the formation of a naval air arm, the Service Aeronautique later that year. The following year saw no less than 30 aircraft and balloons operated. The AM was involved in early experiments in aerial photography and radio transmission. In 1912, the squadron, *l'escadrille*, was created.

On the outbreak of World War I in 1914, the AM had 21 squadrons in France: five with Maurice Farmans, four each with Henri Farmans and Bleriots, two each with Voisins and Deperdussins, and one each with Breguets, Caudrons, Nieuports and REPs. There were another four squadrons in the colonies. Initially, air power was used for reconnaissance, although bombing emerged as observers dropped 90mm shells fitted with fins over the side of aircraft, while aerial combat appeared in 1915. Morane-Saulnier Type C aircraft formed the AM's first fighter squadrons. Rapid expansion under wartime conditions saw the AM suffer 60 per cent casualties, the highest of any Allied service, in the air or otherwise. The return of peace found it with 3,480 combat aircraft in 255 squadrons. These included 1,600 reconnaissance and spotting aircraft in 140 squadrons, 480 bombers in 32 squadrons, and 1,400 fighters in 83 squadrons. Aircraft included Caudron GIII and Maurice Farman AOP aircraft; Caudron GIV, Dorand AR-1 and Voisin reconnaissance aircraft; Breguet-Michelin IV, licence-built Caproni tri-motors and Breguet Br14 bombers; Morane-Saulnier, Spad S7C and S13C, Nieuport 17C-1 Bebe and Caudron RXI fighters.

Peace saw a reduction to 180 squadrons spread across France, Germany, and in the colonies, including Algeria and Tunisia and elsewhere in Africa. In 1920, a squadron was formed in French Indo-China with Breguet Br14 bombers. New aircraft started to appear, including Breguet Br16 bombers

France

ABOVE: *The Dassault Mystere was the manufacturer's first swept-wing design, seeing service with the Armee de l'Air during the mid-1950s. (Dassault Aviation Collection)*

and Nieuport 29C fighters. Peacetime duties consisted mainly of police action in the colonies, but the uprising led by Abd el Krim that started in Morocco in 1925 required reinforcements and continued until 1934. New aircraft continued to enter service, and during the 1920s these included Liore-Gourdou-Leseurre 32, Nieuport-Delage 62C, Spad 81, Wibault, Liore-et-Oliver LeO 20, Amiot 122 and Bleriot 127M fighters; Potez 25 reconnaissance aircraft; Morane-Saulnier MS35, 130 and 138, Hanriot-Dupont 32 and Caudron C59 trainers. Nieuport-Delage 629C and Morane-Saulnier MS225C-1 fighters; Liore-et-Oliver LeO 206 four-engined bomber and Potez 39 AOP aircraft followed during the 1930s.

Post-war, the AM had grown into the 'fifth arm' of the French Army, behind the infantry, cavalry, artillery and engineers. In 1928, it became an autonomous air force, the Armee de l'Air. It included control of six shore-based naval squadrons, including two with Dewoitine D500 fighters. The new service soon suffered neglect, as the uncertain French political situation aggravated the problems already experienced during the Depression, and the lower priority accorded defence expenditure in peacetime. In 1935, events in neighbouring Germany and Italy forced an urgent modernisation and expansion programme, with deliveries of Dewoitine D501 and D520 fighters, Morane-Saulnier MS406 and Bloch MB151 fighters, Farman F221, Bloch MB210 and Liore-et-Oliver LeO 45 bombers, plus Potez and Bloch general-purpose aircraft. The programme soon fell behind, aggravated by the decision to nationalise the French aircraft industry in 1936 and 1937. Aircraft were ordered from the USA, but these could not be delivered in time.

On the outbreak of World War II, most obsolete aircraft were transferred to areas of low risk, although some were used for leaflet dropping. There was a reluctance to mount offensive operations against Germany for fear of retaliation, but once the German invasion started, there were some successes in aerial combat. The Germans invaded the Low Countries in May, 1940, by-passing the French defensive Maginot Line. The Luftwaffe's overwhelming aerial superiority forced French and British forces into retreat, with France surrendering in June. Germany planned to disband the AdlA immediately, but the British attack on the French fleet at Mers el Kebir and Dakar, led to hopes that Vichy France would support the Axis Powers militarily. Units in France were finally disbanded in 1942, after some units in North Africa defected to the Allies following the landings in Algeria and Morocco.

Earlier, many French airmen had managed to escape to the UK, and a Free French Air Force was formed within months. A few aircraft had been flown out of France, but for the most part, the FFAF flew Hawker Hurricane and Supermarine Spitfire fighters, Bristol Blenheim and Martin Maryland bombers, and Westland Lysander army co-operation aircraft. Operations started in 1942, and in 1943 a merger with former Vichy units allowed the Armee de l'Air to be re-established. New equipment during the final years of the war included North American F-51 Mustang, Republic F-47 Thunderbolt, Bell Airacobra, Lockheed Lightning and de Havilland Mosquito fighters and fighter-bombers; Douglas Dauntless, Lockheed Lodestar and Handley Page Halifax bombers, and Douglas C-47 transports.

Post-war saw French forces struggling to quell insurrections in French Indo-China as nationalists attempted to take-over before the colonial power could re-establish itself. Many personnel arrived before their equipment, Supermarine Spitfires, so they flew ex-Japanese Nakajima Ki 43 fighters instead.

In France, the AdlA started to modernise and re-equip itself, standardising on fewer aircraft types to simplify maintenance and training. De Havilland Vampire F1 jet fighters were produced under licence, while the Dassault MD450 Ouragan entered production. As a founder member of the North Atlantic Treaty Organisation, NATO, the country was eligible for American military aid, initially consisting of Republic F-84F/G Thunderjet fighter-bombers and Lockheed T-33A jet trainers. These joined a number of other types in the late 1940s and early 1950s, including Grumman Bearcats, Gloster Meteor NF11 night-fighters, Morane Saulnier MS500s, Nord 1100 (a Bf109 development), Beech C-45s, Bell 47 Sioux and Sikorsky S-55 helicopters. The AdlA had 123,000 personnel in 75 squadrons. Gradually, as the French aircraft

France

ABOVE: *A French Air Force Dassault Mirage 2000D strike aircraft accelerates down the runway. (Armee de l'Air)*

industry re-established itself, French designs - the Dassault Mystere fighter, Sud Vatour fighter-bomber, Nord Noratlas transport and the Morane-Saulnier MS733 Alcyon trainer - came to predominate. At the time of the Anglo-French invasion of Suez in 1956, Mysteres and Thunderjets flew many of the operational sorties.

During the late 1950s and early 1960s, modernisation continued, coupled with a determination to 'fly French'. Morane-Saulnier MS760 Paris and Potez Magister jet trainers replaced the Lockheed T-33As, while the Dassault Etendard IV fighter-bomber entered service, followed by the highly successful Dassault Mirage IIIC interceptor. The export success of this aircraft and the Mirage 5 development was credited by one French air force chief with making a modern air force affordable. Sud Alouette II and III helicopters also appeared during the 1950s. The mid-1960s saw the first appearance of the Mirage IVA bomber to carry the first generation of French atomic weapons and the keystone of the French *Force de Frappe.* As development costs soared, France turned to international collaborative projects, with the Franco-German C-160 Transall transport appearing towards the end of the 1960s to replace the Noratlas. By the early 1970s, only a few foreign aircraft remained. These included 12 Boeing KC-135F tankers, two Douglas Skyraider squadrons based overseas; three Tactical Air Force North American F-100D Super Sabre squadrons; and a few Douglas C-47s and C-54s, Sikorsky H-34 helicopters, and six special duties Canberra B6 jet bombers. Collaborative projects continued, with almost 200 Anglo-French Jaguar strike aircraft entering service during the early 1970s, followed by 130 Franco-German Alphajet trainers during the middle of the decade.

The North Atlantic Council was moved out of Paris to Brussels after France refused to cooperate in the alliance's command structure after the mid-1960s.

Steady contraction also came to affect the AdlA, with personnel falling from 105,000 in 1973 to 60,500 today. Decolonisation meant that overseas commitments fell, but serious difficulties were only encountered in French Indo-China and in Algeria. Despite its lukewarm attitude to NATO, France co-operated in both the Gulf War, as a member of the Coalition Forces established to liberate Kuwait after the Iraqi invasion in 1989, and in NATO operations over the former Yugoslavia during 1999.

France still maintains substantial forces in its former colonies, often with strong AdlA support. This is in addition to UN peacekeeping commitments. Major changes taking place include the move to all-professional armed forces, and French

ABOVE: *For many years, the Dassault Mirage family were the mainstay of many air forces, including, of course, that of France. (T. Holmes)*

commitment to the so-called European Rapid Reaction Force that has grown out of the Franco-German Eurocorps and can be expected to include its own tactical air power and air transport element for rapid deployment.

The Mirage series continued with the development of the first non-delta wing version, the Mirage F1, which was followed by the Mirage 2000 series, including the 2000N designed to deliver nuclear weapons and replace the Mirage IVA. The one major national aircraft project in recent years has been the Rafale, a fighter and fighter-bomber replacement for the Jaguar and much of the Mirage force, but due to budgetary constraints, this programme has been stretched, and the first operational squadron will not be declared ready until 2005. There will be 139 Rafale interceptors, while up to 95 reconnaissance versions may be ordered.

ABOVE: *Every air force needs communications aircraft, such as the SOCATA TBM700. (Armee de l'Air)*

Today, the AdlA is organized in a series of commands. Air Combat Command has six fighter squadrons with 114 Mirage 2000B/C/5F; seven squadrons in the strike role, two of which operate 50 Jaguar A, two have 40 Mirage F1-CT and three have 60 Mirage 2000D. Weapons include ASMP and AS-30/-30L ASM, and Super 530F/D, R-550 Magic I/II and AIM-9 Sidewinder AAM. Strategic nuclear forces include three squadrons with 60 Mirage 2000N, supported by 11 C-135FR and three KC-135 tankers. There are two reconnaissance squadrons with 40 Mirage F1-CR, and a handful of Mirage IVP, converted from the IVA force when replaced by the Mirage 2000N. Electronic warfare uses a squadron of specially modified C-160 Transalls. There are five E-3F Sentry AEW aircraft in a single squadron. Air Mobility Command has 14 transport squadrons, including a long-haul hybrid squadron with two DC-8F and two A310-300; six squadrons with the bulk of upgraded 66 C-160 Transall and 14 Lockheed C-130H/H-30 Hercules for tactical lift. The Transalls are due to be replaced by 50 Airbus A400M transports from 2008. There are seven light transport squadrons, some of which also have SAR or training responsibilities, operating C-160, six DHC-6 Twin Otter, 14 CN235, 14 Falcon 20, four 50 and two 900, and 17 TBM-700, 19 Nord 262 and some of the 43 AS555 Fennec helicopters for communications duties. A further electronic warfare unit operates DC-8s in the ELINT role, while four C-160 Transalls are airborne command posts. There are also five squadrons operating Eurocopter AS332, SA-330, AS555, AS-355 and SA319. In 1999, the first four Europter Cougar Mk2s were introduced for CSAR and special operations. Air Training Command operates Alphajets, TB-30, EMB-312, CAP-10/20/-231, CR-100 and N262 aircraft, but there are also conversion units for the aircraft of the two operational commands.

France

NAVAL AIR ARM/AERONAUTIQUE NAVALE - L'AERONAVALE

Formed: 1945

French naval aviation started just slightly later than that of the army in 1910, when the Service Aeronautique was formed with a Henri Farman and a Voisin seaplane, soon joined by Breguets and Nieuports. Naval exercises during summer 1913 proved the value of the aircraft for reconnaissance, leading to conversion of the cruiser *La Foudre* to a seaplane tender.

During World War I,' two German cargo vessels, *Ann Rickmers* and *Rabenfell*, were seized and used as seaplane tenders. French naval aircraft reconnoitred the Austrian fleet in the Adriatic, and flew reconnaissance missions against Turkish forces in the Middle East for the British Army. Before the end of the war, a platform was built over the forward gun turrets of the battleship *Paris* for take-off trials. Service Aeronautique strength increased to 1,260 aircraft during the war years.

In 1925, the Service Aeronautique was renamed the Aeronautique Maritime, and the French coastline was divided into six Districts Maritimes, later being reduced to four in 1929, amongst which the AM's aircraft were divided. Six shore-based squadrons were lost to the Armee de l'Air in 1928, but overall the French Navy, or Marine Nationale, retained control of its own aircraft, and had already acquired its first aircraft carrier, the converted battleship *Bearn*, in 1925. The ship had earlier, in 1921, been used for further trials with a wooden platform laid over her forward gun turrets. In 1929, a new seaplane tender, the *Commandant Teste*, joined the *Bearn*. A major role became support for ground forces fighting rebels in Morocco. Equipment included shore-based Farman Goliaths, dating from the war, as well as CAMS 37s and 55s, Latham 43s, Gourdou-Leseurre GL-32s and Dewoitine D1s, Levassaeur PL7Bs and PL10Rs, operating from the carrier, while Latecoere 29s and Gourdou-Leseurre GL-810s operated from the seaplane tender. Wibault 74s replaced the D1s, and these in turn were replaced by Dewoitine D373 fighters aboard the *Bearn*. The carrier's other aircraft remained unchanged until after the outbreak of World War II, when they were based ashore while she became an aircraft transport ferrying aircraft from the USA. New seaplanes and flying boats did enter service during the 1930s, including Liore-et-Olivier LeO257, Levasseur PL15, Latecoere 298 and Liore 210 seaplanes, as well as Breguet Bizerta (Short Calcutta), and Liore 70 flying boats.

France's first two purpose-designed aircraft carriers were planned before the out-

ABOVE: *Too small to operate aircraft such as the F-4 Phantom, until the late 1990s France's Aeronavale depended on LTV F-8-E Crusader fighters for operations from its two aircraft carriers,* Foch *and* Clemenceau. *(Vought)*

break of war, but the first of these, *Joffre*, was only a quarter complete when Germany invaded, and the second, *Painleve*, wasn't started. *Joffre* would have been unusual, having a heavily offset flight deck to compensate for a large island, and two hangars despite being just 18,000 tons, indicating little protective armour.

Shore-based aircraft enjoyed some success against German and Italian forces before the fall of France. Some aircraft and their pilots escaped to Britain, but others were in North Africa and found themselves being attacked by the Royal Navy after the Vichy commander refused to surrender. After the Allied landings in North Africa, those AM aircrew who joined the Allies were assigned to the Armee de l'Air.

ABOVE: *Hook down, a Dassault Super Etendard prepares to land on a carrier – the aircraft was developed from the original Etendard when the Jaguar proved to be too heavy for carrier operations. (Dassault Aviation Collection)*

In January, 1945, the French Navy was loaned an escort carrier, the former HMS *Biter*, as an interim measure before the light carrier HMS *Colossus* could be transferred later that year, to be renamed *Arromanches*. The Aeronautique Navale, often referred to as the Aeronavale, was born. A second light aircraft carrier followed in 1951, the Independence-class USS *Langley*, which entered French service on loan as *La Fayette*. These two ships were to prove invaluable during the war in French Indo-China. A third ship, the loaned Independence-class USS *Belleau Wood*, *Bois Belleau* in French service, arrived in 1953, but too late for the war. Carrier-borne aircraft were needed as shore bases had become untenable as Communist forces advanced. Aircraft operated during this period were primarily of US origin. Aboard the carriers, Douglas, Dauntless and Curtiss Helldiver dive-bombers operated alongside Grumman F6F Hellcat and Chance-Vought Corsair fighter-

BELOW: *Given French pressure for a European 'defence identity', a growing need can be seen for strategic transports such as the Airbus A310-300. (Armee de l'Air)*

bombers, and Grumman TBM-3 Avenger anti-submarine aircraft. Ashore, there were Convair P4Y-2 Privateer and Avro Lancaster MR aircraft, although later replaced by Lockheed P2V-6/7 Neptune patrol aircraft, as well as Short Sunderland and captured Dornier Do24 flying-boats, and Grumman JRF-5 Goose and Consolidated PBY-5A Catalina amphibians. The first jets were Sud Aquilan (Sea Venom) fighters. Helicopters were also introduced, with Bell 47, Vertol H-22 and HUP-2, and Sikorsky S-55s entering service. Training used North American SNJ-5s.

By the end of the decade, 75 Breguet Br1050 Alize turboprop ASW aircraft entered service, mainly aboard the carriers, joined by a hundred Dassault Etendard strike aircraft. The early 1960s saw the US carriers returned, and the first two purpose-designed French ships, *Clemenceau* and *Foch*, join the fleet. With their longer angled decks, these ships were able to operate heavier fighter aircraft, and two squadrons of LTV F-8E Crusader interceptors were obtained. Throughout this period *Arromanches* remained in service, operating as a helicopter and training carrier. A helicopter cruiser, *Jeanne d'Arc* also joined the fleet in 1963. Plans to replace the Etendards with Sepecat Jaguar strike aircraft during the mid-1970s had to be abandoned as the aircraft proved too heavy and under-powered for carrier operation, and an up-rated Super Etendard had to be put in service. Nevertheless, the Neptunes were replaced by 40 Breguet Br1150 Atlantique MR aircraft, in a joint venture with Germany. ASW helicopters entering service during the 1970s included 12 Sud Super Frelon, while the Sikorsky S-58 was built under licence. Lighter helicopters for liaison and training included the Alouette II and III. In common with most navies, frigates (often referred to in France as corvettes) and destroyers capable of carrying small helicopters entered service, and for these the Anglo-French Westland-Aerospatiale Lynx shipboard helicopter was developed.

Today, the Aeronavale accounts for 3,500 of the French Navy's personnel. After *Clemenceau* was withdrawn during in the 1997 round of defence cuts, her sister *Foch* remained in service until withdrawn in 2000 when she was sold to Brazil and replaced by the nuclear-powered *Charles de Gaulle*. The F-8s were withdrawn that same year. Many of the 52 surviving Super Etendard remain in service, having been heavily updated during the early 1990s, and use R-550 Magic 2 AAM, but are being replaced with the first of 60 Rafale, the first squadron of which became operation in mid-2001. Both Rafale and Super Etendards operated from the *Charles de Gaulle* in early 2002, with British and US warships as part of a Coalition fleet following the New York and Washington terrorist attacks in September, 2001. Most carrier versions of the Rafale will have two seats for improved air-to-surface capability. Three Grumman E-2C Hawkeye AEW aircraft also operate from the carrier. Ashore, 28

ABOVE: *The Franco-German Alphajet soldiers on in the advanced training role. (Armee de l'Air)*

BELOW: *The new Dassault Rafale. (Dassault Aviation Collection)*

upgraded Atlantique ATL2 aircraft operate in two squadrons. The Marine Nationale still has 37 Lynx HAS4 (FN) helicopters able to operate from 29 frigates and four destroyers, being joined by 27 collaborative NH90 ASW and ASuV helicopters. Other helicopters include 12 SAR and four transport Super Frelon and five SAR and four transport Dauphin II, with 24 Eurocopter AS565MA Panther in course of delivery replacing 30 Alouette III. There are five Nord N262 Fregate transports, and another eleven for training alongside six Dassault Falcon 10MER, which also fill the liaison role. VIP operations and communications duties are covered by 12 Embraer EMB-121AN Xingu. Basic training uses 15 Socata Rallye 100S/100ST and ten CAP10B. A second nuclear carrier is unlikely to be built, as plans for *Foch* to remain in maintained reserve as a standby have already been abandoned.

ABOVE: *France maintains a substantial tactical transport force with more than 60 C-160 Transalls, some of which can also operate in the tanker role, augmenting the force of KC-135s. (Armee de l'Air)*

FRENCH ARMY

BELOW: *The French Army operates Eurocopter Tigers in the attack role. (ALAT)*

AVIATION/AVIATION LEGERE DE L'ARMEE DE TERRE

Formed: 1954

Originally founded to operate light aircraft and small helicopters in support of ground forces, the ALAT proved its worth during the Algerian emergency. After the withdrawal from Algeria, the force continued to increase and during its first twenty years managed to amass a total of a thousand aircraft, of which 600 were helicopters, and an organisation based on a quota of forty aircraft to a division. Heavier helicopters such as the Vertol H21, Sud Super Frelon and SA-330 Puma were added to the Alouette II and III and the small number of Bell 47 Sioux. Fixed-wing aircraft during this peri-

od included Max Holste 1521M Broussard (Bushranger), Piper and Cessna O-1 aircraft, while Nord 3200 and 3400 trainers were also used.

The French Army has been reducing in size over the past twenty years, and is likely to reduce further from its present strength of 169,300 personnel as the decision to abandon conscription takes effect. Few fixed-wing aircraft remain, and the number of helicopters has fallen to around 500. ALAT has 80 collaborative programme Tiger attack helicopters entering service from 2003, with a requirement for up to 215 more. Reconnaissance is provided by 114 Eurocopter SA341M/F Gazelle, with a further 185 SA342M/L1 Gazelle. Four AS-532UL Cougar Horizon helicopters provide electronic surveillance. There are 124 SA330B/H Puma and 64 AS532M Cougar transport and assault helicopters - some SA330B will be replaced by up to 68 European collaborative NH90 TTH helicopters. Training takes place on 18 AS555 Fennec helicopters. The small number of fixed-wing aircraft includes five PC-6B Turbo Porter and two Cessna-Reims Caravan II, as well as eight Socata TBM700, all used on liaison duties.

ABOVE: *The Armee de l'Air is one of the growing number of users of the Embraer EMB-312 Tucano turboprop trainer. (Armee de l'Air)*

Gabon

POPULATION: 1.6 million

LAND AREA: 103,000 square miles, 266,770 sq km

GDP: $6.4bn (£4.5bn), per capita $5,400 (£3,776)

DEFENCE EXPENDITURE: $126m (£88m)

SERVICE PERSONNEL: 4,700 active.

GABON AIR FORCE/FORCE AERIENNE GABONNAISE

Formed: 1960

Originally part of French Equatorial Africa, Gabon became independent in 1960 and received the standard French military gift of three Max Holste 1521 Broussards, an Alouette II helicopter and a C-47. By the early 1970s, this force had reduced to two Broussards, but the force has been restructured since and has acquired combat aircraft, modelling itself on the French pattern. It now has 1,000 personnel. There are nine Mirage 5 aircraft for interception and ground attack duties. An EMB-110P Bandeirante provides MR with two more as transports, with a Lockheed C-130H Hercules plus two civil versions, and one each of CASA CN235, Nord 262C Fregate, ATR 42F and a Bell 412SP helicopter, with a VIP flight consisting of a Douglas DC-8, Falcon 900EX and Gulfstream IVSP. An AS350 Ecureuil is used on communications duties.

The Presidential Guard has a COIN capability, with three Magister and Beech four T-34C-1 Turbo Mentor armed-trainers, as well as VIP Bandeirante, Falcon 900 and AS355 Ecureuil. The small army has five Gazelle armed helicopters and five Pumas.

ABOVE: *Lockheed L 100-30 (Mader-Acaes Collection)*

Georgia

POPULATION: 4.9 million

LAND AREA: 26,900 square miles, 69,671 sq km

GDP: $4.7bn (£3.3bn), per capita $5,289 (£3,698)

DEFENCE EXPENDITURE: $119m (£83m)

SERVICE PERSONNEL: 16,790 active, plus 250,000 reserves.

ABOVE: *The locally-built Sukhoi Su-25 is the mainstay of the Georgian Air Force. (Jane's)*

GEORGIAN AIR FORCE

Formed: 1991

On the break-up of the USSR, the Georgian Air Force was established using aircraft stationed in the country and those of its citizens serving in what had been the Soviet air forces. Compared to many other former Soviet republics, Georgia has the advantage that the Su-25 was manufactured in the country, avoiding the problems that have afflicted many post-USSR air forces. Nevertheless, economic problems are affecting serviceability. Five Su-17s are non-operational. Up to 50 Su-39s, developments of the Su-25s, are supposed to have been ordered for delivery up to 2007. The GAF has 1,330 personnel. Its operational aircraft include seven Su-25/K/UB strike aircraft including a conversion trainer, and three Mi-24 attack helicopters. Six An-2s, ten UH-1s, a gift from the USA, a Mi-2 and four Mi-8/-17 helicopters provide transport. There is a Tu-134 VIP transport. Training uses four L-29 Delfin and four Yak-52/Yak-18T, with ten L-39s decommissioned.

Germany

POPULATION: 82.4 million

LAND AREA: 137,732 square miles, 356,726 sq km

GDP: $1.8tr (£1.26tr), per capita $24,500 (£17,133)

DEFENCE EXPENDITURE: $28.8bn (£20.1bn)

SERVICE PERSONNEL: 308,400 (38% conscript) active, plus 363,500 reserves.

ABOVE: *Both the German Military Air Service and the Navy operated airships during World War I, this is one of the later naval airships, L40, in her shed at Cuxhaven. (RAF Museum)*

GERMAN AIR FORCE/LUFTWAFFE

Re-formed: 1955

German military air power dates from the purchase of a Zeppelin dirigible by the German Military Board in 1907. Further dirigibles followed, and in 1910, the German Army received its first aircraft, five Henri Farmans, five Wrights and an Antoinette. A Naval Air Service was formed in 1911 with two Curtiss seaplanes. In 1912, the army air arm became the Military Aviation Service. Both air arms grew quickly in the couple of years remaining before the outbreak of World War I, in which Germany was one of the Central Powers, allied with the Austro-Hungarian Empire and Turkey. The NAS obtained Rumpler-Etrich Taubes (Doves), Euler biplanes and had Farmans built under licence, having 36 aircraft by 1914. The MAS also favoured the Taube, and versions of this aircraft were built by Albatros, Aviatik, AEG, DFW, Euler, Gotha, LVG and Otto, so that by 1914, half the MAS strength of 250 aircraft were Taubes. Amongst these were the Stahltaube (Steel Dove) built by the Jeannin concern.

At first aircraft were used primarily for reconnaissance, although small bombs were dropped on Paris as early as August, 1914. The Zeppelins proved too vulnerable for reconnaissance duties, and were diverted to bomb targets in London and England's east coast ports, causing the British to divert aircraft from France to home defence. The MAS gained a technical advantage when it adopted the Dutchman Anthony Fokker's invention allowing a synchronised machine-gun to fire through an aircraft's propeller disc, first introduced on the Fokker EI *Eindecker* monoplane in July, 1915. It was to be early the following year before Allied aircraft incorporated such a feature. Tactical developments included the introduction of large formations - known as 'flying circuses' - of fighter aircraft in 1916, the most famous being that led by Baron Von Richthofen, the 'Red Baron'. The 'red' referred to the colour of his aircraft rather than his politics! The NAS also remained active throughout the war, raiding the south-east of England. At sea, the commerce raider *Wolf* used a seaplane for reconnaissance.

German aircraft production soared under the demands of wartime, rising from 1,350 in 1914 and peaking at 19,750 in 1917, before the Allied blockade led to shortages of materials and eventual defeat. The MAS operated Fokker EI, EII, EIII and DIII, and Pfalz DIII and Roland DII fighters; Albatros CIII, Aviatik CII and LVG CIII reconnaissance-bombers; and AEG GIV, Friedrichshafen GIII and Gotha GIV twin-engined bombers. The NAS operated Gotha, Sublatnig, Brandenburg and Friedrichshafen seaplanes. In addition to their own needs, the German services provided aircraft and experienced personnel to other members of the Central Powers. Yet, during the closing stages of the war, despite constant technical improvement in the new aircraft introduced in the final year or so, Allied aerial supremacy grew. At the end of the war, the MAS had 4,000 aircraft in service, and another 15,000 under construction, and 80,000 personnel. Between September, 1915, and September, 1918, two months before the Armistice, the MAS suffered 11,000 casualties and lost 2,000 aircraft.

In June, 1919, the Treaty of Versailles was signed, resulting in the disbanding of both the MAS and NAS in 1920, prohibiting the manufacture of military aircraft and even, until 1926, restricting the size of civil aircraft that Germany could build. The Treaty conditions were circumvented through training pilots in so-called civil gliding schools and at a centre established in the USSR in 1928, while aircraft factories were established in Russia, Sweden and Turkey. A nucleus of wartime pilots was also secretly retained.

After Adolf Hitler came to power in the early 1930s, the Luftwaffe was established in

ABOVE: *The German strategy of* blitzkrieg *or 'lightning war' was well suited to the Junkers Ju87 Stuka dive-bomber. (IWM)*

ABOVE: *Known as the 'flying pencil', the Dornier Do17 was widely used. (IWM)*

Germany

ABOVE: *Although not as famous as the Messerschmitt Bf109, the Focke Wulf Fw190 was a more effective fighter. (IWM)*

March, 1935, as an autonomous air force taking absolute control over all German service aviation. A naval air arm existed at first, but it was merged into the Luftwaffe after World War II began. The new Luftwaffe's initial aircrew and other key personnel were based on a core of wartime veterans, boosted by graduates from the flying schools. Aircraft design and development was achieved through bomber types being developed in the guise of airliners for Deutsche Luft Hansa, the national airline. The Luftwaffe was soon equipped with Heinkel He51 fighters, He45 and 46 reconnaissance aircraft, Junkers Ju52/3M transports, Dornier Do11 and Do23 bombers and Focke-Wulf Fw44 Stieglitz and Arado Ar66 trainers. A research centre was established by the Luftwaffe to evaluate new aircraft, many of which were to play an important part in World War II. Manufacturers were given loans to expand production and as an incentive for engineering firms to move into aviation, of which the most notable example was Blohm und Voss. Within a short time, the young Luftwaffe had 2,000 aircraft.

New aircraft designs were soon feeding into the Luftwaffe's operational units, including the Messerschmitt Bf109 and Me110 fighters, Junkers Ju87 Stuka dive-bombers, and Dornier Do17 and Heinkel He111 bombers. Invaluable operational experience came when the Spanish Civil War broke out, and Germany allied herself with the Nationalists. The forces of General Franco were provided with 20 Ju52/3Mg bomber/transports, and six Heinkel He51 fighters, followed by He70 reconnaissance aircraft and He59 and He60 seaplanes, flown by the 'volunteers' of the German Condor Legion. The transports carried the troops of the Spanish Foreign Legion from North Africa to Spain in one of the major moves in the early part of the war. More up-to-date aircraft followed to counter new Soviet fighters also finding their way to Spain on the Republican side. In 1937, Messerschmitt Bf109B/C fighters replaced the He51s. The following year, He111B and Do17E bombers replaced the Ju52/3Mgs in the bomber role. The Condor Legion returned to Germany in 1939 when the war ended. The conflict had tested aircraft and tactics, and given personnel invaluable combat experience.

The annexation of Austria and the Czech Sudetenland in 1938 and the occupation of Bohemia and Moldavia in 1939, were all assisted by a massive show of Luftwaffe air power that made resistance pointless. The Austrian Air Force was merged into the Luftwaffe. This was followed by a reorganization, replacing the original area groups, *Gruppenkommandos*, with air fleets, *Luftflotten*, one of which was commanded by an Austrian officer. *Luftflotten* were divided into *Luftgou*, air districts, and these were further divided into groups, *Gruppe*, and squadrons, *Staffelnen*. At the outbreak of World War II in Europe in September, 1939, the Luftwaffe had a front-line strength of almost 4,000 aircraft, including 1,300 Bf109 series fighters, 350 Ju87 Stuka dive-bombers and 1,300 Do17 and He111 bombers.

Germany precipitated the war by invading Poland on 1 September, 1939. Aerial supremacy was gained over the Polish Air Force at some cost, with the Poles fighting bravely, but outnumbered and their aircraft outclassed. The Luftwaffe then operated 'hit and run' bombing raids against British naval bases in the east of Scotland, but saw little further action until the invasion of Denmark and Norway the following spring. Over-run, Denmark fell almost immediately, but the Luftwaffe met fiercer resistance in Norway, supported by British and French forces and by RAF and Fleet Air Arm aircraft. Even before the Norwegian campaign ended, the Luftwaffe supported the German Army as it raced through the Netherlands and Belgium and into France, which was largely occupied by the end of June, leaving just the friendly Vichy zone unoccupied. The Luftwaffe bombed the undefended Dutch city and port of Rotterdam, and was left to destroy the retreating British and French forces at Dunkirk, but failed to achieve this before they could be evacuated. These campaigns showed the Luftwaffe's *blitzkrieg* strategy at work, with dive-bombers and medium bombers operating in close co-ordination with ground forces - in complete contrast to the British and American concept of strategic bombing.

The Luftwaffe next attempted to destroy the RAF in the famous Battle of Britain, but failed to do so. This was due in part to the British radar network and to the short range of German aircraft even when operating from France. A switch to bombing the cities gave the RAF a breathing space. The bombing campaign, known as the *Blitz*, was hampered by the lack of a heavy bomber. It too came to a premature end in 1941 as German attentions turned elsewhere to Operation Barbarossa, the invasion of Russia. The Luftwaffe was by this time becoming heavily committed, operating in the Balkans and against Greece and then Crete. Also against Malta, while also supporting German forces in North Africa. Even so, the Luftwaffe had almost 2,000 aircraft available for the invasion of the USSR, with a further 1,000 contributed by Germany's Romanian, Hungarian, Italian and Finnish allies, facing 9,000 Soviet aircraft. The operation opened on 22 June with raids against 66 Soviet airfields, crippling the Red Air Force. Here the Luftwaffe was to suffer the same fate as the ground forces, faced with long distances and unprepared for the worst of the Russian winter. Unusually amongst air forces, the Luftwaffe was responsible for the German paratroop force, but squandered much of this in the ill-conceived invasion of Crete, after which Hitler forbade any further paratroop assaults. Paratroops might have helped Operation Barbarossa and the war in North Africa through an invasion of Malta.

New equipment continued to enter service, with the Luftwaffe benefiting from continued technical improvements. Earlier in the war, the *Blitz* on British cities had demonstrated superior navigational and bomb aiming systems. New aircraft were not neglected, with Focke-Wulf Fw190 fighters, Dornier Do217 bombers, Blohm und Voss reconnaissance flying boats, Focke-Wulf Fw200 Condor and Junkers Ju290 anti-shipping aircraft, and Heinkel He115 seaplanes. A heavy bomber, the Heinkel He177, arrived too late to be truly effective, and was also dangerously unreliable. Towards the end of the war, new weapons were introduced, including the FX radio-controlled missile and the Hs293 glider-bomb, both launched from Do217 bombers. The revenge weapons, the V1 and V2 missiles, were launched from occupied Europe towards British cities, including London, and could be regarded as the predecessors of the cruise missile and ballistic missile respectively. The Messerschmitt Me262 jet fighter entered service in 1944, already too late to stem overwhelming Allied aerial superiority, especially once Hitler insisted that it be used as a bomber. There was also the Arado 234 jet reconnaissance aircraft and the Messerschmitt Me163 Komet rocket-powered fighter, and a few of the unusual twin-engined pusher-puller layout He219 fighters. All too late and in too small numbers.

ABOVE: *Luftwaffe Me.262 preserved in RAF Cosford museum (Jeremy Flack/Aviation Photographs International)*

April, 1945, saw the Luftwaffe grounded by fuel shortages, and its personnel deployed as ground troops. It was disbanded immediately the war ended.

Germany was divided into four Allied Zones of Occupation, and the former capital, Berlin, also divided. In 1949, the three Western zones were amalgamated to form the Federal German Republic, West Germany, while the Soviet zone became the German Democratic Republic, East Germany.

The GDR was the first to re-establish an air force, with an air arm of the People's Police, the Volkspolizei, formed in 1950, using Yakovlev Yak-18, Polikarpov Po-2 and Fieseler Fi156C Storch light aircraft and former Luftwaffe personnel. This became the Luftstreitskrafte in 1955, and once again, 'flying clubs' produced members for the new air force. With the GDR a member of the Warsaw Pact, the USSR supplied MiG-15 fighters and MiG-15UTI and Yak-11 and Yak-18 trainers, and these were joined by MiG-17s, Ilyushin Il-14M and Tupolev Tu-104 transports by 1960, with Mi-1 and Mi-4 helicopters following later.

The Luftwaffe reformed in 1955. West Germany became a member of NATO and qualified for US military aid. Its first aircraft were 30 Piper Cub trainers, which arrived in 1956, and were followed by 30 North American T-6 Harvard trainers and some Lockheed T-33A jet trainers. US aid included 450 Republic F-84F Thunderstreak fighter-bombers and 100 RF-84F Thunderflash reconnaissance-fighters, as well as Canadair CL-13B Sabre 6 and Fiat F-86K Sabre fighters. Hunting Pembroke, de Havilland Heron, Douglas C-47 and licence-built Nord Noratlas provided a strong transport element. Potez Magister jet trainers and Piaggio P149D trainers, as well as Bell 47 Sioux, Vertol H-21H and H-25, Sikorsky H-34, Saro Skeeter and Bristol Sycamore, and Hiller UH-12C helicopters, and the indigenous Do27 liaison aircraft completed the support element.

This left a divided Germany with two air

Germany

ABOVE: *The Luftwaffe has more than 200 Panavia Tornados for the attack, reconnaissance and ECM roles. (BMVg)*

ABOVE: *The Luftwaffe has almost 150 McDonnell Douglas F-4F Phantoms, up-graded to extend their operational life until they can be replaced by Eurofighter 2000 from 2003 onwards. (BMVg)*

forces facing each other across a frontier throughout the Cold War. The Luftwaffe was not allowed into West Berlin, itself well inside East Germany.

The Luftwaffe continued to develop during the 1960s, with licence-built Lockheed F-104G Starfighters and Fiat G91 ground attack aircraft replacing the F and RF-84s and the Sabres. The German version of the Starfighter was a multi-mission aircraft giving a heavy pilot workload, and as a result of many accidents became nicknamed the 'Widow-Maker'. The next major aircraft into the inventory was the McDonnell Douglas RF-4 Phantom which replaced the RF-104 version of the Starfighter, and in 1973, F-4s were also introduced. Joint projects between Germany and France saw the Alphajet advanced trainer and the C-160 Transall transport. Dornier built large numbers of Bell UH-1H helicopters under licence. In the mid-1970s, the Anglo-German-Italian multi-role combat aircraft, the Panavia Tornado, entered Luftwaffe service, replacing the Starfighters, while the Fiat G91s had earlier been sold to Portugal.

Similar developments took place in East Germany, with the Luftstreitkrafte receiving successive marks of MiG aircraft, including the MiG-19, MiG-21, MiG-23 and continuing up to the MiG-29.

German reunification came in 1990. The first move was for the Luftstreitkrafte to be merged into the Luftwaffe, although the older Soviet-era aircraft were quickly retired with many of their personnel, leaving the Luftwaffe with just a small number of MiG-29s to convert to NATO standards. Over the past thirty years, the two air forces have moved from a combined personnel strength of 125,000 to the present figure of 73,300, while average flying hours are 150 a year, probably the bare minimum to main-

ABOVE: *The Luftwaffe's first operational mission since it re-formed in 1955 was participation in Operation Allied Force over Kosovo - here a Tornado takes off from Piacenca in Italy. (BMVg)*

tain standards.

The 19 MiG-29As and four MiG-29UB conversion trainers, modified to NATO standards and with updated navigational equipment, were sold to Poland in 2002.

Today, the Luftwaffe is the second largest NATO air force, and the largest in Western Europe. Its structure is divided into an Air Force Command, itself divided into two air commands, North and South, and a Transport Command. AFC has 193 Panavia Tornado IDS in five strike wings with a total of ten squadrons, while another wing has two squadrons with 35 ECM and reconnaissance Tornado. Fighter and interceptor operations are handled by four wings with eight squadrons operating updated 145 F-4E Phantom IIs. Conversion training also takes 27 of the total Tornado strength and 23 Phantom. Transport Command has three wings: Four squadrons (one of which is a conversion unit) operate 83 Transall C-160; four squadrons (one of which is a conversion unit) operate 82 Bell UH-1H Iroquois in the transport and SAR roles. A special mission wing has an assortment of aircraft, including seven Airbus A-310, VFW-614, and for VIP duties, six Canadair CL-601, two L-410S and three AS-532U2 Cougar helicopters. In addition to the front-line types in the OCU units, training also takes 37 T-37B and 40 T-38A Talon jets. As a result of absorbing the Luftstreitkrafte, the Luftwaffe has AIM-9 Sidewinder, AA-8, AA-10 and AA-11 AAM, while ASM include AGM-65 Maverick and AGM-88A HARM. There are Hawk, Roland and Patriot SAM.

ABOVE: *German re-unification and the merger of the East and West German air forces brought MiG-29 interceptors into the Luftwaffe. (BMVg)*

The Luftwaffe is currently introducing 30 European NH90 TTH helicopters that will replace some UH-1H. During the next few years, it will introduce up to 180 Eurofighter 2000 Typhoon jets in the inter-

Germany

ABOVE: *The Marinflieger operates British-built Sea King 41s (S-61) for SAR. (BMVg)*

ceptor and attack roles, replacing the Phantoms and possibly also the MiG-29s. Although 73 Airbus A400M transports were ordered to replace the recently refurbished Transalls from 2010, it seems that the order will be cut to 48 aircraft, despite Germany's involvement in the European Rapid Reaction Force and the implied support for strategic mobility. Germany has a record of declaring an interest in far more aircraft in collaborative programmes than it eventually introduces to service.

GERMAN NAVAL AIR ARM/MARINEFLIEGER

Formed: 1957

While German naval aviation existed during World War I, it was disbanded after the war. The German Navy, or Kriegsmarine, maintained a small air arm during the 1930s, with shore-based He59 and He60, Dornier Do18 flying boats and Heinkel He115 seaplanes, as well as aircraft aboard battleships and commerce raiders. This force was merged into the Luftwaffe after World War II started. Rivalry between the two services over control of naval aviation fatally delayed the completion of an aircraft carrier, *Graf Zeppelin*, launched in 1938, while work on a second ship, *Peter Strasser*, started but was abandoned. These ships would have operated Messerchmitt Bf109 fighters and Junkers Ju87 Stuka dive-bombers. *Graf Zeppelin*'s catapults were donated to the Italian Navy for their planned carrier. Post-war, *Graf Zeppelin* was seized by Soviet forces, and is believed to have capsized and sunk.

The Marineflieger was formed in 1957 for shore-based air operations over the Baltic and the North Sea. Early equipment included Armstrong-Whitworth Sea Hawk strike aircraft, Fairey Gannet ASW aircraft,

BELOW: *The Marineflieger operates Westland Lynx 88/88A from its frigates. (BMVg)*

Hunting Pembroke transports and Bristol Sycamore helicopters. During the 1960s, the Sea Hawks were replaced by licence-built Lockheed F-104G Starfighters, with 72 aircraft in four squadrons, and the Gannets by the Franco-German Breguet Atlantic, while Sikorsky CH-34 helicopters were also introduced for SAR, before being replaced in this role by 22 Westland S-61N Sea King helicopters. Other aircraft operated included eight Grumman HU-16 amphibians, 20 Dornier Do28 Skyservant liaison aircraft, as well as some of the smaller Do27, and Potez Magister jet trainers.

During the 1970s, the Starfighters were replaced by more than a hundred Anglo-German-Italian Panavia Tornado in the anti-shipping strike role, but in 1994, more than half of these aircraft were transferred to the Luftwaffe.

The arrival of six Type 122 frigates, designed jointly with the Netherlands, during the early 1980s, saw the Marineflieger acquire its first shipboard helicopters, ASW and ASuV Westland Lynx.

Today, the Marineflieger accounts for 4,200 of the German Navy's personnel strength of 26,600. It did not gain any significant increase in strength through the reunification of Germany, since the East German Navy was primarily a coastal defence force. There are 49 shore-based Tornado IDS, carrying AIM-9 Sidewinder and Roland AAM, and HARM ASM. While the original force of 40 Atlantics is now down to 14 aircraft on MR and another four on SIGINT duties, although these have been updated to extend their life until 2010. There are 22 Sea Lynx, capable of carrying Sea Skua anti-shipping missiles and upgraded to Mk88A standard, which will be replaced by 38 European NH90 helicopters in 2015, after first replacing 21 Sea Kings in the SAR role. Four Dornier Do228LM/LT cover liaison and pollution control. A project exists to replace the Atlantics in a joint venture with Italy, possibly using a modified transport airframe. Kormoran ASW missiles are also deployed.

ABOVE: *The Franco-German Breguet Atlantic continues in German naval service for maritime-reconnaissance following a major up-grade. The next priority is to develop a successor, probably in conjunction with Italy. (BMVg)*

GERMAN ARMY AVIATION/HEERESFLIEGER

Formed: 1957

The Heeresflieger was founded in 1957 to provide liaison, communications and AOP facilities for the new Federal German Army. Battlefield transport followed later using Bell UH-1D Iroquois, of which 200 were in service at one time, and Sikorsky CH-53 heavy lift helicopters. North American OV-10Zs were operated at one time on target-towing duties. The move into operating

BELOW: *The German Army's Heeresflieger has its own heavy lift capability with almost 100 Sikorsky CH-53G helicopters. (BMVg)*

attack helicopters came with more than two hundred MBB PAH-1s, essentially BO105P/M equipped with HOT anti-tank missiles, while another 96 BO-105 operated in the observation and liaison roles.

Today, the Heeresflieger is undergoing substantial change. The European Tiger attack helicopter is entering service, with the aircrew trained at a joint Franco-German OCU at Luc en Provence in France. Originally, 212 were planned to replace the PAH-1, but this figure was reduced in a defence review and currently just 80 are on order. Initially the Tiger will use the HOT missile, but this may be replaced later by the European Trigat missile if this project survives. From 2004, the UH-1H will be replaced by the European NH90 TTH, with up to 120 delivered for replacement on a one-for-one basis. These changes may leave a number of PAH-1 in service, as well as the BO105 AOP and liaison machines. The 96 CH-53G includes 20 that have been upgraded with long-range tanks, EW equipment and NVG compatible cockpits. Training uses 15 Eurocopter EC635 (EC135).

ABOVE: *The Heeresflieger BellUH-1D iroquois are being replaced by the European NH90TTH helicopter. (BMVg)*

BELOW: *More than 200 Tiger attack helicopters are entering Heeresflieger service, giving the force a significant increase in its combat ability. (BMVg)*

Ghana

POPULATION: 20.8 million

LAND AREA: 92,100 square miles, 238,539 sq km

GDP: $10.7bn (£7.2bn), per capita $2,400 (£1,678)

DEFENCE EXPENDITURE: $96m (£86.4m)

SERVICE PERSONNEL: 7,000 active.

GHANA AIR FORCE

Formed: 1959

Ghana became an independent member of the British Commonwealth in 1957, but the Ghana Air Force was not founded until 1959, with Israeli and Indian assistance and a gift of Hindustan HT-2 trainers. A Piper Super Cub was also obtained. RAF personnel were seconded in 1960, when de Havilland Canada Chipmunk trainers were also provided, followed later that year by DHC-2 Beaver light transports. Early plans were to provide a transport and communications force, so DHC-3 Otters and DHC-4 Caribou were also introduced. Soviet attempts to influence Ghana involved the supply of Antonov An-12 and Ilyushin Il-18 transports in 1963, with Mil Mi-4 helicopters. These aircraft were returned to the USSR when the Nkrumah dictatorship was overthrown. Eventually, Westland S-55 Whirlwind and then S-58 Wessex helicopters were obtained, while during the late 1960s, six Aermacchi MB326F armed-trainers and Short Skyvan light transports were introduced.

Ghana has in recent years been active in supporting UN and Organisation of African Unity operations. The emphasis is still on transport and communications, although there is a COIN element using MB339A armed jet trainers obtained during the late 1970s to augment the ageing MB326s. There are just 1,000 personnel in the GhAF. There are 19 combat aircraft, all armed-trainers, including five MB326K/E, four MB339F, two L-29 and 12 L-39s. There are two Fokker F-27-400M Troopship with a VIP F-27-600 and a VIP F-28-3000, four refurbished Short Skyvan 3M and four Britten-Norman BN-2T Turbine Islander transports. Helicopters include two VIP AB412, as well as two SA319 Alouette III and two A109 for communications.

Greece

POPULATION: 10.7 million

LAND AREA: 50,534 square miles, 132,561 sq km

GDP: $113bn (£79bn), per capita $14,624 (£10,226)

DEFENCE EXPENDITURE: $5.6bn (£3.9bn)

SERVICE PERSONNEL: 159,170 (60% conscripts) active, plus 291,000 reserves.

HELLENIC AIR FORCE

Formed: 1931

Greece occupies an important strategic position offering control over the Balkans and eastern Mediterranean, in which air power plays a vital part. The importance of air power has been further heightened by the long-running feud with neighbouring Turkey, including territorial disputes and tension over the island of Cyprus and parts of the Aegean Sea. Greece itself was occupied by Turkey until 1829.

The Royal Hellenic Army established an air squadron in 1912 using four Farman biplanes during the Balkan Wars of 1912-13. The Royal Navy provided assistance for the Royal Hellenic Navy to form an aviation service in 1914, using Farman and Sopwith seaplanes. The two air arms were merged into the Hellenic Air Service in 1916, operating against German and Turkish bases in Bulgaria before re-dividing into separate air arms in 1917, becoming the Royal Hellenic Naval Air Service and the Royal Hellenic Army Air Force. Post-war, both services were strengthened using surplus British and French aircraft. The 'Royal' prefixes were dropped when Greece became a republic in 1924. To counter the effects of the Depression, a Greek Air Force Fund was established and several towns raised funds to purchase aircraft, with Salonika buying 25. New aircraft for the HNAS included Avro 504N trainers, Hawker Horsley torpe-

ABOVE: *The Hellenic Air Force's A-7H Corsairs are likely to remain in service longer than expected, as financial problems are delaying the Eurofighter Typhoon purchase. (LTV)*

Greece

ABOVE: *The Lockheed-Martin F-16 equips two of the Hellenic Air Force's six fighter squadrons. (Jane's)*

do-bombers, Greek-built Blackburn Velos (Dart) torpedo and reconnaissance seaplanes, and Armstrong-Whitworth Atlas general-purpose biplanes. French aircraft predominated, with Breguet Br19A and 19B reconnaissance-bombers and Morane-Saulnier trainers.

The two air arms were unified once again in 1931, this time into an autonomous Hellenic Air Force, becoming the Royal Hellenic Air Force with the restoration of the monarchy in 1935. Before the outbreak of World War II in Europe in 1939, new aircraft included PZL P-24 and Bloch MB151 fighters, Bristol Blenheim, Fairey Battle and Potez 63 bombers, Henschel Hs126 AOP aircraft, Avro Anson reconnaissance-bombers, and Dornier Do22 and Fairey IIIF seaplanes. Many of these types - including the Battle and Blenheim, and in the offensive role, the Anson - were to prove poor performers elsewhere. Even though the RHAF managed to repel an Italian invasion in 1940, Greece was overrun by German forces in 1941. Some personnel managed to escape to Egypt where fighter squadrons were formed with the help of the RAF using Hawker Hurricanes and a bomber squadron was equipped with Blenheims. Later, these were replaced with Supermarine Spitfire fighters and Martin Baltimore bombers.

Greece was liberated in 1944, but immediately Communists tried to seize power and the ensuing civil war lasted for five years. The RHAF returned home to engage in COIN operations bringing its wartime equipment, plus Anson, Airspeed Oxford and Douglas C-47 transports, Curtiss SB2C-4 Helldiver dive-bombers, Stinson L-5 liaison aircraft and de Havilland Tiger Moth and North American Harvard trainers. The civil war left the RHAF and the Greek economy in a much-weakened condition. Greek membership of NATO in 1952 opened the way for assistance under the US Mutual Defence Aid Programme. Personnel were trained in the USA and Germany. Substantial numbers of aircraft were delivered, including 80 Canadair F-86D Sabre Mk2/4 fighters, 250 Republic F-84G Thunderjet fighter-bombers and RF-84F Thunderflash reconnaissance-fighters, and Canadair T-33A jet trainers. These were soon joined by additional C-47s, North American T-6G Texan trainers, and the RHAF's first helicopters, Sikorsky H-19D Chickasaws. Tension between Greece and Turkey, another NATO member, in the late 1960s saw a temporary cessation of arms deliveries, resumed in 1970 with Convair F-102 Delta Dagger interceptors.

A military coupe in 1967 eventually led to a republic. The 'royal' prefix was dropped.

Many Thunderjet and Thunderflash aircraft remained in service with the Delta Dagger force. However, this was soon augmented with two squadrons each with 18 Lockheed F-104G Starfighter interceptors and four Northrop F-5A Freedom Fighter squadrons. Additional transports included Nord Noratlas and Fairchild C-119G Packet, while a SAR squadron of Grumman HU-16 Albatross amphibians was also operated.

The tendency for the USA to interrupt arms supplies whenever tension rose between Greece and Turkey led to attempts to diversify supplies. While LTV A-7 Corsairs were added during the 1970s, followed by F-4E Phantoms, the HAF also added Mirage F1G and, later, Mirage 2000EG interceptors as well as later obtaining F-16C/D Fighting Falcon interceptors. Lockheed C-130 Hercules were introduced, but other transports included Dornier Do28s and NAMC YS-11s, with Canadair CL215 and, later, CL415 amphibians being obtained for fire fighting and MR, and many Agusta-Bell helicopters.

Despite Greek sympathies with the Greek-Cypriot authorities in Cyprus, the HAF could do little to counter the Turkish invasion of Cyprus in 1973, since the only direct route to Cyprus lay over Turkish territory.

The HAF has 30,170 personnel, almost a third more than 30 years ago. It is divided into a Tactical Air Force and an Air

Training Command. There are six fighter squadrons, two equipped with F-16C, two with upgraded F-4E and two with Mirage 2000EG/BG. Fighter ground-attack squadrons total eleven, and of these two have LTV A-7H and two A-7E, two have F-16CG/DG, two F-4E, one F-5A/B and two Mirage F-1CG. There is also a reconnaissance squadron with RF-4E Phantom. There are 75 upgraded F-16C/D, being joined by another 34 F-16C and 14 F-16D, while 34 Mirage 2000 are being joined by another 18 in 2003, allowing many of the 89 A-7E/H and TA-7C Corsair to be withdrawn. Upgrading has been applied to 36 of the 62 F-4E/RF-4E. There are also 53 RF-5A/F-5A/B Tiger. Missiles used include AIM-7 Sparrow, AIM-9 Sidewinder, R-550 Magic 2, Super 530D and AIM-120 AMRAAM AAM, as well as AGM-65 Maverick and AGM-88 HARM. Greece was interested in ordering up to 90 Eurofighter Typhoons, but a firm decision has been delayed beyond 2008 due to the cost of the country mounting the Olympic Games in 2004.

The RHAF has three transport squadrons operating five C-130B and ten C-130H, which are to have their lives extended while purchases of C-130J are considered. But meanwhile, 12 Alenia/Lockheed-Martin C-27 Spartans have been ordered, with options on three more, a C-47, eight Do28 and two YS-11s starting in 2004. There is a VIP RJ-135, while four RJ-145 are used for AEW. MR is handled by six P-3B Orion, with joint navy/air force crews, aided by some of the ten CL-215 and five CL-415 fire-fighting amphibians. Other fire-fighting aircraft include PZL M-18A Dromader. The helicopter force includes ten AB205A for SAR and transport, and a JetRanger and four AB212 on VIP duties. Training uses seven Bell 47G, also used for liaison and agricultural duties, while there are four fixed-wing training squadrons. Two with 45 T-6s which replaced T-41As and T-37B/Cs in 2002, two operate T-2E Buckeye that will be replaced by new advanced jet trainers once funding allows. There are Nike Hercules, Patriot PAC-3, Skyguard, Sparrow, Crotale and SA-15 SAM.

HELLENIC NAVY AIR ARM/ AEROPORIA ELLENIKI POLEMIKOU NAFTIKOU

Formed: 1975

The Greek naval air arm was founded in 1975, initially with four Alouette IIIs for liaison duties, although these were soon supplemented by 16 Agusta-Bell AB212s for shipborne operation as the first of six Dutch-designed Kortenaer-class frigates entered service during the early 1980s. More recently, the Sikorsky S-70B Aegean Hawk has entered service, armed with the Penguin missile, and this also operates from frigates, with ten out of the 12 currently in service able to operate helicopters. A number of personnel serve with the air force's Orions.

Currently, 250 out of the Navy's 19,000 personnel are directly involved in naval aviation. There are ten S-70B Aegean Hawk in the ASW and ASuV roles that can also provide SAR, while eight AB212 also undertake these roles with another two on electronic warfare duties. Two Alouette III remain in service on SAR, liaison and training.

HELLENIC ARMY AIR ARM/ELLENIKI AEROPORIA STRATOU

Formed: 1975

Greek Army aviation was originally founded for liaison, communications, and transport. It acquired six heavy lift Meridionali-built Boeing CH-47 Chinooks during the late 1970s, by which time it also had eight Bell AH-1S HueyCobras for the anti-tank role. The main force at this time comprised 50 Agusta-Bell 204 and 205 helicopters, with a small liaison and AOP force of Bell 47G.

Currently, the air arm is spearheaded by 24 Boeing AH-64 Apache, plus another four purchased in 1999 from the US Army and stored. There are 16 CH-47D (with the original machines upgraded by Boeing in 1992), 12 UH-60 Black Hawk, 30 UH-1H Iroquois and 80 AB205A for the transport and utility roles. A Beech C-12A King Air 200 operates in the VIP role, with an AB212. Three Aero Commanders are used on communications and survey work. Training uses 26 Nardi-built Hughes 300C, 20 Cessna U-17A (Cessna 180), and five OH-13H. Up to 24 additional attack helicopters are planned for 2004 onwards, followed by 50 medium transport helicopters to replace the UH-1H, with the choice between the NH90 TTH, additional UH-60, Cougar or Agusta-Westland EH101.

ABOVE: *The Dassault Mirage 2000EG equips another two Hellenic Air Force fighter squadrons, spreading procurement to avoid occasional US delays in authorising aircraft purchases. (Jane's)*

Guatemala

POPULATION: 11.5 million

LAND AREA: 42,042 square miles, 108,889 sq km

GDP: $14.8bn (£10.3bn), per capita $4,500 (£3,147)

DEFENCE EXPENDITURE: $155m (£108m)

SERVICE PERSONNEL: 31,400 (70% conscript) active.

GUATEMALAN AIR FORCE/FUERZA AEREA GUATEMALTECA

A French aviation mission in 1919 with Avro 504 trainers marked the start of military aviation in Guatemala, forming an air unit within the Army. This became the Cuerpo de Aeronautica Militar in 1929 with a miscellany of French aircraft. Standardisation occurred in 1937, with Boeing P-26A fighters, Waco general-purpose aircraft and Ryan STM-2 trainers. During World War II, Guatemala allowed the Allies to use bases, and in return received Douglas C-47 transports, Boeing-Stearman PT-17 Kaydet, North American AT-6 and Vultee BT-15 trainers. An American aviation mission in 1945 provided a squadron of North American F-51D Mustang fighters. Guatemala joined the Organisation of American States in 1948, but did not receive new aircraft until ten Douglas B-26 Invader bombers and a number of Lockheed T-33A jet trainers were delivered in 1960. Later, the T-33As were joined by Cessna A-37B Dragonfly armed jet trainers, Pilatus PC-7 armed turboprop trainers, and Basler Turbo 67 turboprop conversions of the C-47 for COIN.

ABOVE: *A Guatamalam Cessna A-37B soldiering on as an armed jet trainer. (Mader-Acaes Collection)*

The FAG has some 700 personnel, but receives basic logistic support services from the Army. Aircraft serviceability is believed to be no better than 50 per cent. Seven PC-7s still operate alongside eight remaining A-37Bs. The Turbo 67s can be used as transports, with two Fokker F-27-100/400, an F-27-400M Troopship, and seven IA-201 Arava. A handful of Cessna T210 Centurion, U206 Stationair and a Piper Navajo operate on communications, with four Beech Super King Air. Helicopters include 12 Bell 212/412, four 206B JetRanger, three 206L LongRanger, four UH-1H Iroquois and three VIP Sikorsky S-76 Spirit. During 1999, five Enaer Pillan T-35B trainers were obtained from the Chilean Air Force, although one was lost, augmenting six Cessna T-41D Mescalero and five R172K trainers.

Guinea

POPULATION: 7.6 million

LAND AREA: 94,927 square miles, 245,861 sq km

GDP: $3.9bn (£2.7bn), per capita $992 (£693)

DEFENCE EXPENDITURE: $58m (£40m)

SERVICE PERSONNEL: 9,700 (75% conscript) active.

GUINEA AIR FORCE/FORCE AERIENNE DE GUINEA

Formed: 1958

The former French Guinea became independent in 1958 and almost immediately left the French Community. Close links with the USSR led to military aid being provided, including MiG-17 and MiG-21 fighters, Antonov transports and Mil helicopters. Today, the FAG has just 800 personnel. Its combat aircraft are no longer operational, other than a Mi-24 attack helicopter obtained in 1998 from the Ukraine. There are four Antonov An-14, an An-24 and two Ilyushin Il-18 transports, two Alouette III and an AS350 Ecureuil on liaison, with an SA342K Gazelle for VIP use. Two MiG-15UTI trainers are also believed to be out of service.

Guinea-Bissau

POPULATION: 1.2 million

LAND AREA: 13,948 square miles, 36,125 sq km

GDP: $340m (£237m), per capita $1,100 (£769)

DEFENCE EXPENDITURE: $6m (£4.2m)

SERVICE PERSONNEL: 9,250 active.

ABOVE: *Mil Mi-8 (Mark Wagner/Aviation -Images.com)*

GUINEA-BISSAU AIR FORCE/FORCE AERIENNE DE GUINEA-BISSAU

Formed: 1974

Formerly Portuguese Guinea, Guinea-Bissau became independent in 1974, beset with economic problems and internal unrest, including an Army mutiny in 1998. Internal conflict re-ignited in 1999. The small air force has 100 personnel with two MiG-17F and a MiG-15UTI, probably no longer serviceable, and has withdrawn MiG-21s from use. Transport is provided by a Falcon 20 and an An-24, with an Alouette II and two Alouette III for liaison. The coastguard has an FTB337 (Cessna 337) for off-shore patrol.

Guyana

POPULATION: 868,000

LAND AREA: 83,000 square miles, 214,970 sq km

GDP: $800m (£559m), per capita $3,400 (£2,378)

DEFENCE EXPENDITURE: $7m (£4.9m)

SERVICE PERSONNEL: 1,600 active, plus 1,500 reserves.

GUYANA DEFENCE FORCE AIR CORPS

Formed: 1967

British Guiana became independent in 1967 and established a small Guyana Defence Force, of which 100 personnel are in the air corps. It is purely a transport force, necessary for jungle-covered terrain. At one time five Mi-8 helicopters were operated, but only two remain in service. Two refurbished Short Skyvan 3M and a BN-2A Defender are operated with a Bell 206B JetRanger and a 412.

Haiti

POPULATION: 8.4 million

LAND AREA: 10,700 square miles, 27,713 sq km

GDP: $3.3bn (£2.3bn), per capita $1,100 (£769)

SECURITY EXP: $49m (£34.3m)

SERVICE PERSONNEL: none officially active.

HAITIAN AIR CORPS/CORPS D'AERIEN D'HAITI

Formed: 1943

The Haitian Air Corps was founded in 1943 with US assistance within the Army to carry mail between the main towns. After the end of World War II, Haiti became a member of the Organisation of American States in 1948, and received North American F-51D Mustang fighters, and Beech, Boeing, Fairchild and Vultee trainers. Douglas C-47 and Beech C-45 transports arrived later, as did North American T-6G Texan and T-28A Trojan trainers. In 1994, with US help a military dictatorship was replaced by a civilian administration after an invasion. The armed forces and police were disbanded prior to first an interim police force and then a new National Police Force being formed.

Honduras

POPULATION: 6.6 million

LAND AREA: 43,227 square miles, 111,958 sq km

GDP: $5.8bn (£4.1bn), per capita $2,300 (£1,608)

DEFENCE EXPENDITURE: $95m (£66m)

SERVICE PERSONNEL: 8,300 active, plus 60,000 reserves.

HONDURAS AIR FORCE/FUERZA AEREA HONDURENA

Formed: 1948

The Honduras Army formed an Air Corps after the end of World War I using two surplus Bristol F2B fighters for AOP and liaison duties. Between the two world wars, some Boeing 95 mail planes were obtained for use as light bombers, but little progress was made. In return for support in safeguarding the Panama Canal during World War II the USA provided Beech, Boeing, Fairchild, North American, Ryan, Vultee and Waco trainers as well as Beech C-17 and Noorduyn Norseman transports.

Further US military aid followed Honduran membership of the Organisation of American States in 1948, including Lockheed P-38 Lightning and Bell P-63 Kingcobra fighters, followed by North American F-51D Mustang and Republic F-47D Thunderbolt fighter-bombers, and Beech C-45 and Douglas C-47 transports. The fighter-bombers were replaced by Vought F4U Corsair fighter-bombers in 1958, when Douglas C-54 transports, Lockheed T-33A jet trainers and Cessna liaison aircraft entered service. The first helicopters, three Sikorsky H-19, followed. Later, Beech AT-11 and North American T-6G Texan trainers were added.

Jet combat aircraft followed during the 1970s, with COIN Cessna A-37B Dragonfly, and ex-Israeli Dassault Super Mystere fighter-bombers. Later, Northrop F-5E Tiger II fighter-bombers and Lockheed C-130 Hercules were added.

Modernization has been delayed by economic difficulties. The FAH has 1,800 personnel. Plans to reactivate the remaining 11 Super Mysteres have been abandoned. Two FGA squadrons include one with 13 Cessna A-37B and another with 11 F-5E Tiger II, Shafrir AAM-equipped, plus two conversion trainer F-5F. There are 11 Embraer EMB-312 Tucano armed-trainers. Two Lockheed C-130A Hercules and one C-130D, seven Douglas C-47 and two IAI-201 Arava transports operate alongside three VIP aircraft, a Lockheed L-188 Electra, an IAI-1124 Westwind and a Piper Cheyenne. Communications aircraft include a Turbo Cheyenne, Cessna 310, 401 and five 185s, a Commander 690 and two Piper Navajo. Helicopters include nine Bell 412EP, five UH-1H, four MD500D and three TH-55 (Hughes 269) Osage. Training uses two Cessna 180 and five T-41D Mescalero, with four CASA C101BB Aviojet likely to be sold.

Hungary

POPULATION: 10 million

LAND AREA: 35,912 square miles, 93,012 sq km

GDP: $47bn (£32.9bn), per capita $8,528 (£5,963)

DEFENCE EXPENDITURE: $793m (£554m)

SERVICE PERSONNEL: 33,810 active, plus 90,300 reserves.

ABOVE: *MiG-21 'Fishbed' (Andrews/Aviation Photographs International)*

HUNGARIAN AIR FORCE

Formed: 1949

Hungary was formerly part of the Austro-Hungarian Empire, one of the Central Powers during World War I, becoming an independent republic in 1918. As a former belligerent, Hungary was not allowed an air force under the Treaty of Versailles. Nevertheless, in 1936, a small military air arm came into existence using Fiat CR32 fighters, Meridionali Ro37 reconnaissance aircraft and Heinkel He46 AOP aircraft, with Hungarian-designed Weiss WM13 trainers. During 1938, German assistance was provided, including advisers, Junkers Ju86D and Heinkel He70 bombers, and Bucker Bu131 Jungmann trainers. The following year, Fiat CR42 fighters, Caproni Ca135 and C310 bombers and Nardi trainers joined the now fast-growing air arm,

Hungary

ABOVE: *Saab built Grippen (Jane's)*

retitled the Hungarian Army Air Force.

In 1940, Hungary entered World War II on the side of the Axis Powers. Hungarian units were sent to the Russian front, assisting in Operation Barbarossa. German-built aircraft entered service in substantial numbers, including Messerschmitt Bf109 fighters, Junkers Ju87D Stuka dive-bombers and Ju88A bombers, while there were also Italian Reggiane Re2000 fighters. Many aircraft were built under licence in Hungary, even for the Luftwaffe! Under the gruelling conditions of the Russian Front and the massive Soviet counter-attacks, Hungary and its armed forces suffered badly during the closing stages of the war, surrendering early in 1945.

Post-war, Hungary was forbidden more than a few fighter and transport aircraft until a Communist government took control in 1949, and Soviet military aid commenced, creating a new Hungarian Air Force, the MHRC. At first, Yakovlev Yak-9 fighters, Lisunov Li-2 (C-47) transports and a variety of trainers were supplied. The first jets, MiG-15 fighters, arrived during the early 1950s with obsolete Ilyushin Il-10 ground-attack aircraft and Tupolev Tu-2 bombers. These were followed later by MiG-17 fighters and Ilyushin Il-28 jet bombers, Antonov An-2 transports, and additional Li-2s, Mil Mi-1 and Mi-4 helicopters, Yakovlev Yak-11, -18 and MiG-15UTI trainers. Hungary became a member of the Warsaw Pact in 1955, but attempted to break away from Soviet influence in 1956, with Hungarian and Soviet forces opposing each other. The USSR quelled the revolution. In its aftermath the MHRC was stood down for some years, without new aircraft until the late 1960s, when MiG-21 interceptors, MiG-19 fighter-bombers, and Sukhoi Su-7B ground-attack aircraft were introduced. These aircraft and the remaining MiG-17s equipped 12 squadrons, with one Ilyushin Il-28 bomber squadron. Ilyushin Il-14 transports joined the earlier An-2 and Li-2 and Mi-1 and Mi-4 helicopters. Later MiG-29 interceptors, Mi-24 attack helicopters, Mi-8 and, more recently, Mi-17 transport helicopters were supplied. The main training aircraft became the Aero L-39, with a number of Yak-52, while the MiG-21 force was upgraded.

The break-up of the USSR and the collapse of the Warsaw Pact caused a major reorientation of Hungarian foreign and defence polices. Despite having the Soviet Bloc's strongest economy, the cost of new equipment has seriously hampered modernisation. Hungary became a member of NATO in 1999, and has since struggled to upgrade MiG-29s to become compatible with NATO. The MiG-21s were retired in 2000, as were Mi-2s and Yak-52s. Leased Gripens are entering service, but at just 14 aircraft, this still leaves the MHRC short of adequate numbers of modern combat aircraft.

The MHRC has 7,500 personnel, having contracted considerably in recent years. It has one squadron with 14 Gripen and another with 12 MiG-29A/UB, with at least as many in store. There is an attack helicopter wing with 20 Mi-24D and 10 Mi-24V, many of which are stored. There are 15 Mi-8 and five Mi-17 transport helicopters, with another four Mi-8 as VIP transports and one as a command post; two EW Mi-17 are in store. Transport is provided by four An-26s with four Zlin 43 communications aircraft. Training uses 19 L-39ZO obtained from the Luftwaffe which had gained them on German reunification.

Iceland

POPULATION: 283,000

LAND AREA: 39,758 square miles, 102,846 sq km

GDP: $9bn (£6.3bn), per capita $27,000 (£18,881)

SECURITY EXP: $19m (£13.3m)

PARAMILITARY PERSONNEL:120 active.

ICELANDIC COAST GUARD

A member of NATO, Iceland does not maintain standing armed forces, due to the very small population and the presence of US forces. The Icelandic Coast Guard has three fishery protection vessels, augmented by aircraft including a single MR Fokker F-27-200 Friendship, a Eurocopter AS332L Super Puma, a SA365N Dauphin II and an AS350B Ecureuil. Based at the international airport at Reykjavik, these can be deployed aboard the fishery protection vessels.

India

POPULATION: 1,029 million

LAND AREA: 1,262,275 square miles, 3,268,580 sq km

GDP: $471bn (£329.4bn), per capita $1,900 (£1,329)

DEFENCE EXPENDITURE: $14.7bn (£10.3bn)

SERVICE PERSONNEL: 1,263,000 active, plus 535,000 reserve.

ABOVE: *The Indian Air Force's Folland Gnats, built under licence in India, were so successful in aerial combat with Pakistan that they were nick-named the 'Sabre Slayer'. (IAF)*

INDIAN AIR FORCE

Formed: 1933

Indian military aviation dates from 1933 when the Indian Air Force was formed under RAF control, initially operating just four Westland Wapiti general-purpose aircraft on army co-operation duties. It was not until 1940 that this force reached squadron strength, but considerable expansion occurred during World War II. Westland Lysanders replaced the Wapitis in 1941. The IAF expanded to seven Hawker Hurricane fighter squadrons and two Vultee Vengeance dive-bomber squadrons. Pilots were trained mainly in Australia and Canada under the Empire Air Training Scheme, but a number were trained in India and the UK. Supermarine Spitfire fighters arrived in 1944. Throughout the war, the IAF operated mainly as a tactical air force in close support of ground forces. The 'Royal' prefix was granted in recognition of the IAF's wartime achievements in 1945. At the end of the war, a squadron was deployed to Japan as part of the Commonwealth Occupation Force.

ABOVE: *The Dassault Ouragan entered Indian service in 1953 - these belonged to No29 'Scorpion' Squadron. (IAF)*

Indian divided into India and Pakistan on independence in 1947. The RIAF was also divided. India's share was seven fighter squadrons with Supermarine Spitfires, and Hawker Hurricanes and Tempests, and a Douglas C-47 transport squadron. Additional Spitfires and Tempests were delivered following independence, and Hindustan Aircraft Industries refurbished redundant Consolidated B-24 Liberator bombers to provide a small bomber force. These were followed by the first jets, de Havilland Vampire F3 jet fighters and FB9 fighter-bombers. De Havilland Devon (Dove) light transports were also added, while Percival Prentice trainers were built in India pending production of the Hindustan HT-2 trainer. The 'Royal' prefix was dropped in 1950 when India became a republic within the British Commonwealth.

Over a hundred Dassault Ouragan jet fighter-bombers were bought in 1953, and the following year 26 Fairchild C-119G Packet transports were introduced. In 1955, the IAF acquired Vickers Viscount 700 and, its first Soviet aircraft, Ilyushin Il-14 transports, as well as Auster AOP9 aircraft. More than a hundred Dassault Mystere IVA fighters were ordered in 1956, while licence-production of the Folland Gnat light fighter started. The IAF continued rapid expansion throughout the late 1950s, introducing Hawker Hunter F56 fighters and T7 trainers, English Electric Canberra B8 jet bombers, de Havilland Vampire T55 jet trainers, de Havilland DHC-3 Otter transports, Bell 47G Sioux and Sikorsky S-55 helicopters, often in substantial numbers. North American T-6G Texan trainers were built under licence.

Throughout the 1960s and for the two decades that followed, Indian defence procurement relied heavily on Soviet armaments. Substantial defence forces were developed because of border clashes with Pakistan, often breaking into warfare, as well as several border clashes with Communist China. Licence-production of the MiG-21 interceptor started and the Hindustan HK-24 Marut fighter-bomber and Kirshak AOP aircraft entered production. The Gnat returned to production after proving to be particularly successful, earning itself the nickname of 'Sabre Slayer', during one battle with Pakistan. The Indo-Pakistan War of 1971 saw East Pakistan break free from West Pakistan with Indian assistance to become the new state of Bangladesh. Antonov An-12 transports, Mil Mi-4 helicopters and de Havilland Canada DHC-4 Caribou transports were also introduced during the 1960s. Within the first quarter century after independence, the IAF grew to 90,000 personnel. Throughout this period, the IAF also provided AOP cover for the Indian Army.

ABOVE: *One of the Indian Air Force's English Electric Canberra B8 jet bombers, in the air before delivery. (IAF)*

ABOVE: *Given the large distances to be covered, and often difficult terrain, transport has always been an important function for the Indian Air Force, this is an Antonov An-12. (IAF)*

The 1970s saw the IAF continue its preference for Soviet equipment. It bought the Anglo-French Jaguar strike aircraft, building this under licence, including a number for shore-based anti-shipping duties. The 1980s saw the IAF acquire Dassault Mirage 2000 interceptors, joining MiG-23, a small number of MiG-25, and, later MiG-27. The helicopter force also grew rapidly, with Mil Mi-8 and, more recently, Mi-17 transport helicopters, as well as Mi-24 and Mi-35 attack helicopters. The Alouette III was built in India as the Chetak, but an indigenous helicopter is now in production, the ALH. A substantial transport force includes the An-32 and Il-76. Recently, MiG-29A interceptors have been bought from Russia, and Sukhoi Su-30MKI strike aircraft and interceptors built under licence.

Despite occasional border incidents, there have been fewer border clashes between India and Pakistan in recent years - the most recent being in 1999 when the IAF shot down two Pakistani MR aircraft that had strayed into Indian airspace. India's defence budget rose by 30 per cent between 2000 and 2001 and by a further 13 per cent between 2001 and 2002. The number of frontline squadrons will grow from 39 to 57 by 2020, with sophisticated aircraft such as the Su-30MKI. Meanwhile, the MiG-21 fighters are being upgraded to MiG-21-93 standard. Existing Jaguars are being upgraded while additional aircraft are being built as well as Mirage 2000s. There have been delays in placing an order for BAe Hawk advanced jet trainers, but this is likely to include a mix of 24 used aircraft to ensure early delivery, 36 imported new aircraft, and licence-production in India of the remainder.

India

ABOVE: *A long-serving member of the Indian Air Force, many MiG-21s have been upgraed. (IAF)*

ABOVE: *A more recent arrival has been the MiG-29. (IAF)*

ABOVE: *The Sukhoi Su-30 is built under licence in India. (IAF)*

Currently, the IAF has 150,000 personnel, with an annual average flying hours of 150. It is organised into five regional air commands: Central based on Allabad; Western based on New Delhi; Eastern at Shillong; Southern at Tiruvettipuram, and South-Western at Gandhinagar. There are 21 fighter squadrons: four of which operate 66 MiG-21FL/U; ten have 169 MiG-21bis/U; one has 26 MiG-23MD/UM; three with 64 MiG-29; two with Mirage 2000H/TH, and one with eight Su-30Ks. The 18 ground-attack squadrons include one with eight Su-30K; three with 53 MiG-23BN/UM; four with 88 Jaguar S(I); six with 147 MiG-27 and four with 69 MiG-21MF/PFMA. A single shore-based maritime attack squadron operates six Jaguar S(I) able to deliver the BAe Sea Eagle anti-shipping missile. There are no bomber squadrons as such. Reconnaissance is provide by a squadron of eight Canberra PR-57-/67 and one with eight MiG-25R/U. These squadrons are supported by four Canberra B(I) 58 for ECM, but which also double-up on target towing duties with another two Canberra TT-18. There are two ELINT Boeing 707 and two 737, as well as four AEW BAe748, and six Ilyushin Il-78 tankers. For tactical support of ground forces, there are 20 Mi-24 and 40 Mi-35 attack helicopters in three squadrons.

The air transport force has also grown, and currently accounts for another 12 fixed-wing squadrons. The most widespread aircraft is the Antonov An-32, known as the Sutlej, with 105 in six squadrons. There are two squadrons with 45 Do228; two with 28 BAe748 and two with 25 Il-76, known as the Gajraj. Another 11 squadrons operate helicopters in the transport role, with a total of 73 Mi-8 and 50 Mi-17, and another 40 on order, with ten Mi-26 for heavy lift. VIP duties fall to a single squadron with two Boeing 737-200, seven BAe748 and six Mi-8. Up to 150 ALH light helicopters are also planned.

This substantial force has extensive training support, using conversion trainer versions of the combat aircraft, denoted by the 'U' suffix for MiG and Su types, and by 'TH' for the Mirage, although most of the two-seat MiG-21s are being withdrawn. In addition, there are 28 BAe748, many of which are navigational trainers; 14 two-seat Jaguar B(I); 20 Hunter F-56 and 18 T-66; nine MiG-29UB; 120 Kiran I and 56 Kiran II; 88 HPT-32; 44 TS-11 Iskara; and for helicopter training, 20 Alouette III, or Chetak; two Mi-24 and two Mi-35. Eight Super Dimona motor-gliders were introduced in 2000.

Missiles in use include the AS-7 Kerry; AS-11B anti-tank weapon; AS-12 and AS-30; Sea Eagle and AM 39 Exocet; and AS-17 Krypton in the ASM role. AAM include AA-7 Apex; AA-8 Aphid; AA-10 Alamo; AA-11 Archer; R-550 Magic; Super 530D. SAM missiles are used by 38 squadrons equipped with a total of 280 launchers, including Divina (SA-2), Pechora (SA-3), SA-5 and SA-10.

ABOVE: *Two IAF Mil Mi-35 attack helicopters in flight. (IAF)*

ABOVE: *The Indian Air Force buys equipment from East and West, whenever possible building aircraft under licence. The Anglo-French Sepecat Jaguar IS is one example of this. (Hindustan Aeronautics)*

ABOVE: *The Jaguar IB is mainly used for conversion training. (Hindustan Aeronautics)*

INDIAN NAVAL AVIATION

Formed: 1950

Indian Naval Aviation first emerged in 1950 with the creation of a Fleet Requirements Unit for communications and target towing duties, equipped with Short Sealand amphibians, five Fairey Firefly target-tugs and three Hindustan HT-2 trainers. It was not until 1961 that the Majestic-class light carrier, HMS *Hercules*, was refitted and transferred to India as INS *Vikrant*. A total of 35 Armstrong-Whitworth Sea Hawk fighter-bombers was acquired to equip two squadrons, with ten Breguet Br1050 Alize ASW aircraft, six Westland SH-3D Sea King and ten Alouette III helicopters. The carrier was able to carry 16 Sea Hawks, four Alize and two Alouette at any one time. Eventually, the number of Sea Kings rose to 31, with later Mk42A/B versions able to carry both AS torpedoes and Sea Eagle anti-shipping missiles.

Aircraft from *Vikrant* played an important part in the Indo-Pakistan War of 1971, as a result of which East Pakistan broke away to become independent as Bangladesh. Sea Hawk fighter-bombers flew from the carrier

India

ABOVE: *The Indian Navy also uses Kiran trainers for aspiring Sea Harrier pilots. (Hindustan Aeronautics)*

to attack Chittagong and Cox's Bazaar in the then East Pakistan, helping to overwhelm Pakistani air defences in the final days of the conflict.

In the mid-1980s, *Vikrant* was joined and then replaced by the former HMS *Hermes*, renamed *Viraat*. *Viraat* came with a 'ski-jump' to operate the Sea Harrier FRS51 V/STOL fighters and T60/T4 conversion trainers acquired at the same time. *Viraat* has just completed a major refit to extend her operational life to beyond 2005. She is being joined by the former Soviet *Admiral Goshkov*, acquired for the bargain price of just US $400,000 (£260,000), but needing a refit which, with the cost of 46 MiG-29K aircraft, will cost US1.8 billion (£1.2 billion). Renamed INS *Baku*, it is likely that she will operate alongside *Viraat* until 2010, when another carrier is planned.

Today, 5,000 of the Indian Navy's 53,000 personnel are involved in naval aviation. There are 23 Sea Harrier FRS51 in two squadrons, and being joined by 46 MiG-29K, likely to operate in at least three squadrons. Shore-based aircraft include eight long-range Tupolev Tu-142M and five Ilyushin Il-38 for MR duties, with 24 Dornier Do228 on transport and offshore MR, operated alongside 13 BN-2A/B Defender, some of which are being upgraded with turboprop engines. INA is leasing four Tu-22 bombers for a long-range shore-based strike capability. Helicopters include 31 Sea King Mk42/A/B/C for ASW, ASUW, SAR and transport, with 12 Kamov Ka-28, five Ka-25 and seven Ka-31 handling ASW, usually from frigates. The Kamov helicopters will be supplemented, and a few might be replaced, by at least 40 ALH helicopters in the ASW and light transport roles. Sea King replacement may use either Russian aircraft or Cougar Mk2s, but tight budgets might mean that secondhand aircraft are acquired. Training uses three Harrier T60 and two T4, as well as 12 Kiran I/II and eight Deepak, while 50 Chetak (Alouette III) operate in the training and communications roles.

The Indian Navy controls the operations of the Indian Coast Guard, which has an additional 8,000 personnel. Coastal patrol is provided by three squadrons operating 36 Do228, while another squadron has 15 Chetak helicopters that are being replaced by up to 40 ALH. There is also a single Fokker F-27 Fellowship.

INDIAN ARMY AVIATION CORPS

India has the fourth largest army with 1,100,000 personnel. The Aviation Corps is relatively young, having taken over AOP and liaison from the IAF in the 1980s, and still relatively small. An attack capability is developing with the introduction of 150 Lancers, an armed version of the Cheetah helicopter, for COIN operations in the Himalayas. Currently, 120 Chetak (Alouette III) and 40 Cheetah (Lama) are also operated on observation and liaison duties, and will be replaced by the Indian ALH (Advanced Light Helicopter).

Indonesia

POPULATION: 216.2 million

LAND AREA: 736,512 square miles, 1,907,566 sq km

GDP: $160bn (£112bn), per capita $4,000 (£2,797)

DEFENCE EXPENDITURE: $1.5bn (£1bn)

SERVICE PERSONNEL: 297,000 (inc 50%+ conscripts) active, plus 400,000 reserves.

INDONESIAN NATIONAL DEFENCE-AIR FORCE

Formed: 1949

Indonesian military aviation preceded independence in 1949 as an aviation division of the People's Peace Preservation Force and was established in December, 1945, following Japanese surrender and before Dutch forces could return in spring, 1946. The new air arm had just five Indonesian pilots and used abandoned Japanese aircraft. These were soon put out of action by a Dutch air attack. India provided assistance before and immediately after independence, when Dutch assistance was also provided for what had become the Indonesian Air Force, AURI. Former Dutch aircraft included North American F-51D/K Mustang fighter-bombers and B-25 Mitchell bombers; Convair PBY-5A Catalina amphibians; Lockheed 12A transports; Piper L-4J Cubs, Auster IIIs and Aiglets for training and AOP duties. India provided Hindustan HT-2 trainers. Indonesia extends for more than 3,000 miles end-to-end with many islands, making the Catalinas important. Lockheed 12As also provided air services before the national airline, Garuda, started operations in 1950.

In 1955, IAF personnel were seconded to the AURI to help reorganise it. At this time, aircrew were being trained in the USA, UK and the Netherlands, with the first jets, de Havilland Vampire fighter-bombers, being supplied by the UK that same year. Indonesia began to be drawn into the Soviet sphere of influence during the late 1950s, and in 1958, Czechoslovakia provided 60 MiG-17 fighters, a number of MiG-15 fighter-bombers, and 40 Ilyushin Il-28 light bombers, while pilots were trained in Egypt. Western aircraft continued to enter service, including DHC-3 Otters, Cessna 180s, Grumman HU-16 amphibians, Fairey Gannet AS4 ASW aircraft, and T5 trainers, Lockheed C-130B Hercules transports and Sikorsky S-58 helicopters.

By 1960, Indonesian policy had become aggressive. An invasion of West Irian, formerly Dutch New Guinea, in 1962 brought Indonesian and Dutch forces into conflict, although the territory was ceded to Indonesia the following year. This period also marked a confrontation with the UK, Australia, New Zealand and Malaysia over the creation of the Federation of Malaysia, through which Malaya and Singapore were to become independent from the UK, lasting until 1966. Western supplies were cut off, so the USSR supplied 35 MiG-19 and 15 MiG-21 fighters, with 25 Tupolev Tu-16 bombers, Ilyushin Il-14 and Antonov An-2 transports. AURI's operations throughout the Malaysian confrontation were largely unsuccessful, and the end coincided with the downfall of the then dictator and president, Sukarno.

In 1975, Indonesia seized the Portuguese colony of East Timor. East Timorese attempts to gain independence and the use of the AURI against them caused controversy over arms sales to Indonesia. When, under UN auspices, East Timor did finally become independent in 1999, the AURI, now with its present title of Indonesian National Defence-Air Force, TNI-AU, had no major role to play. Meanwhile,

ABOVE: *The Indonesian Air Force uses 20 Beech T-34C Turbo Mentor trainers - although this pilot is obviously flying solo! (Raytheon)*

Indonesian foreign relations had become less focussed on Russia. Western aircraft types were ordered once again, including Northrop F-5E/F, McDonnell Douglas A-4 Skyhawk, BAe Hawk trainers and attack aircraft, a few North American OV-10 Bronco COIN aircraft, and later, Lockheed F-16A interceptors. More modern variants of the C-130 Hercules were also introduced.

Economic problems affecting the Far East have meant that the TNI-AU has contracted in recent years, and now has just 27,000 personnel, about half the figure for the early 1970s, although reorganization of Indonesia's armed forces has meant that the paratroops have now passed to the army. In 1998, orders for Mi-17 helicopters and eight Sukhoi Su-30K were cancelled. Upgrades to the F-16 and F-5 aircraft have been kept to the necessary minimum. Indonesia now builds aircraft, including a joint utility transport venture with CASA of Spain, the CN235. In future, the TNI-AU will be needed increasingly for COIN, as

ABOVE: *A more up-to-date Indonesian Air Force trainer is the BAE Hawk. (BAE SYSTEMS)*

Indonesia

the widely disparate population of this sprawling nation seeks greater self-determination. It could also have a considerable role in disaster relief.

The TNI-AU operates seven F-16A plus three F-16B in the training role, as well as eight F-5E plus two F-5F in the training role, as an interceptor force. There are 14 A-4E/J Skyhawk in the attack role, plus three TA-4H for training. The BAe Hawk force, delivered between 1992 and 1996, with a further batch suspended in 2000 due to sanctions, includes 30 Mk209 single-seat attack aircraft, plus seven Mk109 and 14 Mk53 two-seat aircraft capable of being used for advanced jet training or light strike duties. The combat element is completed by five OV-10F Bronco COIN aircraft, although another seven are in store. There are three Boeing 737-2X9 Surveiller in the MR and transport role, as well as three CN235 MR aircraft, with another 18 operating as transports. There are two KC-130B tankers, and a further 22 C-130B/H/H-30 and civil L-300-30 in the transport role. Transport aircraft also include seven Fokker F27-400M Troopship and three F28-1000/2000, as well as a VIP Boeing 707-320C, and ten NC212-100/200 Aviocar. A Skyvan 3M is used for aerial surveys. Seven Cessna 401/402s are used for communications duties. There are 16 licence-built NAS332 Super Puma helicopters, with two for VIP duties, five for CSAR, and another nine for transport. There are still ten S-58T, a single Bell 412, and seven BO105CB transports. Training aircraft include 20 T-34 Turbo Mentor, two Cessna 172, with a further four of the T-41 version for liaison duties, as well as 39 AS-202. Ten Bell 47G Sioux are being replaced in 2003 by 12 EC120s, when the first of up to 20 KT-1s basic trainers will also be delivered.

INDONESIAN NATIONAL DEFENCE-NAVY

At one time, the Indonesian Navy operated MiG-19 and MiG-21 interceptors, but in recent years it has become more focussed on MR, transport and communications, SAR and ASW duties, although longer-range MR is left to the air force. Its development has been affected both by the state of the economy and by sanctions. Before sanctions were applied, ex-Australian Army Nomads were introduced for a variety of roles, including transport and MR. The CN235 collaborative venture transport aircraft has also been introduced, but six MPA versions ordered in 1995 had still not been delivered by 2000.

Currently, about 1,000 personnel out of a total 40,000 are directly involved with the air arm. There are two ex-Australian DHC-5 Buffalo transports, and 35 Nomad for MR and transport duties. The main SAR helicopter is the NAS332L Super Puma, of which there are four. It is believed that all of nine Wasp HAS1 used for ASW from frigates are grounded. There are ten NBO105CB and four Indonesian-built Bell 412 helicopters in the light transport role. Fixed-wing transport aircraft include six CN235 and six NC212, also available for MR, with four Commander 100 for communications and training. A variety of training types are in use, including six Piper Tomahawk and four Seneca, two Beech Bonanza and a TB9 Tampico, as well as two Alouette II helicopters.

INDONESIAN NATIONAL DEFENCE-ARMY

Indonesian Army Aviation is centred around transport and liaison types, with no attack capability. In recent years, the TNI-AD has also been affected by the economic difficulties, but a slow recovery meant that four out of eight Mi-17V postponed in 1997 were delivered in 2000. The army is considering the CH-70 for the transport role, but might opt for licence-built Super Pumas. There are 30 Bell 205A and 28 locally-assembled Bell NB412SP, with fixed-wing aircraft including four NC212 and three DHC-5D Buffalo. Liaison duties are covered by 17 Indonesian-built NBO105CBs, as well as an Alouette III, plus fixed-wing aircraft, including a BN-2A Islander, two Commander 680 and 18 PZL Wilga 32. Training is provided by 15 Schweizer 300C.

Iran

POPULATION: 68.3 million

LAND AREA: 627,000 square miles 1,626,520 sq km

GDP: $99bn (£69.2bn), per capita $7,400 (£5,175)

DEFENCE EXPENDITURE: $7.5bn (£5.2bn)

SERVICE PERSONNEL: 513,000 (40% conscripts) active, plus 350,000 reserves.

ISLAMIC REPUBLIC OF IRAN AIR FORCE

Formed: 1932

At one time possessing the most modern and powerful air force in the Middle East, Iranian military aviation dates from 1922, when the Iranian Army created an Air Office. A Junkers F13 was purchased and a German pilot hired. The following year, an Aero A30 was bought for AOP duties, followed in 1924 by four DH4s and DH9s as well as additional F13s. The first Iranian pilots were trained in France during 1924, when the Air Office became the Iranian Air Force, remaining part of the Army.

In 1932, the IAF became the Imperial Iranian Air Force, a separate service. Modernization started, with 18 de

Iran

Havilland Tiger Moth trainers, followed by 12 Hawker Fury fighters in 1933, and later by Hawker Hart light bombers and 32 Audax AOP aircraft, while instructors were seconded from the Royal Swedish Air Force. Although 38 Hawker Hurricane fighters and some Curtiss 75A Hawk fighter-bombers were ordered in 1938, just two Hurricanes reached Iran before deliveries were suspended following the outbreak of World War II in Europe. In 1941, Iran was invaded by British and Soviet forces to guarantee an overland supply route to Russia and prevent Iran joining the Axis powers. During the war, the IIAF stagnated, except for the delivery of a small number of Avro Ansons and some additional Tiger Moths after Iran declared war on Germany in 1943.

ABOVE: *The then Imperial Iranian Air Force operated Hawker Hurricanes during World War II and for some time afterwards, with additional aircraft delivered post-war. This is a rare shot of the two-seat trainer version. (BAE SYSTEMS)*

Post-war, the IIAF eventually received 34 Hurricane fighters. These were soon joined by the first of many American aircraft, including Republic F-47D Thunderbolt fighter-bombers, Douglas C-47 transports, Piper L-4 AOP aircraft, and North American T-6G Texan and Boeing-Stearman PT-13 Kaydet trainers. Until 1950, all aircrew were trained in the USA or West Germany. As a founder member of the Baghdad Pact (later renamed the Central Treaty Organisation), Iran received American military aid throughout the 1950s, receiving Republic F-84F Thunderstreak and North American F-86F Sabre fighter-bombers, and Lockheed T-33A trainers. During the 1960s and early 1970s, these aircraft were supplemented and to some extent replaced by Northrop F-5A and McDonnell Douglas F-4D Phantom fighter-bombers. Lockheed C-139E Hercules transports were also introduced, while helicopters included 16 Meridonali-built CH-47 Chinook, 40 Bell UH-1D and 40 Agusta-Bell 205, as well as 100 206A, 16 Sud Super Frelon and a number of Kaman HH-43B Huskie. Short Tigercat SAM were also deployed. Many helicopters were operated on behalf of the Army and the Navy, although these were developing their own air arms. The paramilitary Imperial Iranian Gendarmerie also had five Augusta-Bell 205 helicopters. By this time, there were 17,000 personnel.

The 1970s saw continued expansion while the IIAF kept pace with technological developments, using the rapid rise in the price of oil to buy sophisticated aircraft - 79 Grumman F-14 Tomcat interceptors - and Northrop F-5E Tiger II fighter-bombers. Sea King ASW helicopters were acquired for naval use, with Sea Stallions for minesweeping and MR Lockheed P-3F Orions. The Hercules and Chinook fleets increased. In 1979, the Shah was deposed and an Islamic republic declared, with a change of name to the Islamic Republic of Iran Air Force. The revolutionary fervour of the new regime, including storming the US embassy and the taking of its staff hostage led to a complete breakdown in the relationship between Iran and the West, allowing the USSR to become Iran's main supplier. Naval and army air power finally moved from air force control at this time.

Iran's Islamic fundamentalism also raised tensions with its neighbours. This led to what some historians describe as the 'First Gulf War', when from 1979 and throughout the 1980s, Iran and Iraq were at war, and while the intensity of operations fluctuated, conflict was often heavy. The outcome of the conflict was inconclusive, with Iraq receiving some support from the West, and Iran moving closer to the USSR. Soviet aircraft were introduced, including Sukhoi Su-24Mk strike aircraft, Ilyushin Il-76 and Antonov An-74 transports, and in 1990, MiG-29A interceptors. The conflict ended in 1989, with no clear outcome. A complete change in relations with Iraq appeared to come following that country's invasion of Kuwait in 1990. During the 'Second Gulf War' that followed, Iraq sent many aircraft to Iran for safety from Allied air attacks, with between 100 and 120 aircraft flown into Iran, which promptly retained the aircraft as reparations for earlier losses!

Over the last ten or twelve years, relations with Iran have remained difficult for both the Western nations and its neighbours, while the break-up of the USSR has also resulted in a less close relationship with Russia. Although tentative steps have been made towards an easier relationship with the rest of the world, US pressure on the CIS still resulted in a reluctance to provide up-to-date equipment for Iran, forcing the country to turn to China. Chinese aircraft include the Shenyang F-6 (MiG-19), F-7N (MiG-21), and a number of transport aircraft.

Today, the IRIAF is one of the few air forces to have increased in size substantially

Iran

ABOVE: *Iranian Air Force's C-130H Hercules (Jeremy Flack/Aviation Photographs International)*

over the past thirty years with 45,000 personnel, although about a third are engaged in air defence. Serviceability of aircraft is a growing problem; with many aircraft cannibalised for spares to keep others flying. Attempts to normalize relations with the West will depend on Iran's attitude to any future regime in neighbouring Afghanistan. Mainstays of the interceptor force are 35 MiG-29A, with another five -29UB conversion trainers, some of which are ex-Iraqi aircraft, as well as 20 surviving F-14As and 24 F-7N, in five squadrons. On FGA duties there are nine squadrons, with four operating 66 F-4D/E, four with 60 F-5E/F and one with a mixture of 24 Su-24MK and up to seven former Iraqi Su-25Ks. There are 18 F-6s in a single squadron. One squadron operates up to 15 RF-4E. There are three Lockheed P-3F Orion MR aircraft. Transport includes five squadrons with a variety of aircraft, including 23 Lockheed C-130E/H Hercules plus two reconnaissance RC-130H. There are ten Antonov An-74 and 12 Ilyushin Il-76, plus a single Il-76 Adnan AEW aircraft; 14 Harbin Y-7 and nine Y-12; 11 Boeing 747-100/200, including four for tanker duties, and ten Fokker F-27-400M/600 Troopship/Friendship. Smaller transport aircraft include ten PC-6 Turbo Porter. VIP duties are handled by five Falcon 20/50, and a Lockheed Jetstar, while communications aircraft include three Aero Commander 690. There are five Agusta-Bell 212 and 20 214A/C helicopters in the transport role. Apart from the conversion trainers in the OCU, training takes 26 Beech Bonanza, 15 Embraer EMB-321 Tucano, 45 PC-7 Turbo Trainer, and 25 MFI-17B Mushshak. Five Do228 are used on aerial survey work. Hawk, Stinger, Rapier and Tigercat SAM are still believed to be deployed, as well as SA-7, while AAM missiles include a mixture of Western and Soviet types, including AIM-7 and AIM-9, AA-8, AA-10 and AA-11.

REVOLUTIONARY GUARD CORPS AIR FORCE

Formed: 1979

The Islamic revolution in Iran channelled the energies of its supporters into the Revolutionary Guard, initially while evaluating the loyalty and commitment of members of the former Imperial armed forces. This force has since acquired an air arm, in some cases using aircraft taken from the armed forces and, given the paucity of information from the country, there may be some double accounting of equipment. The Revolutionary Guard has 125,000 personnel, of which 100,000 are ground forces, 20,000 naval and 5,000 marines, but little is known about air arm numbers other than that they are commanded by an officer of Brigadier General (equivalent to Air Commodore) rank. Aircraft are believed to include 20 PC-7 used both on COIN and for training; 20 Mi-8AMTSH helicopters; and a fixed-wing transport force with six An-74 and 20 Shahed-5s (CASA-IPTN CN212).

ISLAMIC REPUBLIC OF IRAN NAVY

Although the IRIAF still operates fixed-wing MR aircraft, the Iranian Navy has control of its helicopters, without warships capable of operating these. Some 2,000 of the total of 18,000 naval personnel are engaged in naval aviation. There are two Sikorsky RH-53D Sea Stallion minesweeping helicopters and ten SH-3D ASW helicopters, with six Agusta-Bell AB212AS also for ASW, five AB205 for transport and ten AB206 JetRanger for liaison. Fixed-wing aircraft include four VIP Falcon 20E and eight Aero Commander 500/690 liaison aircraft.

ISLAMIC REPUBLIC OF IRAN ARMY

The Iranian Army has its own aircraft, although serviceability is likely to be low given the fact that most aircraft are of Western origin. The 40 CH-47C Chinooks are now down to around 20, and possibly no more than 70 of the original 100 AH-1J Cobra attack helicopters are still operational. There are 100 Agusta Bell AB-214A/C, 20 AB212 and 10 AB205A-1 in the transport role, with 40 AB206A/B JetRanger on observation and communications duties. There are two VIP Falcon 20Es, five communications Aero Commander 690 and ten liaison Cessna 185.

Iraq

POPULATION: 22.3 million

LAND AREA: 169,240 square miles, 435,120 sq km

GDP: $15.4bn (£10.8bn).

DEFENCE EXPENDITURE: $1.4bn (£979m)

SERVICE PERSONNEL: 424,000 (mainly conscript) active, plus 650,000 reserves.

IRAQI AIR FORCE

Formed: 1931

Originally founded as the Royal Iraqi Air Force in 1931 to coincide with the return of the first Iraqi pilots from training in the UK, the RIAF initially had five de Havilland Gipsy Moth trainers, to which another four were soon added. In 1932, four Puss Moths fitted with bomb racks arrived to be deployed on 'air control', suppressing rebel uprisings. Further expansion followed in 1934, with additional Puss Moths as well as de Havilland Tiger Moths and Dragon Rapides and Hawker Nisrs. Breda Ba65 fighter-bombers were delivered in 1937 with Savoia-Marchetti SM79B bombers, while during 1938-40, Gloster Gladiator fighters, Avro Anson and Douglas DB-8A bombers were delivered, as well as additional Dragon Rapides and Dragonfly light transports. Initially, Iraq was unaffected by World War II, until a German-inspired uprising in 1941 saw the RIAF in conflict with the RAF, losing most of its aircraft.

Re-equipment started in 1946 with 30 Hawker Fury fighter-bombers. In 1948, four Bristol Freighters, some de Havilland Doves and Auster AOP6 and T7 aircraft were added. Iraq became a founder member of the Baghdad Pact and was able to acquire further British equipment throughout the early and mid-1950s. New aircraft included 20 DHC-1 Chipmunk T20 trainers in 1951; the first jets, 12 de Havilland Vampire FB52 fighter-bombers and six T55 trainers in 1953; and the first helicopters, two Westland Dragonflies (S-51), in 1955. Two de Havilland Heron light transports arrived in 1956, followed by a squadron of Hawker Hunter F6 fighters in 1957. In 1958, the royal family was assassinated, a republic declared and the RIAF dropped the 'royal' prefix. Iraq left the Baghdad Pact, which was re-named the Central Treaty Organisation, and the IAF turned to the USSR for help.

The USSR wasted little time - the first MiG-15 fighter-bombers arrived in October, 1958, followed by Russian instructors and Ilyushin Il-28 jet bombers. The aid continued into the next decade, with MiG-17, MiG-19 and MiG-21 fighters and interceptors, Antonov An-12 transports and Mil Mi-1 and Mi-4 helicopters. Soviet supplies deliveries were halted after defections by MiG-21 pilots, with their aircraft, during 1965 and 1966. Once again, Iraq turned to the UK for supplies, buying Hawker Hunter FGA9, FR10 and T66/9, Westland Wessex (S-58) helicopters and BAC Jet Provost T52 jet trainers. Soviet supplies resumed following the Arab-Israeli War of 1967, including Sukhoi Su-7B ground-attack aircraft. Iraq became a substantial air power in the region, with 60 MiG-21, 50 MiG-19 and MiG-17, 50 Su-7B, 36 Hunter FGA9, ten Il-28 and eight Tu-16, as well as a wide variety of transport aircraft, with the Antonov An-2, An-12 and An-24, and Ilyushin Il-14. Training aircraft included Yakovlev Yak-11 and Yak-18, and Aero L-29 Delfins. Guided missiles were also obtained from the USSR, starting with the SA-2 SAM.

The IAF continued to expand, more than doubling in personnel over the next thirty years. Iraq used its considerable oil wealth to purchase more sophisticated Soviet equipment. MiG-23 and then MIG-25 aircraft were delivered, while a number of Chengdu F-7 (MiG-21) were also obtained. The start of the war with neighbouring Iran in 1979, which lasted throughout the 1980s,

BELOW: *Latest addition to the Iraqi Air Force is the MiG-29, but just ten are in service. (Jane's)*

Iraq

saw a slight improvement in relations with the West, and the IAF was able to obtain helicopters from France, the USA and Germany, and Dassault Mirage F1EQ attack aircraft.

Following the 'First Gulf War', Iraq turned its attention to its much smaller neighbour Kuwait, long been claimed as an Iraqi province. An Iraqi threat to Kuwait in 1961 had been countered by the deployment of British aircraft carriers and marines. In 1990, Iraq invaded. Under the auspices of the United Nations, a coalition of nations assembled forces in the area to first counter any threat to Saudi Arabia and then to liberate Kuwait from Iraqi occupation. The coalition ground assault was preceded by an air campaign, which the IAF attempted to counter for no more than a day or so, with many of its aircraft, including all the Tu-16s, either shot down or destroyed on the ground. Many surviving aircraft were sent to neighbouring Iran for safety, where the aircraft, between 100 and 120 in number, were seized as reparations.

Since the end of the 'Second Gulf War', Iraq has been banned from flying military aircraft north of the 36th parallel and south of the 33rd parallel, known as the 'no fly zones'. This is to protect Kuwait, and also the Marsh Arabs in the south and the Kurdish population in the north. Initial strict economic sanctions were later relaxed to allow Iraq to sell oil to earn currency for food and medical supplies. Revenue from the oil sales has been used mainly to upgrade Iraq's anti-aircraft defences, prompting repeated strikes by US and British aircraft against air defence radar and missile sites. The IAF's state of readiness is very low, with one defector reporting that no more than 75 aircraft are serviceable. Other estimates believe that perhaps half of the aircraft are operational, although the figures for helicopters may well be much lower.

The IAF has 35,000 personnel. Flying hours are believe to be around 100 hours per year for senior pilots, but just 20 hours for junior pilots. Interceptors include ten MiG-29, twelve MiG-25, 60 MiG-23ML and BN (the latter in the attack role), 40 Chengdu F-7 and 36 MiG-21PFM/MF. Attack aircraft include some of the MiG-25BN, 60 Dassault Mirage F1EQ and F1BQ trainers, and 46 Sukhoi Su-7/-20/-22/-25. There are also 20 Mi-25 attack helicopters. There is an Ilyushin Il-76 AEW aircraft. The transport force includes three Antonov An-12 tanker/transports and two An-26s. There is a substantial helicopter force, mainly devoted to supporting ground forces. There are 80 Mi-8/-17 transport helicopters, and in the same role are ten Bell 214ST and two Mi-6, while ten SA312 Super Frelon are used for anti-shipping operations. There are 20 SA342L Gazelle and 40 BO105C used for attack, communications and training duties, while 15 Alouette III are also for communications. VIP duties take up some of the 20 SA330 Puma transport helicopters and three AS61TS. Ten BK117A are used for SAR. Training uses 50 L-39BZ and 40 EMB-312 Tucano, as well as 20 PC-7, ten PC-9, and 18 AS202.

Little is known about the SAM force, but ASM missiles include AM-39, AS-4, AS-5, AS-9, AS-11, AS-12 and AS-30L as well as the C-601. AAM missiles include AA-2, AA-6, AA-7, AA-8 and AA-10, as well as R-530 and R-550.

Ireland

POPULATION: 3.8 million

LAND AREA: 26,600 square miles, 68,894 sq km

GDP: $97.9bn (£68.5bn), per capita $25,085 (£17,541)

DEFENCE EXPENDITURE: $698m (£488m)

SERVICE PERSONNEL: 10,460 active, plus 14,800 reserves.

ABOVE: *Ireland remained neutral during World War II, but increased the strength of the Irish Air Corps before hostilities broke out. Here are Gloster Gladiator biplane fighters and a Westland Lysander army co-operation aircraft. (Irish Air Corps)*

IRISH ARMY AIR CORPS

Formed: 1922

Most of Ireland became independent of the UK in 1922, and immediately the Irish Army Air Corps was formed to provide air cover for ground forces on internal security duties as a civil war raged between the Government and those opposed to the treaty with Britain. Early equipment included Avro 504K trainers, Bristol F2B and Martinsyde F4 Buzzard fighters, and DH9 bombers. The IAAC took over former RAF bases. De Havilland Tiger Moths were purchased for training, and by 1929, the Air

Corps had 160 personnel. A variety of aircraft were introduced, all in small numbers, before World War II, including Avro 621, 626, 631 and Ansons, Gloster Gladiators, a Fairey IIIF, Miles Magisters and a de Havilland Dragon Rapide. Although Ireland remained neutral throughout the war, the Air Corps attained a strength of three squadrons, one each with Gloster Gladiator fighters, Avro Anson patrol aircraft and Supermarine Walrus amphibians, as well as a small number of Vickers Vespa army co-operation aircraft. During the war, the Gladiators were replaced by Hawker Hurricanes and the Vespas by Westland Lysanders, supplied by the UK against the possibility of a German invasion.

Post-war, replacement and modernisation was slow, as Ireland remained neutral throughout the Cold War. Throughout the 1950s, the Air Corps operated Supermarine Spitfire fighters and twin-seat trainers, de Havilland Dove light transports and Vampire jet trainers, as well as DHC-1 Chipmunk basic trainers. A number of Percival Provost piston-engined trainers were also acquired. The first helicopters, three Alouette III, were not introduced until 1966. A small number of SIAI SF260WE armed-trainers were later acquired both for training and for anti-terrorist operations, while the growing importance of fisheries protection resulted in the acquisition of two CASA CN235MP. Four SA365F Dauphin helicopters were acquired for naval liaison and SAR. Recently, helicopters have been chartered from commercial operators to enhance Ireland's SAR capability.

The Irish Air Corps has had the unusual distinction of being investigated by management consultants, who have recommended greater standardisation and the creation of its own SAR and medium lift transport capability, possibly leading to an order for up to four helicopters such as the Merlin or an equivalent. It is also likely to buy an additional CN235MP and possibly some additional aircraft in the BN-2 Defender category to replace the Cessna 172s. The present 15 helicopters are also likely to be rationalized into fewer types.

The Air Corps has 860 personnel. There are two VIP aircraft, a Beech Super King Air 200 and a Gulfstream IV. There are seven remaining SF260WE for COIN and training, with two fishery protection CN235MP and a single BN-2T Defender and an SA355 Squirrel for Garda (police) support. Army support is provided by six Cessna FR172H/Ks. Naval liaison is provided by five SA365F Dauphins. There are seven SA316B Alouette III covering a variety of roles, including SAR, and communications, and two SA342L Gazelle for training.

ABOVE: *A small number of SIAI SF260WE are used by the Irish Air Corps for training, and could be used for COIN if the need arose. (Irish Air Corps)*

BELOW: *Ireland's Aerospatiale Alouette III helicopters are used on a wide variety of tasks, including SAR. (Irish Air Corps)*

Israel

POPULATION: 6.3 million

LAND AREA: 7,993 square miles, 20,850 sq km

GDP: $107bn (£748bn), per capita $19,200 (£13,426)

DEFENCE EXPENDITURE: $9.5bn (£6.6bn)

SERVICE PERSONNEL: 163,500 (60% conscript) active, plus 425,000 reserves.

ABOVE: *IAF F-4 following upgrade to F-4/2000 configuration. (Michael J. Gething)*

ISRAEL DEFENCE FORCE - AIR FORCE

Formed: 1951

Part of the unified Israel Defence Force - the Air Force dates from 1951. The origins of Israel's military aviation date from before the founding of the state. While the country was still under a League of Nations mandate, the Zionist Haganah underground movement formed the Sherut Avir which used Auster and Taylorcraft light aircraft for AOP duties, obtained by cannibalising aircraft abandoned by the British Army during World War II. After independence in 1948, the Sherut Avir briefly became the Chel Ha'avir. The need for more potent aircraft followed an attack by Egyptian Spitfires in 1949, and again the Chel Ha'avir was forced to cannibalise abandoned aircraft, Spitfires and de Havilland Mosquito fighter-bombers found on former RAF bases. Later, some 300 war-surplus aircraft of these two types were purchased, plus some Avia C210 fighters. These were followed by Boeing B-17G Fortress heavy bombers, Curtiss C-46, Douglas C-47 and C-54 transports, and Boeing-Stearman PT-17 Kaydet, North American T-6 Harvard and Avro Anson and Airspeed Oxford trainers, and in 1951, by some former Swedish North American F-51D Mustang fighter-bombers. Some of the transports were converted to bombers for raids on Cairo, the Egyptian capital, and some of the Harvards were adapted for ground-attack duties. The Chel Ha'avir attracted experienced pilots from a large number of air forces.

The IDF-AF was formed in 1951. The first jets, 14 Gloster Meteor F8 fighters, were obtained in 1953, and were soon joined by six Meteor NF13 all-weather fighters and T7 trainers. In 1955, 30 new Dassault Ouragan fighter-bombers were delivered, and joined by 45 ex-Armee de l'Air Ouragans. Eight Nord 2501 Noratlas transports entered service, with 41 Fokker S-11 Instructors. When Egypt received new equipment in 1955, Israel ordered 24 Canadair-built CL-13 Sabre 6 and 24 Dassault Mystere fighters, and increased the Mystere order to 60 aircraft when the Sabres were embargoed. During the Suez crisis of 1956, the IDF-AF gained aerial superiority for the loss of eleven aircraft against a numerically superior Egyptian force. The Mustangs had remained in service on ground-attack duties, but their water-cooled engines were vulnerable to ground fire, so they were replaced by licence-built Potez Magister trainers during the late 1950s. Twenty-four Sud Vatour II light bombers and a number of Dassault Super Mystere interceptors were also obtained at this time.

The 1960s saw 60 Dassault Mirage IIICJ fighter-bombers enter service. These were to prove highly successful in the June, 1967, war with Israel's Arab neighbours, when the IDF-AF virtually wiped out the Egyptian and Jordanian air forces. A further 50 were ordered, but the French government embargoed the order. This marked a growing reliance on the United States for arms supplies, but it also led to Israeli development and production of a Mirage derivative, the Kfir. US aircraft introduced during the late 1960s and early 1970s included 50 McDonnell Douglas F-4E Phantom fighter-bombers and 85 A-4E/M Skyhawk strike aircraft. The IDF-AF had also acquired a substantial number of helicopters by this time, including 25 Agusta-Bell 205, 20 Alouette III and 12 Super Frelon, as well as eight Sikorsky CH-53 Sea Stallion and 15 H-34 Choctaw.

The easy victory in 1967 had lulled Israel into a false sense of security with Arab territory as a buffer zone. In October, 1973, Arab forces once again attacked Israel, selecting 6 October, Yom Kippur, a Jewish day of fasting when Israeli forces would be at their most vulnerable. The Arabs had also learnt lessons from their earlier defeat, using heavy air strikes to neutralise the IDF-AF, and having heavy SAM missile defences within an interlocking air defence system based on SA-2, SA-3, SA-6, SA-7 and SA-9 missiles. The IDF-AF was familiar only with the SA-2, but found that evading this brought aircraft into the range of the other missiles. Subsequently, the IDF-AF was to admit to losing 115 aircraft in the war, although US sources have suggested that the true figure was nearer 200. To counter the improved Arab defences, the IDF-AF had to attack in greater strength, sending a squadron of aircraft at a time rather than a flight of four, and using stand-off or 'smart' weapons, such the US-supplied Walleye. The war ended with a cease-fire on 22 October, although some Israeli operations continued until 24 October, before stopping under US pressure.

While US military aid has fluctuated, overall it has proved constant. The easing of tension between Israel and Egypt during the 1980s meant a less active existence for the IDF-AF, but tension has continued between Israel and Syria, with neighbouring Lebanon often being used as a battleground between the two nations or forces supported by them. IDF-AF operations have included attacks on terrorist camps inside Lebanon, but more usually have consisted of helicopter support for attacking ground forces. Equipment has kept pace with technical developments, with older aircraft upgraded using Israel's growing aerospace expertise. Airborne early warning support came with the delivery of four Grumman E-2C Hawkeye aircraft, superseded by Israel's own Phalcon radar mounted in Boeing 707s. The IDF-AF became a major customer for the F-15 series of interceptors, following this with large numbers of Lockheed F-16s. An attack helicopter capability is based on the AH-64 Apache, known as the Peten.

The IDF-AF was not involved during the Gulf War of 1991, with Israel under strong US pressure to remain neutral for fear of losing the support of the Arab nations involved in the anti-Iraqi coalition. More recently, the IDF-AF has not been heavily involved with clashes between Israeli security forces and Palestinain militants operating from the Left Bank of the Jordan or from Gaza.

The IDF-AF has 37,000 personnel, rising to 57,000 on mobilisation. Large numbers of aircraft are in storage, either as reserves or awaiting disposal. Combat aircraft are often given extensive upgrades. There are 12 squadrons for fighter and strike operations: two have 50 F-4E-2000 upgraded Phantoms and 20 F-4E; two have 73 F-15A/B/C/D; one has 25 F-15I Ra'am; and seven have 171 F-16A/C, with 66 F-16B/D mainly for use in conversion training. There are also 25 A-4N in a ground-attack squadron, although another 80 are in storage. A squadron provides reconnaissance with ten RF-4E. Many of the Phantoms will remain in service until 2010, largely due to their role in delivering the Popeye stand-off weapon, and there are more than 100 in storage. There are six AEW Boeing 707 Phalcon, with another three 707 ELINT aircraft, and other 707s on tanker duties. ELINT is also handled by ten IAI-201 Arava, with another nine on transport duties. Three Gulfstream IVs are on order for command and control and for SIGINT duties. Attack helicopters include 52 AH-64A/D Peten (Apache), 30 Hughes 500MD, 21 AU-1G and 36 AH-1F Tsefa (Cobra), while in early 2002, some of the CH-53s were armed with up to eight laser-guided Nimrod anti-tank missiles. MR is handled by three IAI-1124 Seascan versions of the Westwind. Transport aircraft include five 707s, of which three also operate as tankers; 22 C-130E/H/KC-130, including four tankers, and 12 C-47. Helicopters include eight AS-565 Atalef and two SA-336 for ASW. Transport helicopters include 38 CH-53D, upgraded in Israel, 60 UH-60A/L Yanshuf, and 54 Bell 212, some of which are available for SAR, while more than 40 Bell 206/L JetRanger/LongRanger are used for communications and training. Communications aircraft also include 12 Dornier Do.28 and a number of Beech King Airs, including the U-21 version. Training aircraft include 40 Super Magister, which may be replaced by an aircraft in the Hawk or Pilatus/Beech PC-9 category, 35 PA-18 Super Cub and 16 Beech Queen Air. Hawk SAM are deployed, while the IDF-AF uses AIM-7 Sparrow, AIM-9 Sidewinder, AIM-120 AMRAAM, R-530, Shafir, Python III and IV AAM, and a wide range of ASM.

ABOVE: *IAI Kfir C2 (Michael J. Gething)*

Italy

POPULATION: 57.2 million

LAND AREA: 116,280 square miles, 301,049 sq km

GDP: $1.1tr (£769bn), per capita $23,436 (£16,388)

DEFENCE EXPENDITURE: $21bn (£14.7bn)

SERVICE PERSONNEL: 230,600 (44% conscript) active, plus 65,200 reserves.

ABOVE: *The Aermacchi MB339 is a popular advanced trainer and light strike aircraft, found in air forces throughout the world and, of course, in Italy. (Aermacchi)*

ITALIAN AIR FORCE/ AERONAUTICA MILITARE ITALIANA

Formed: 1945

The Italian Army formed an aeroplane company during the Italo-Turkish War in 1911, buying Bleriot XIs, Etrich Taubes, Maurice Farman S-11s and Nieuports for reconnaissance duties. These aircraft were credited with making the first bombing attack. In 1912, the Battagliore Aviatori, or Aviation Battalion, was formed; being renamed the Military Aviation Service by the end of the year. That same year, the Servizio d'Aviazione Coloniale, Colonial Aviation Service, was also formed. The MAS changed its name to the Military Aviation Corps, Corpo Aeronautico Militare, in 1914.

Italy was one of the Allies in World War I. The CAM rose from 70 aircraft at the outbreak of the war to 1,800 at the end. Wartime aircraft included Nieuport 17C-1 Bebe and 110, Spad SVII, Hanriot HD-1 and Macchi M14 fighters; Caproni Ca33, Ca40 and Ca46, and Macchi M7 and M8 bombers; and Ansaldo SVA4, SVA5, SVA9 and SVA10, Savoia-Pomilio SP3 and SP4, and Fiat R2 reconnaissance aircraft. Post-war, the CAM's strength dropped sharply, with little new equipment.

Benito Mussolini's assumption of power in 1923 marked a revival for Italian military aviation. The CAM became the autonomous Regia Aeronautica. Steps were taken to increase its strength, and boost morale through prestigious events, the most notable being a mass flight of 24 Savoia-Marchetti SM55X flying boats to New York and back in 1933. This was the same year that the RA reached a strength of 1,200 aircraft: 37 fighter squadrons operated Fiat CR20s and CR30s; 34 bomber squadrons with Caproni Ca73s and Ca101s; 37 reconnaissance squadrons operated Romeo Ro1s, Caproni Ca97s and Fiat R22s. Flying-boat squadrons were equipped with Savoia-Marchetti SM55X, while transport squadrons used Caproni Ca101s, Ca111s and Ca133s, and Savoia-Marchetti SM81s. The RA was involved in Italys invasion of Abyssinia (Ethiopia) in 1935, and was accused of bombing as well as reconnaissance and transport. There was no Ethiopian air defence of any consequence. In 1936, the Spanish Civil War started, and the RA sent a strong contingent to fight alongside German and Nationalist forces.

Italy entered World War II in June, 1940, shortly before the fall of France, as an ally of Germany. At the outset, the RA had circa 3,000 aircraft, of which some 400 were obsolete or obsolescent types based in the African colonies. The RA was active mainly in the Mediterranean, bombing Malta, supporting ground forces in North Africa and in Italy's invasions of Yugoslavia and Greece. A token force of 75 Fiat BR20M Cicogna bombers and 50 CR42 and CR50 fighters was sent to Belgium for operations over England. The RA also provided aircraft for Operation Barbarossa, the German invasion of the Soviet Union, suffering heavy losses. A number of German aircraft entered RA service during the war, includ-

ABOVE: *The Marina Militare Italiana was one of the launch customers for the Agusta-Westland EH101 helicopter, replacing Agusta-built ASH-3D/H Sea Kings. (Agusta)*

ing Junkers Ju87 Stuka dive-bombers, and, later, Messerchmitt Bf109F and Me110G fighters and Dornier Do217 bombers. Daimler-Benz water-cooled engines replaced Italian air-cooled engines in a number of aircraft types, including Fiat CR52 and CR55 fighters, improving their performance by reducing drag.

Italy capitulated on 8 September, 1943, splitting the RA as many units joined the Allies, becoming the Italian Co-Belligerent Air Force. Those left became the Aviazone della Republica Sociale Italiano.

Post-war, the Aeronautica Militare Italiano was formed as an autonomous air service. Initially, it used surviving wartime aircraft as well as Supermarine Spitfire and Bell P-39 Airacobra fighters, Martin Baltimore bombers and Douglas C-47 transports. As a former member of the Axis alliance, the 1947 Peace Treaty dictated that only 200 out of a permitted maximum of 350 aircraft could be combat types. These restrictions were removed when Italy became a member of the North Atlantic Treaty Organisation in 1949.

A major re-equipment programme started once Italy joined NATO. US military aid included 80 Lockheed P-38J Lightning fighters, 100 Beech C-45 transports and a number of Stinson L-5 liaison aircraft. In 1950, the first jets, de Havilland Vampire FB5 fighter-bombers arrived; Fiat and Aermacchi built this type under licence. US equipment predominated throughout the 1950s, with Republic F-47D Thunderbolt fighters being joined and then replaced by F-84F Thunderstreak fighter-bombers and RF-84F Thunderflash reconnaissance-fighters. Lockheed PV-2 Harpoon MR aircraft, Grumman S2F-1 Tracker ASWt, Grumman SA-16A Albatross amphibians, and North American T-6G Texan and Lockheed T-33A jet trainers entered service. De Havilland Vampire NF54 night fighters and Canadair-built F-86 Sabres were introduced in 1955. Helicopters arrived during the late 1950s, including Sikorsky UH-19 and SH-34J. Agusta started building Bell helicopters, starting with the 47G Sioux, under licence.

Aermacchi MB326 trainers replaced the Texan, while the 1960s also saw Lockheed F-104G Starfighter and Fiat G91 fighters enter service, with Fairchild C-119G Packet and Fiat G222 transports. Italy also entered collaborative projects, joining the UK and Germany to produce the Panavia Tornado, and Brazil to produce the AMX attack aircraft. Italy ordered the Tornado as a strike and reconnaissance aircraft, but recently it has leased F3 interceptor versions from the RAF while awaiting the Anglo-German-Spanish-Italian Eurofighter 2000. Many older aircraft have been extensively upgraded, including the Starfighter, Tornado, AMX and Atlantic.

The AMI has fallen from 73,000 personnel to 55,350 over the past thirty years, and may fall further as conscription ends. There are five fighter squadrons, four operating up to 86 upgraded F-104ASA-M Starfighters, and two with 24 Tornado F3, although these and the F-104s are being replaced in 2003 with 30 F-16A/B, leased until 2010 when they will be replaced by up to 121 Eurofighter 2000s. Strike squadrons include four with 89 Tornado IDS/ECR, receiving an MLU, and four with 104 AMX, with half a squadron in each case devoted to reconnaissance duties, in addition to an AMX reconnaissance squadron. Two squadrons of

ABOVE: *The MMI uses 50 Agusta-Bell 212s for ASW, often operating from frigates and a cruiser. (Agusta)*

ABOVE: *The Italian Army's Air Cavalry uses militarized Augusta A109A and A109CM (nearest the camera) Hirundo helicopters for attack and observation, supporting the A129 attack helicopters. (Agusta)*

Italy

ABOVE: *An A129 Mangusta of the Air Cavalry with rocket pods. Most A129s do not have the chin-mounted gun seen here. (Agusta)*

Atlantic MR aircraft are operated under naval control, and may be replaced in a joint programme with Germany. An electronic warfare squadron operates a mixture of aircraft, including G222VS, PD-808, P-180 and P-166DL-3, and AWACs aircraft may be acquired in the near future. Transport includes two squadrons with 33 G222, although one is being replaced by 12 Alenia C-27J Spartans, and one with 22 C-130J Hercules, with some of the older C-130H being converted to tanker duties. There are 16 Airbus A400M on order for 2014, and four Boeing 707 tanker transports are being replaced by 767s by 2004. A substantial VIP fleet includes two A319CJ, two Falcon 900EX, 18 P180 Avanti and four Falcon 50s, as well as two SH-3D. Training uses 38 SIAI SF260AM and 73 MB339A, many of which are having an MLU, with another 14 attack MB339CD and an aerobatic display team with up to 17 MB339PAN. SAR is provided by 40 S208M, 33 HH-3F and 36 AB212, while 50 NH500E are used for both liaison and training. Ten NH90 TTH are on order for transport and CSAR.

ITALIAN NAVAL AVIATION/MARINA MILITARE ITALIANA

Until the post-war period, all aviation was in the hands of the air force, or Regia Aeronautica. During World War II, the liner *Roma* was converted to an aircraft carrier, re-named *Aquila*, and intended to operate up to 50 Reggiane Re2001 aircraft. Equipped with catapults intended for the *Graf Von Zeppelin*, *Aquila* was completed as Italy capitulated, fell into German hands, was bombed by the Allies, and attacked by human torpedoes before being scuttled.

Post-war, the MMI, or *Marinavia*, was established as a helicopter force, with Italian law dictating that the navy could not operate fixed-wing aircraft. After operations with Agusta-Bell 47G Sioux, the force expanded to include Sikorsky SH-3D and SH-34J, Agusta A106 and Agusta-Bell 204B helicopters. Rapid expansion resulted from the introduction of helicopter cruisers, including the *Andrea Doria*, *Caio Duilio*, and the larger *Vittorio Veneto*. Experiments with Harrier V/STOL jet fighters aboard the first of these cruisers in the early 1970s, encouraged the Italian Navy to introduce its first aircraft carrier, the *Guiseppe Garibaldi*, during the late 1980s. The law was changed to allow the navy to operate fixed-wing aircraft, although the IMMI uses the USMC's AV-8B rather than the Sea Harrier. Italy has joined the US JSF programme with plans for 20-24 aircraft, and plans a second aircraft carrier, a new *Andrea Doria*, for service in 2007.

Currently, some 2,500 of the navy's 38,000 personnel are involved in aviation. In addition to the carrier and one remaining helicopter cruiser, there are four destroyers and 16 frigates able to carry helicopters. There are 16 AV-8B Harrier II and two TAV-8B conversion trainers. Helicopters include 31 ASH-3D/H Sea Kings for ASW and transport, some of which have been replaced by the first 20 of a planned requirement for Westland Agusta Merlins, with four providing carrier-borne AEW capability. There are 50 AB212 ASW helicopters aboard the cruiser and escort vessels, which will be replaced by up to 46 NH90.

ITALIAN ARMY AVIATION - 'AIR CAVALRY'/CAVALLERIA DELL'ARIA

Formed during the post-war period, Italian Army aviation was known as the CAALE until 2000, when the title Air Cavalry was adopted. Early aircraft included 150 Piper L-18/21 and Cessna O-1E light aircraft, replaced by Aerfer AM-3C and Savoia-Marchetti SM1019 liaison aircraft during the 1970s, as well as 125 Agusta-Bell 47G/J Sioux and 70 204B helicopters. The CAALE was one of the first export customers for the CH-47C Chinook, operating 26 licence-built versions during the early 1970s.

In more recent years, a combat ability was acquired, initially using armed versions of the Agusta A109 helicopter, followed by the A129 Mangusta attack helicopter. The CAALE saw service in Somalia, deploying the Mangusta there on UN operations and, more recently, in Albania, Bosnia and Macedonia.

Currently, the Air Cavalry has 45 A129 Mangusta, which are being upgraded, but requires at least 15 more, and has cascaded its 28 A109A onto observation duties, supporting the A129 with roof-mounted sights. Reconnaissance is provided by 115 AB206A2 JetRanger. Transport is provided by 38 CH-47C Chinook, 80 AB205A/B and 25 AB412, while there is a requirement for up to 60 NH90 TTH helicopters. Three P180 Avanti are used on VIP duties. There are also three Do2228 light transports.

Jamaica

POPULATION: 2.6 million

LAND AREA: 4,411 square miles, 11,424 sq km

GDP: $6.9bn (£4.8bn), per capita $3,500 (£2,447)

DEFENCE EXPENDITURE: $50m (£35m)

SERVICE PERSONNEL: 2,830 active, plus 953 reserves.

JAMAICA DEFENCE FORCE AIR WING

Formed: 1963

Part of an integrated defence force, the Jamaica Defence Force Air Wing came into existence in 1963, with a Cessna 185, later joined by two Bell 47G Sioux helicopters and a DHC-6 Twin Otter. It has 140 personnel, and operates two BN-2A Defender, a Cessna 210 and a Beech King Air 100, three Bell 412EP, three 212 and four 206B JetRanger, and six UH-1H Iroquois, and four Eurocopter AS335N Squirrel.

Japan

POPULATION: 127 million

LAND AREA: 142,727 square miles, 370,370 sq km

GDP: $4.7tr (£3.3tr), per capita $24,600 (£17,202)

DEFENCE EXPENDITURE: $45.6bn (£31.9bn)

SERVICE PERSONNEL: 239,800 active, plus 47,400 reserves.

ABOVE: *A captured World War II Imperial Japanese Navy Aichi D3A1 dive-bomber, known to the Allies as 'Val'. (FAAM)*

JAPANESE AIR SELF-DEFENSE FORCE

Formed: 1954

Japanese military aviation dates from 1911 and the founding of both the Japanese Army Air Force and the Japanese Navy Air Force. The JAAF started with three Henri Farman and two Wright biplanes, an Antoinette and a Bleriot monoplane. The JNAF had two Maurice Farman and two Curtiss seaplanes. These aircraft were soon joined by Japanese-designed Tokogawa 1 and Sei Model 1 and 2 biplanes, licence-built Maurice Farmans, a Nieuport and a Rumpler Taube for the JAAF, with Otari and Ushioku biplanes and a Bleriot for the JNAF. Japan sided with the Allies during World War I, but saw little action other than the occupation of the German mandated port at Tsingtao on the Chinese mainland, although a few Japanese pilots flew with the French Aviation Militaire. Towards the end of the war, the JNAF received a number of Short reconnaissance-seaplanes, Sopwith seaplane-fighters, and Deperdussin seaplane-trainers, in addition to Yokosuka Model A seaplanes.

After the Armistice, the foundations were laid for a Japanese aircraft industry using the three main industrial giants, Mitsubishi, Nakajima and Kawasaki. A French aviation mission visited the country to advise on the structure and future of the JAAF, while the JNAF received similar aid from a British naval mission. Both missions had a considerable influence on Japanese aviation, and fostered interest in the aircraft carrier. The JAAF was completely re-equipped with Spad S13C and licence-built 20C, Nieuport 24C and 29C fighters, as well as 50 ex-RAF Sopwith 1 1/2-Strutter and Pup fighters; Breguet Br14B, Farman F50 and licence-built F60 Goliath bombers; Salmson SA-2 reconnaissance aircraft; Nieuport 81E, 83E and 24C, and Hanriot and Caudron C6 trainers, as well as the Nakajima Type 5 advanced trainer. The JNAF received Gloster Sparrowhawk Mars II and III and Mitsubishi Type 10 shipboard fighters, with reconnaissance and training versions of both; Short F5 America, Schreck FBA17 and Tellier flying-boats; Avro 504K and 504L trainers; plus Sopwith, Airco, Vickers, Blackburn and Supermarine Types. A Gloster Sparrowhawk Mars IV fighter was used in trials flying from a platform built over a forward gun turret of the battleship *Hamishiro* in 1922, leading to the conversion of an oil tanker into the first Japanese aircraft carrier, the *Hosho*, that same year. The ship entered service in 1923 with Mars IV and Mitsubishi Type 10 fighters.

The late 1920s saw two more carriers for the Imperial Japanese Navy, the *Akagi*, a converted battlecruiser, entered service in 1928, and the *Kaga*, a converted battleship, in 1929. Aircraft were developed or built under licence specifically for carrier operations, including Nakajima A1N1 Type 3 (Gloster Gambit) fighters, Mitsubishi B2M1 Type 89 (based on a Blackburn design) naval bombers and the Mitsubishi C1M2

Japan

ABOVE: *In desperation, the Japanese resorted to Kamikaze suicide attacks on Allied warships, but many never reached their target, as in the case of this Yokosuka D4Y3A* Suisei, *or* Comet. *(IWM)*

reconnaissance aircraft. Seaplanes and flying boats continued to enter service, including Yokosuka E1Y1 and Aichi Type 2 (Heinkel HD25) seaplanes, and Hiro H1H1, H1H2 and H2H1 flying boats. During the early 1930s, these aircraft were followed by Nakajima A2N1 shipboard fighters and advanced trainers, E4N1 and Kawanishi E5K1 reconnaissance seaplanes, and Hiro H3H1 flying boats.

The JAAF meanwhile re-equipped with Mitsubishi Type 87 light bombers, Kawasaki Type 87 (Dornier F) heavy and Type 88 reconnaissance-bombers. These were followed during the early 1930s by Nakajima Type 91 and Kawasaki Type 95 fighters, Mitsubishi Type 92 (Junkers G38) and Type 93, and Kawasaki Type 93 bombers, and Nakajima Type 94 AOP aircraft.

In 1931, Japan invaded Manchuria, which became protectorate of Manchouko. This was followed in 1932 by action against Shanghai, with the JNAF attacking from two carriers. This was the start both of Japanese territorial expansion that was to lead to World War II in the Pacific, and of major expansion of the armed forces. Expansion had to be achieved without British or French assistance once they realized the threat posed by Japanese ambitions.

Japanese aircraft designations were based on the year of the reign of the emperor, so that 7-Shi meant the seventh year of the Showa reign. In 1932, two aircraft of the 7-Shi range of prototypes, the Hiro G2H1 Type 95 naval bomber and Kawanishi E7K1 Type 94 reconnaissance aircraft were put into production. These were followed by the 9-Shi range of prototypes in 1934, including the Mitsubishi A5M1 Type 96 fighters and G3M1 Type 96 land-based long-range bomber and the Aichi D1A2 and Nakajima B4Y1 Type 96 carrier-borne bombers, the Watanabe E9W1 and Aichi E1A1 Type 96 reconnaissance seaplanes, and the Kawanishi H6K1 Type 97 four-engined flying-boat, which all entered service around 1936. These aircraft were followed by Nakajima C5M1 Type 98 reconnaissance aircraft, and Hiro H5Y1 Type 99 reconnaissance flying boats.

During this period, the JAAF also selected and introduced new aircraft. Kawasaki Type 95 and Nakajima Ki27a and Ki27b Type 97 fighters; Mitsubishi Ki30 and Ki21, and the Kawasaki Ki48 Type 98 bombers; Nakajima Ki34 Type 97 transports; Tachikawa Ki36 AOP aircraft; Mitsubishi Ki15 and Ki51 Type 97 reconnaissance aircraft; and Ki51b ground-attack aircraft.

Operations in China resumed in 1937, with the JAAF and JNAF outnumbering Chinese forces by more than five to one. During the winter of 1937-38, Japanese forces advanced rapidly across China, although, outside the range of escorting fighters, Japanese bombers were subjected to devastating Chinese fighter attack. New aircraft continued to enter service, with the most notable being the Mitsubishi A6M2 Type O shipboard fighter, the famous Zero. New aircraft carriers had been entering service, giving Japan a strong naval air arm. By the end of 1941, China was virtually defeated.

On 8 December, 1941, 353 JNAF aircraft operating from six aircraft carriers attacked the US naval base at Pearl Harbor, Hawaii, achieving complete surprise and inflicting major losses. The attack was a tactical success, but a strategic blunder. It brought the United States into the war, but failed to make Pearl Harbor unusable while the US Pacific Fleet's aircraft carriers were safely at sea. Japanese forces then swept into Hong Kong, and into Thailand and down the Malay Peninsula to take Singapore. The Dutch East Indies were taken, and Japanese forces advanced into Burma, eventually being held in check both there and in New Guinea. JAAF and JNAF aircraft were present wherever they were needed, aiding the fastest and best co-ordinated advances in military history. The Vichy French government allowed the Japanese to use bases in French Indo-China. On 10 December, 1941, Japanese shore-based aircraft found and attacked the new British battleship HMS *Prince of Wales* and the elderly battlecruiser HMS *Repulse* off the coast of Malaya, finding the ships without air cover in the first instance of warships at sea being sunk in an aerial attack. On 5 April, 1942, aircraft from the carrier *Soryu* found and sunk the British heavy cruisers HMS *Dorsetshire* and HMS *Cornwall*. A few days later, on 9 April, they sunk the small British carrier HMS *Hermes* and her escort, the Australian destroyer HMAS *Vampire*. Darwin in Australia and bases in Ceylon were bombed.

The US Pacific Fleet was quick to move to the offensive. Carrier-based bombers attacked Tokyo and other large cities, and while the damage was slight, the blow to Japanese confidence was immense, as was the damage to the reputation of the JAAF. The Battle of the Coral Sea was inconclusive, a Japanese plan to seize Midway Island resulted in the Battle of Midway on 4 June, 1942, in which the Imperial Japanese Navy lost all four carriers - *Akagi, Kaga, Hiryu* and *Soryu* - assigned to the operation. The IJN was never to recover from this crushing defeat.

Wartime aircraft operated by the JNAF included developments of the Zero fighter, including seaplane versions. Kawanishi H8K2 Type 2 flying boats, Nakajima J1N1 Type 2 fighter-reconnaissance aircraft, Aichi D4Y1 and D4Y2 and Mitsubishi G4M2 bombers were also used. Like the JNAF, the JAAF was also heavily dependent on pre-war designs and their developments, but new aircraft included the Nakajima Ki44 Type 2 and Ki84 Type 4, and Kawanishi Ki4 and Ki61 Type 3 fighters, Nakajima Ki49 Type 0 and Mitsubishi Ki67 Type 4 bombers, and Kokusai Ki49 Type 0 AOP aircraft.

In desperation towards the end of the war, Kamikaze (Divine Wind) suicide groups were formed by both services. Initially, standard combat aircraft were used, but Oka (cherry blossom) piloted bombs were also developed. These had to be carried close to the target by heavy bombers before being released; the concept was flawed because of the vulnerability of the mother aircraft. Suicide bombers proved effective against smaller warships, including escort carriers, but often failed to penetrate intense AA fire and were useless against the armoured decks of the new British carriers. Throughout the war, the relative lack of technical progress by the Japanese, compared with the Germans, was a major weakness. As US forces advanced across the Pacific towards Japan, bases became available for USAAF heavy bombers, and the JAAF found that its aircraft were completely unable to reach these aircraft to shoot them down. Again the solution lay in suicide attack, with pilots using aircraft stripped of armament to 'ram' the bombers. The fire bombing of Japanese cities, the Allied advance towards Japan, and finally the dropping of the first atomic bombs on Hiroshima and Nagasaki in August, 1945, led to Japanese surrender and occupation by the Allies.

Japan's armed forces were disbanded after the surrender. In 1950, a National Police Reserve Force, a paramilitary organisation, was formed under US sponsorship to relieve the pressure on the occupation forces during the Korean War. This was followed in 1954 by the creation of land, air and sea 'self-defence' agencies, effectively re-establishing an army, air force and navy in all but name. These were formed with US assistance, largely because of Japan's proximity to Communist North Korea and China.

ABOVE: *Another US aircraft in service with the JASDF was the North American F-86F Sabre. (JASDF)*

ABOVE: *JASDF Mitsubishi T-2 trainers. (JASDF*

LEFT: *Many F-4EJ Phantoms remain in Japanese service. (JASDF)*

Initially, the Japanese Air Self-Defence Force operated North American T-6G Texan and Beech T-34 Mentor trainers, with a few Lockheed T-33A jet trainers and Curtiss C-46 Commando navigational trainers. The first students were mainly wartime veterans seeking refresher and conversion training. They started training in 1955, and by the end of the year the first North American F-86F Sabre fighter squadron was operational. Within three years, the JASDF had almost 300 Sabres, 35 C-46 Commando transports, and some 300 trainers, with many of the Sabres and T-33s built in Japan. In 1959, Japanese-designed Fuji T-1 jet trainers were in service, followed in 1960 by Japanese-built Lockheed F-104J Starfighters. McDonnell Douglas F-4EJ Phantom fighter-bombers were also built in Japan and entered service in 1970, by which time personnel numbers had reached 40,000 and the JASDF's first SAM missiles, three Nike-Ajax battalions, were operational. Japanese designed aircraft entering service included military versions of the NAMC YS-11 transport. Most aircraft were licence-built US designs. Helicopters were also introduced, mainly for SAR, including Sikorsky H-19 and S-62, and Vertol 107. In 1972, NAMC C-1 tactical jet troop transports entered service, and these were followed a couple of years later by the Mitsubishi T-2 supersonic trainer. The Japanese-designed Mitsubishi F-1 strike aircraft entered service in the 1980s, followed by McDonnell Douglas F-15J Eagle interceptors.

Japanese neutrality and their sensitivity over military expansion has meant that the country avoided the post-war regional conflicts, and has had only minimal involvement in UN peace-keeping. The JASDF is a formidable force today, with 45,400 personnel. Latest aircraft into service has been the Mitsubishi F-2A/B, a Japanese development of the Lockheed F-16 with a composite wing, of which 130 are being delivered as F-1 replacements: Some are also likely to replace a number of the 135 F-4EJ and RF-4EJ Kai Phantom interceptors and reconnaissance aircraft which equip two fighter

ABOVE: *A formation of Kawasaki C-1 transports. (JASDF)*

ABOVE: *A variety of tasks falls to the NAMC YS-11s, including VIP transport, ECM and radar calibration. (JASDF)*

squadrons, a ground-attack squadron and a reconnaissance squadron. There are 153 F-15J and 38 F-15DJ Eagle interceptors, in eight out of the ten fighter squadrons. Aggressor training is provided by a squadron with ten F-15DJ. There are ten Grumman E-2E Hawkeye and four Boeing 767s in an AEW squadron. ECM are provided by two squadrons with Kawasaki EC-1 and NAMC YS-11. Four Boeing 767 tankers were ordered in 2002 for delivery in 2006. Transport aircraft are operated in three squadrons with 16 Lockheed C-130H Hercules, up to 26 C-1, and a number of YS-11, while a fourth squadron has two VIP Boeing 747-400 and some YS-11. The C-1s are likely to be replaced by a new type in the near future, while a large business jet or regional airliner type will replace the YS-11s. Four flights operate 17 Boeing CH-47J Chinook heavy lift helicopters, some of which can be used for SAR, on which 30 U-125 (Hawker 800) are operated, also handling radar calibration, replacing Mitsubishi MU-2J/S. There are 22 UH-60J Black Hawk and 22 Kawasaki-built CH-46 helicopters. Training aircraft include 22 Fuji T-1A/B, 41 Mitsubishi T-2 and 43 Fuji T-3 basic trainers, now being augmented by 21 T-7 variants, as well as 60 T-4 and ten T-400 (Mitsubishi Diamond), operated as 12 squadrons.

There are six SAM groups with 24 Patriot squadrons, while ASM missiles include the ASM-1 and ASM-2, with AIM-7 Sparrow and AIM-9 Sidewinder AAM.

JAPANESE MARITIME SELF-DEFENSE FORCE

Formed: 1954

The Japanese Maritime Self-Defence Force was founded in 1954 primarily as an anti-submarine force with US assistance. The initial equipment included four Bell TH-13 helicopters and some North American SNJ-6 trainers. The following year, 20 Grumman TBM-3W-2 and TBM-3S Avengers, 17 Lockheed PV-2 Harpoon, ten Convair PBY-6A Catalina, four Grumman Goose amphibians, and Sikorsky S-51 helicopters, were being operated. The rapid expansion was helped by the recall of many wartime JNAF veterans. In 1956, the first of 42 Japanese-built Lockheed P2V-7 Neptunes started to replace the Harpoons, and in 1957, the Avengers were replaced by 60 Grumman S-2A Tracker ASW aircraft.

Licence-manufacture of US aircraft continued, but also allowed some development, with 46 Kawasaki P-2J MR aircraft being produced as turboprop developments of the Neptunes which they replaced in JMSDF service during the early 1970s. A Japanese design was the Shin Meiwa PS-1 turboprop flying-boat, which at one time had 36 aircraft equipping three squadrons, and with a few of this aircraft's US-1 development still in service on SAR duties. Longer-range MR eventually passed to the Lockheed P-3C Orion. As with most navies, helicopters can be operated from escort vessels, and this has meant a considerable expansion of the helicopter force in recent years, with some destroyers able to carry three or four helicopters. The role of the JMSDF has developed beyond anti-submarine operations in recent years, and it now has the *Osumi*, officially classified as an LST, capable of carrying 330 troops, but with a through flight deck, effectively making it a small helicopter carrier.

Currently, some 12,000 of the JMSDF's 42,600 personnel are directly involved with the air arm. This has 96 Lockheed P-3C/OP-3 Orion MR aircraft spread over ten squadrons, one of which is assigned to training. Another seven EP-3C, NP-3C and UP-3D Orion are assigned to ELINT and radar calibration duties. SAR uses ten Shin Meiwa US-1/1A flying boats, as well as nine Sikorsky S-61A Sea King and 12 UH-60J Seahawk helicopters, while Gulfstream Ivs are being introduced for anti-piracy patrols. Ten MH-53J Sea Stallion helicopters are used for minesweeping. There are 39 HSS-2B Sea King helicopters for shipboard ASW, but some at least of these will be replaced

Japan

ABOVE: *The Japanese Air Self-Defence Force has 30 Hawker 800s, designated U-125 by the JASDF, for radar calibration and rescue - this is a rescue-configured U-125A. (Raytheon)*

by up to 60 SH-60J Seahawk. ASW training uses six YS-11T and another four YS-11M transports, with 35 King Air being used for transport and training. Another aspect of training is provided by five Learjet U-36A target tugs. The main training type is the T-5 (KM-2D) while13 OH-6DA Cayuse helicopters are also in use.

JAPANESE GROUND SELF-DEFENSE FORCE

Formed: 1954

The Japanese Ground Self-Defence Force was created in 1954 out of the National Police Reserve, which included a number of pilots trained from 1952 onwards by the US Army to fly Piper L-21 and Stinson L-5 Sentinel AOP aircraft and who formed the nucleus of the JGSDF's air element. Helicopters were soon introduced, including Japanese-built versions of the Vertol 107, or CH-46, Bell UH-1 Iroquois and 47 Sioux, Hughes OH-6A and Sikorsky S-62, although the earlier H-19 Chickasaw was supplied direct from the US. During the late 1960s and early 1970s, a massive expansion programme saw the number of helicopters rise from 290 to 400. Attack helicopters were introduced later, Fuji-Bell AH-1 Cobras.

The JGSDF has formed an air mobile brigade and in 2000 introduced the first of up to 200 Kawasaki OH-1 armed scout helicopters, replacing the large force of OH-6D/J Cayuse. It will be replacing its 86 AH-1F Cobra helicopters with ten AH-64D Longbow Apaches in 2004, with another 50 planned. Transport is provided by 45 CH-47J Chinook, with a handful of CH-46 still operational. There are eight UH-60A and 170 UH-1H/J Iroquois. Three AS332L Super Puma are used for VIP transport. Communications duties are handled by 16 Mu-2 and three Beech Super King Air fixed-wing aircraft.

BELOW: *A JASDF Boeing CH-47J Chinook, mainly used for transport and SAR. (JASDF)*

Jordan

J O R D A N

POPULATION: 6.9 million

LAND AREA: 36,715 square miles, 101,140 sq km

GDP: $7.6bn (£5.3bn), per capita $3,200 (£2,238)

DEFENCE EXPENDITURE: $520m (£363m)

SERVICE PERSONNEL: 100,240 active, plus 35,000 reserves.

ROYAL JORDANIAN AIR FORCE

Formed: 1956

Originally established by the UK as the Arab League Air Force in 1949 following the first Arab-Israeli War, initially as a transport unit within the army using two elderly de Havilland Rapide biplanes. Seconded RAF personnel filled key posts. During its first year, the ALAF received two de Havilland Tiger Moth and four Percival Proctor trainers. In 1950, training of Arab personnel began, using Auster light aircraft for this and for AOP. Three Turkish MKEK4 Ugur trainers were operated briefly before being replaced by DHC-1 Chipmunks. During the early 1950s, four de Havilland Dove light transports, a Vickers Viking airliner and a Handley Page Marathon communications and VIP aircraft entered service, joined in 1956 by a Vickers Varsity, nine de Havilland Vampire FB9 fighter-bombers. Two T55 trainers, a gift from the British government in 1955, were the first jets, joined by ex-Egyptian FB52s in 1956. Three ex-RAF North American T-6 Harvard trainers were delivered in 1955.

The ALAF became the separate Royal Jordanian Air Force in 1956. Jordan underwent a brief flirtation with Egypt, but with Saudi support the country returned to her former alliance with the UK. Twelve Hawker Hunter F6 fighters, a Westland Scout and four Alouette III helicopters, with additional Doves and a Beech Twin Bonanza were delivered during the late 1950s and early 1960s. Almost all of the aircraft were lost during the June, 1967, Arab-Israeli War, when Jordan also lost the West Bank of the River Jordan. Re-equipment took place with British, American and Pakistani help, the latter providing four loaned North American F-86F Sabres. By the early 1970s, the RJAF had grown to 2,000 personnel. It was operating two squadrons with a total of 36 Lockheed F-104A Starfighters and a squadron with 18 Hawker Hunter FGA9 fighter-bombers, as well as Westland Whirlwind (S-55), four Alouette III helicopters, Douglas C-47 and de Havilland Dove and Heron transports. Short Tigercat SAM had been acquired for airfield defence. Pilots were trained in the USA.

Jordan occupies a difficult strategic position. The country did not take part in the October, 1973 Yom Kippur War, although it did tie down considerable Israeli forces sim-

ABOVE: *The Royal Jordanian Air Force grew out of the Arab League Air Force - one of its first aircraft was this Auster for AOP and liaison duties. (RJAF)*

ABOVE: *The Royal Jordanian Air Force was one of many operators of the Hawker Hunter fighter-bomber. (RJAF)*

ABOVE: *Few air forces managed without the legendary Douglas C-47, known in the UK as the Dakota. Even the Communist Bloc air forces had the aircraft as the Li-2. (RJAF)*

ABOVE: *The Lockheed F-104 Starfighter was a potent addition to the RJAF inventory. (RJAF)*

BELOW: *Dassault Mirage F-1C attack aircraft are part of the RJAF's current frontline strength. (Dassault Aviation Collection)*

Jordan

ABOVE: *Lockheed C-130 Hercules deploy paratroopers. (RJAF)*

ABOVE: *CASA C212 handle a variety of tasks, including light transport. (RJAF)*

ABOVE: *The Royal Jordanian Air Force provides air support for the Jordanian Army, including Bell AH-1F Cobra anti-tank helicopters. (RJAF)*

ply by mobilising. It did not take part in the Gulf War of 1991, but offended moderate Arab and Western opinion by siding with Iraq.

The Hunters were replaced by Northrop F-5E/F Tigers and the F-104s by Dassault Mirage F1B/C/E interceptors, while substantial tactical air mobility came with a large number of Bell UH-1H/L Iroquois helicopters. Anti-tank helicopters were also introduced in the form of the Bell AH-1F Cobra. Over the past 30 years, the RJAF has grown considerably, and now has 15,000 personnel. It introduced Lockheed F-16 interceptors during the winter of 1997-98, leasing these from the USAF. Current plans are to standardise on this aircraft, ideally acquiring another 70-80 by 2007.

There are 12 F-16A and four F-16B fighters in a single fighter squadron, with another squadron operating 25 Mirage F-1CJ/BJ. Four strike squadrons include three operating a total of 50 Northrop F-5E/F and one operating 15 Mirage F1-RJ. A transport squadron operates eight Lockheed C-130B/H Hercules and two leased CN-235, while two C212 Aviocar are used for transport and surveying. A VIP Royal Flight operates a Lockheed TriStar 500, a Canadair Challenger 604 and a Grumman Gulfstream IV, as well as three Sikorsky S-70 helicopters and an Alouette III. Attack helicopters are 20 Bell AH-1F Cobra, currently being upgraded, fitted with TOW ASM. Transport helicopters include 54 Bell UH-1H/L Iroquois and ten AS332M-1 Super Puma. Training uses 17 BAe Bulldog, although these may soon be replaced, as well as 13 C101CC Aviojet and six Hughes 500D helicopters. The RJAF operates three BO105CBS for the Jordanian police. AIM-9 Sidewinder, R-530 and R-550 AAM are deployed, with TOW anti-tank missiles and AGM-65D Maverick ASM.

Kazakstan

POPULATION: 16.1 million

LAND AREA: 1,048,070 square miles, 2,778,544 sq km

GDP: US $18.2bn (£12.7bn), per capita US $3,700 (£2,587)

DEFENCE EXPENDITURE: $364m (£255m)

SERVICE PERSONNEL: 64,000 (inc many conscripts) active.

KAZAKSTAN AIR FORCE

Formed: 1991

Kazakstan became independent of the former USSR in December, 1991. Equipment for the new Kazakstan Air Force was acquired by taking Soviet equipment stationed at air bases within the country, not all of it suitable for the KAF. Former Soviet Tupolev Tu-95MS Bears with a nuclear-strike capability abandoned at the airbase at Semipalatinsk were returned to Russia in exchange for new MiG-29s and Su-27s. The new state also found itself with a number of SS-18 ICBM in silos, these were removed and the silos destroyed in 1996. Kazakstan now has a significant air defence and ground attack capability, although it is not known how many aircraft are in storage. Flying hours are claimed to be around 100 annually, far more than in most former USSR states, and the country is a member of the CIS joint air defence plan.

Currently, the KAF has 19,000 personnel. It has one fighter regiment with 40 MiG-29, and one with 43 MiG-31 and 16 MiG-25. For the FGA role it has three regiments, one with 14 Su-25, one with 25 Su-24 and one with 14 Su-27. A reconnaissance regiment has 12 Su-24. Many older aircraft are in storage, including more than 100 MiG-23, and possibly available for sale, including MiG-23, 25, 27 and 29, and some Su-27. There are also up to 42 Mi 24 attack helicopters. Transport is provided by a variety of Antonov and Tupolev types, of which only the An-26, of which there are 14, is available in significant numbers. Helicopters include around 60 Mi-8, as well as six Mi-6 and 24 Mi-26. A Boeing 757-200 and a Falcon 900 provide VIP transport. Training uses 12 L-39 and four Yak-18 aircraft. Missile defences include upwards of 100 launchers for SA-2, SA-3, SA-4, SA-5, SA-6 and S-300. ASM missiles include AS-7, AS-9, AS-10 and AS-11, while AAM are mainly AA-6, AA-7 and AA-8.

Kenya

POPULATION: 30.5 million

LAND AREA: 224,960 square miles, 582,646 sq km

GDP: $10.7bn (£7.5bn), per capita $1,500 (£1,049)

DEFENCE EXPENDITURE: $313m (£219m)

SERVICE PERSONNEL: 24,400 active.

ABOVE: *Bae Hawk series 50. (Michael J. Gething)*

KENYA AIR FORCE

Re-formed: 1994

Kenya became independent from the UK in 1963, with British assistance in establishing national defence forces, including six DHC-1 Chipmunk trainers. The emphasis was on transport and communications using DHC-2 Beaver and DHC-4 Caribou, with an Aero Commander 500 for VIP duties. A limited combat capability came with the delivery of six BAC167 Strikemaster armed trainers in 1971. BAe Bulldogs replaced the Chipmunks in 1972. During the late 1970s, the KAF received Northrop F-5E Tigers for interception and ground attack, with DHC-5 Buffalo transports. The KAF was involved with an attempted *coup d'etat* in 1982, afterwards being disbanded and placed under army control. Reformed afterwards using loyal elements, it was renamed 'Kenya 1982 Air Force' to avoid any confusion with the disgraced former service. The present title was re-adopted in 1994. COIN remains important for the KAF, although there has been relatively little organized insurrection, and the Strikemasters have been replaced by BAe Hawk Mk52 and Embraer Tucano Mk51 trainers also capable of fulfilling the attack role. A new task has been anti-poaching patrols to preserve the country's wildlife using Hughes 500MD helicopters, which can also be armed.

Currently, the KAF has 3,000 personnel. It has five F-5E Tiger II interceptors, capable of using AIM-9 Sidewinder AAM, and two F-5F trainers, while it can also use nine Hawk Mk52, 11 Tucano Mk51 and many of the 32 Hughes 500MD helicopters in the attack role. There are eight DHC-5 Buffalo, expected to remain in service until 2005 or 2006, as well as three DHC-8-100 Dash 8, 12 HAMC Y-12 and six Dornier Do28D Skyservant transports, with a VIP Fokker 70, as well as 12 SA330 Puma transport helicopters. TOW anti-tank and AGM-65 Maverick missiles are used.

Korea, Democratic People's Republic of (North)

POPULATION: 24.5 million

LAND AREA: 46,814 square miles, 121,248 sq km

GDP: $15bn (£10.5bn), per capita $1,000 (£699)

DEFENCE EXPENDITURE: $2.1bn (£1.5m)

SERVICE PERSONNEL: 1,082,000 active, plus 4,700,000 reserves.

KOREAN PEOPLE'S ARMY AIR FORCE

Formed: 1953

Korea was occupied by Japan from early in the twentieth century until the end of World War II, when the north was occupied by Soviet troops and the south by US troops. Almost immediately after Soviet occupation in 1945, a North Korean Aviation Society was formed, and in 1946 this became the Korean People's Army Aviation Division, becoming the Korean People's Armed Forces Air Corps in 1948. Considerable Soviet assistance was received, including Yakovlev Yak-9 fighters, Ilyushin Il-10 ground-attack aircraft and Yak-18 trainers, with most of these aircraft reaching North Korea between 1948 and June, 1950, when North Korean forces invaded the south. Towards the end of 1950, MiG-15 jet fighters started to enter service, bearing the brunt of aerial action throughout the Korean War.

The end of the war in 1953 saw Korea divided into two nations, with North Korea becoming the Democratic People's Republic of Korea. The title of the Korean People's Army Air Force was also adopted at the time. Aircraft included MiG-15, Yak-9 and Lavochkin La-9 fighters, Tupolev Tu-2 bombers and Ilyushin Il-10 ground-attack aircraft, as well as Yak-11 and Yak-18 trainers. Ilyushin Il-28 jet bombers followed, and in 1957, MiG-17s were delivered, including

Korea, Democratic People's Republic of (North)

a number of the Chinese-built version, the Shenyang F-4, with Antonov An-2 and Lisunov Li-2 (C-47) transports. The first helicopters, Mi-1s, arrived at this time and were soon joined by Mi-4s. A few MiG-19s were delivered before MiG-21 interceptors were supplied in 1966.

Despite attempts at easing the situation, the two Koreas continue an extreme version of the Cold War stand-off, aggravated by the very poor state of the North Korean economy, especially compared with South Korea. North Korea spends massively on defence, with conscription lasting for up to ten years. Despite economic problems, KPAAF personnel numbers have virtually trebled over the past 30 years from 30,000 to 86,000. Much of the equipment is obsolescent since the collapse of Communism and the break up of the Soviet Union mean that it no longer enjoys Soviet aid. Aircraft have been acquired from air forces in the newly-independent former states of the USSR, including MiG-21s from Kazakstan. The main emphasis is on attack, and this is reflected in the helicopter units with MiG-24 attack helicopters, with relatively few transport types. The most modern aircraft are MiG-29s and Su-25s, both acquired in 1988, although numbers of the former have been boosted in recent years.

The KPAAF has 35 MiG-29A in an air defence regiment, with another five for training, while there are 36 Su-25K in a ground-attack regiment, with another four aircraft for training. Three regiments operate 107 J-5 (MiG-17); four have 159 J-6 (MiG-19); four have 130 J-7 (MiG-21); one has 46 MiG-23; while one has 18 Su-7. There are 20 Mi-24 attack helicopters. There are 300 Antonov An-2 or Harbin Y-5 light transports whose role is to infiltrate snipers in any future conflict. Apart from 12 An-24s, the remainder of the fixed-wing transport force consists of a varied selection of Antonov, Ilyushin and Tupolev types with each in small numbers. The most numerous helicopter is the Mi-2, with 140, followed by the Hughes 500D/E, of which there are 86, with just 15 Mi-8 and Mi-17, and 48 Mi-4. Training uses 170 Yak-18, 10 CJ-5, 30 MiG-15UTI, known as the FT-2 in Korea, and 12 L-39 Albatros. Average flying hours are believed to be around 20, and doubtless much less for units other than those with the latest MiG-29A and Su-25K aircraft.

Korea, Republic of (South)

POPULATION: 47.3 million

LAND AREA: 38,452 square miles, 99,591 sq km

GDP: $457bn (£319bn), per capita $15,000 (£10,490)

DEFENCE EXPENDITURE: $12.8bn (£9bn)

SERVICE PERSONNEL: 683,000 (25% conscript) active, plus 4,500,000 reserves.

ABOVE: *A South Korean Navy Westland Super Lynx. (GKN Westland)*

REPUBLIC OF KOREA AIR FORCE

Formed: 1949

The Republic of Korea was formed in 1948, with the Republic of Korea Air Force, RoKAF, formed the following year. The first pilots were Koreans who had flown with the Japanese forces during World War II. Plans for combat aircraft were over-ruled on the grounds that Korea was not of strategic significance, with the first aircraft mainly liaison and AOP types, including Piper L-4s, as well as North American T-6 Harvard trainers. North Korea invaded the south in June, 1950, starting the Korean War in which the United Nations forces, led by the United States, attempted to protect South Korea from first North Korean and then Chinese attack. The RoKAF started to receive combat aircraft, with the first being North American F-51D Mustang fighters.

After the Korean War ended in an uneasy armed truce, US military aid continued at a high level. In 1956, the first jets, North American F-86F Sabre fighters, entered operational service, followed by Lockheed T-33A jet trainers in 1957. In 1965, Northrop F-5A/B fighter-bombers were introduced, and in 1970, McDonnell Douglas F-4D Phantoms. By the early 1970s, the RoKAF's personnel strength had reached 23,000. Modernisation since has taken the form of Northrop F-5E/F Tiger IIs, followed later by Lockheed F-16s, but

Korea, Republic of (South)

many older types remain in service, including the now ageing F-5As and F-4 Phantoms. In recent years, the ability to fund upgrades and replacements suffered during the Asian economic crisis of the late 1990s, which delayed development of the indigenous T/A-50 advanced trainer and light fighter programme, a joint venture between Lockheed Martin and KAI. Meanwhile, fighters are being assessed as a replacement for both the F-4s and F-5s, with initial deliveries due to start in 2004.

The RoKAF has grown over the past 30 years to 63,000 personnel. Fighter and ground-attack formations are in seven tactical wings. Two operate 100 Lockheed Martin F-16B and 50 F-16C, with 20 F-16D conversion trainers; another 20 F-16C/Ds are being added. Three have 145 Northrop Grumman F-5E, with 35 F-5F, which may be upgraded to extend their life beyond 2010; another two have 130 F-4D/E. A reconnaissance group has 18 RF-4C and five RF-5A. Close support is provided by 27 Cessna A-37B Dragonfly, with forward air control by seven Rockwell OV-10D Bronco and ten Cessna O-2A, as well as 20 O-1A Bird Dog AOP aircraft. Eight Hawker 800s provide reconnaissance and ELINT. Airlift capability includes 12 Lockheed C-130H/H-30 Hercules, 20 CN235-100/200, and six CH-47D Chinook helicopters, with a Boeing 737-300, two BAe748 and three Bell 412 for VIP duties. SAR is provided by a helicopter squadron with five Bell UH-1D/H. There are 16 BAe Hawk Mk67, 30 leased Northrop T-38 Talon, 25 Cessna T-37, 15 T-41B Mescalero and 35 T-33A Shooting Star trainers. The T-37s and T-41s are being replaced by the first of up to 110 KT-1 Woong Bee, followed by 95 T-50, which will replace the T-38 Talons from 2005, and which may be joined by up to 50 A-50. Missiles include AIM-7 Sparrow, AIM-9 Sidewinder and AIM-120B AMRAAM AAM, with AGM-65A Maverick and AGM-88 HARM ASM, and Nike Hercules, Hawk, Javelin and Mistral SAM.

Future equipment plans may be affected by attempts at rebuilding relations with North Korea.

REPUBLIC OF KOREA NAVY

The Republic of Korea Navy initially started using aircraft for liaison duties, before acquiring Grumman S-2 Trackers for shore-based MR. The introduction of warships with helicopter capability led to the introduction of the Westland Lynx. The Trackers were replaced by eight Lockheed P-3C in 1996 and the RoKN is believed to be looking for eight additional aircraft. The Westland Lynx were upgraded to Super Lynx standard in the late 1990s after an order was placed for new Super Lynx in 1997. Today the RoKN has eight Orion, 24 Super Lynx Mk99 for ASW and ASuW, as well as 25 ASW MD500M. Communications duties are undertaken by two Bell 206B JetRanger and three Cessna Caravan II.

REPUBLIC OF KOREA ARMY

Initially, the Republic of Korea Army relied upon the RoKAF for its close support needs, but a significant transport and anti-tank capability has been developed. The main combat capability is with 60 Bell AH-1J/S/F Cobra and 45 TOW-equipped MD500 helicopters, with another 130 AOP MD500s. There are 18 CH-47D Chinook heavy lift helicopters, as well as 20 Bell UH-1H Iroquois, with up to 138 Sikorsky UH-60P Black Hawk and 12 BO105 helicopters entering service. Three AS332L Super Puma are used for communications duties. Up to 60 new attack helicopters may be ordered shortly.

Kuwait

POPULATION: 2.1 million

LAND AREA: 9,375 square miles, 24,235 sq km

GDP: $33.4bn (£23.4bn), per capita $15,000 (£10,490)

DEFENCE EXPENDITURE: $3.3bn (£2.3bn)

SERVICE PERSONNEL: 15,500 (some conscripts) active, 23,700 reserve.

KUWAIT AIR FORCE

Re-formed: 1991

The Kuwait Air Force was formed with RAF assistance as an extension of the Security Department in 1960, initially for liaison and communications with five Auster AOP aircraft, some de Havilland Doves and a Heron. Threatened by its neighbour Iraq, which claimed Kuwait as an Iraqi province, it quickly acquired sophisticated aircraft. Six Hawker Hunters and six Agusta-Bell 204s were followed by six BAC167 Strikemaster armed-trainers, and then by 12 BAC Lightning interceptors, with two conversion trainers. These aircraft were replaced in due course and expansion continued during the late 1970s and 1980s including Dassault Mirage F1CK interceptors and McDonnell Douglas A-4KU Skyhawk ground attack aircraft.

Plans were laid for the acquisition of F-18C Hornets and Short Tucano trainers when, on 2 August, 1990, Iraqi forces invaded. The small country was quickly overrun, but the KAF managed to resist for up to 48 hours before the remaining aircraft were flown to bases in neighbouring Saudi Arabia. About 40 aircraft, including helicopters, were caught in the Iraqi invasion. The KAF became the Free Kuwait Air Force, and was able to join the Coalition Forces in the 1991 Gulf War, Operation Desert Storm, in the air campaign that preceded the ground assault to liberate Kuwait,

Kuwait

losing at least one aircraft in the campaign. As Kuwait was liberated, retreating Iraqi forces destroyed captured Kuwaiti aircraft.

The Kuwait Air Force was reformed, and aircraft ordered before the invasion were soon delivered, with the first of 16 Tucanos arriving in Kuwait in October, 1991, and the first Hornets delivered early the following year. Altogether, 40 Hornets were delivered, including eight F/A-18D twin-seat aircraft, and surviving Mirage F1s were refurbished. The Skyhawks were sold to Brazil. Other new aircraft included BAe Hawk 64 armed jet trainers.

The KAF has 2,500 personnel. There are 32 F/A-18C and eight F/A-18D Hornets, which are being upgraded while Kuwait is interested in the F/A-18E/F Super Hornet for 2005/6. The Mirage F1 force is for sale. There are 12 Hawk 64 and 16 Short Tucano, for training and close support. There are 16 Boeing AH-64D Longbow Apache attack helicopters, equipped with Hellfire anti-tank missiles, displacing a similar number of Eurocopter SA342K Gazelle onto AOP and police duties. Three AS532AF Cougar operate in the anti-shipping role. Three Lockheed L-300-30 Hercules are being replaced by six C-130J Hercules II. There are eight SA330H Puma transport helicopters. For VIP use, there is a DC-9 and a MD-83. Additional AH-64Ds may be bought.

Kyrgyzstan

POPULATION: 4.7 million

LAND AREA: 73, 861square miles, 191,299 sq km

GDP: $1.3bn (£909m), per capita $2,300 (£1,608)

DEFENCE EXPENDITURE: $31.2m (£21.8m)

SERVICE PERSONNEL: 9,000 (many conscripts) active, plus 57,000 reserves.

REPUBLIC OF KYRGIZIA AIR ARM

Formed: 1991

Formerly a member of the USSR, Kyrgyzstan became independent in 1991, creating its own air force from aircraft and helicopters belonging to the former Central Soviet Air Force Training School. The potential to continue training students for other former USSR states seems to have been missed, possibly due their funding problems. The country is a member of the CIS joint air defence treaty. The original equipment included almost 70 L-39 jet trainers, Mi-8 and Mi-24 helicopters, and up to 50 MiG-21 aircraft in various states of repair.

Today, there are 2,400 personnel. Operational aircraft include 24 L-39 Albatros, two An-12 and two An-26 transports, with 11 Mi-24 attack and 19 Mi-8 transport helicopters. The 50 MiG-21 are all believed to be unserviceable. In 1999 Russia offered Su-24 and Su-25 ground-attack aircraft to deal with unrest in Central Asia. Air strikes were launched against rebels in September, 2000. SA-2, SA-3 and SA-4 SAM launchers are available in small numbers.

Laos

POPULATION: 5.6 million

LAND AREA: 88,780 square miles, 231,399 sq km

GDP: $1.7bn (£1.2bn), per capita $2,800 (£1,958)

DEFENCE EXPENDITURE: $19.7m (£13.8m)

SERVICE PERSONNEL: 29,100 active.

LAOS PEOPLE'S LIBERATION ARMY AIR FORCE

Formed: 1975

Laos was formerly part of French Indo-China and became an independent member of the French Union in 1949. In 1953, the country was invaded by communist Viet Minh forces from North Vietnam, and although these withdrew, the country was then subjected to repeated attempts to overthrow the government by communist Pathet-Lao forces. American assistance brought the Laotian Army Aviation Service into existence in 1954, initially for counter-insurgency and AOP duties. The first aircraft included 20 Cessna O-1 Bird Dog AOP aircraft, ten Douglas C-47, three DHC-2 Beaver and four Aero Commander 520 transports. Sikorsky S-55 helicopters and North American T-6G Texan trainers followed later, with T-28D armed-trainers for COIN operations. In 1960, the Royal Lao Air Force title was adopted.

Unrest in neighbouring Cambodia and Vietnam, coupled with US withdrawal from Vietnam, continue to destabilise Laos, and

Laos

in 1975, the government was overthrown and a republic declared, with the RLAF becoming the Laos People's Liberation Army Air Force. Soviet military aid was received, including MiG-21PF interceptors and Mi-8 helicopters. Despite a defence pact between Laos and Russia in 1997, it seems that the MiG-21 force is no longer serviceable. An attempt to return the aircraft to service with an overhaul failed as the structural life of the aircraft had expired. Russia supplied Mi-17s in 1998 and Kamov Ka-32T helicopters in 2000.

The LPLAAF has 3,500 personnel. The 29 MiG-21PF/U are unserviceable, but nominally equip two squadrons. There are two Yak-40 VIP transports, three Antonov An-2 and three An-24, but three HAMC Y-12 and five Y-7 are shared with the national airline, Lao Aviation. Helicopters include a single Mi-6 heavy-lift machine, as well as nine Mi-8s and 12 Mi-17, and six Kamov Ka-32T Helix.

Latvia

POPULATION: 2.3 million

LAND AREA: 25,590 square miles, 66,278 sq km

GDP: $7.2bn (£5bn), per capita $7,219 (£5,048)

DEFENCE EXPENDITURE: $72m (£50.3m)

SERVICE PERSONNEL: 6,500 (includes 35% conscripts) active, plus 14,400 reserves.

REPUBLIC OF LATVIA AIR FORCE

Formed: 1994

During World War II, Latvia was annexed by the USSR. The Latvian Air Force was formed in 1994 after the last Soviet forces departed. It is mainly a communications and liaison force as plans to acquire combat aircraft have been thwarted by a lack of funds. The RLAF has just 210 personnel and operates four Antonov An-2 and two Let L410Ups, as well as eight Mi-2 and a single Mi-8 helicopters. Five PZL Wilga are operated on behalf of the National Guard. Up to six transport helicopters may be purchased, depending on funding.

Lebanon

POPULATION: 3.1 million

LAND AREA: 3,400 square miles, 8,806 sq km

GDP: $16bn (£11.2bn), per capita $6,800 (£4,755)

DEFENCE EXPENDITURE: $564m (£394m)

SERVICE PERSONNEL: 71,830 (31% conscript) active.

LEBANESE AIR FORCE/FORCE AERIENNE LIBANAISE

Formed: 1949

Lebanon was under Turkish rule until the end of World War I, then administered by France under a League of Nations mandate until independence in 1943. The Force Aerienne Libanaise was formed with British assistance for internal security duties. The first aircraft were two Percival Prentice trainers, but these were soon joined by three Savoia-Marchetti SM79 tri-motor transports, a de Havilland Dove and some DHC-1 Chipmunk trainers in 1950. Former French air bases were used, and by 1955, de Havilland Vampire FB52 fighter-bombers and T55 jet trainers, North American T-6 Harvard and additional Chipmunk trainers were in service, with an Aermacchi MB308 communications aircraft. AOP duties were filled by some ex-Iraqi de Havilland Tiger Moths. The FAL's combat capability grew throughout the 1950s, with five Hawker Hunter F6 fighters and five FB9 fighter-bombers. During the 1960s, twelve Dassault Mirage IIIC fighter-bombers were introduced, while Potez Super Magisters took over the advanced jet training role, and BAe Bulldogs replaced the Chipmunks as primary trainers.

Civil war erupted in the Lebanon during the mid-1970s, with Israeli and Syrian intervention, during which some aircraft were destroyed, often on the ground. Some equipment continued to arrive throughout this period, mainly helicopters. The return of peace during the late 1990s saw Bell UH-1H helicopters donated by the USA, but the remaining Mirage IIIs were sold to Pakistan.

The FAL has 1,000 personnel, with just one combat squadron of five Hawker Hunter F9, with a sixth aircraft for training, although three SA342L Gazelle have an anti-tank capability. There are 24 Bell UH-1H and nine Agusta-Bell 212 helicopters, nine SA330L Puma and six Alouette III in the transport role, with two SE3130 Alouette II for communications. VIP transport is a Falcon 20. Five BAe Bulldog trainers are being sold, leaving five Super Magisters as the only training aircraft. The Gazelles and Pumas are believed to be in need of major attention.

Lesotho

POPULATION: 2.2 million

LAND AREA: 11,716 square miles, 30,344 sq km

GDP: $730m (£510m), per capita $2,400 (£1,678)

DEFENCE EXPENDITURE: $30m (£21m)

SERVICE PERSONNEL: 2,000 active.

LESOTHO DEFENCE FORCE - AIR WING

Formed: 1978

The former Basutoland in southern Africa, land-locked Lesotho became independent within the British Commonwealth in 1966, but an air wing for the defence force was not formed until 1978. Early aircraft included a Cessna 182 and a Westland-built Bell 47G Sioux helicopter, later joined by BO105 helicopters. Serious unrest in 1998 was quelled by the arrival of troops from South Africa and Mozambique, who were withdrawn in 1999. The status of the air wing is uncertain, but it is believed to still have two CASA C212 Aviocar transports and the Cessna, two Bell 412SP/EP and an Agusta-Bell 412, a 47G and two BO105S.

Liberia

POPULATION: 3.3 million

LAND AREA: 43,000 square miles, 99,068 sq km

GDP: $450m (£315m), per capita $600 (£420)

DEFENCE EXPENDITURE: $25m (£17.5m)

SERVICE PERSONNEL: up to 6,000 active, plus 6-8,000 reserves..

LIBERIAN ARMY AIR RECONNAISSANCE UNIT

In 1847, Liberia became the first independent African state. Its long history was relatively stable until a presidential assassination in 1990 sparked off internal strife. At the time, the small army had an air arm for reconnaissance and communications duties, but all the aircraft were destroyed. There are plans to rebuild the armed forces, retaining an integrated defence force, but meanwhile the air arm is believed to have some 300 personnel. Pre-1990 equipment included two DHC-4 Caribou, a VIP Falcon 20, two IAI-101B Arava, two Cessna 172, a 185 and a Caravan 1.

Libya

POPULATION: 5.6 million

LAND AREA: 679,358 square miles,1, 759,537 sq km

GDP: $38bn (£26.6bn), per capita $6,200 (£4,335)

DEFENCE EXPENDITURE: $1.2bn (£839m)

SERVICE PERSONNEL: 76,000 (55% conscript) active, plus 40,000 reserves.

LIBYAN ARAB REPUBLIC AIR FORCE

Formed: 1959

The Libyan Air Force was formed in 1959 with two training aircraft supplied by Egypt, accompanied by two Auster AOP6s provided by the UK. In 1963, the USA provided a Douglas C-47 and two Lockheed T-33A jet trainers as part payment for the use of Libyan bases. Later, seven Northrop F-5A fighter-bombers were supplied. A major arms deal with the UK, including SAM was scrapped after the monarchy was overthrown and a revolutionary government installed in 1970, breaking the links with the UK and USA. This led to arms procurement from France and the USSR. Substantial numbers of Mirage V and F1 aircraft were matched by MiG-21, MiG-23 and MiG-25, as well as Su-20, Su-22 and Su-24, Tu-22A bombers and Mi-24 attack helicopters, transport and training aircraft. The new regime engaged in border disputes with many of its African neighbours.

The last Soviet aircraft were 15 Su-24 in early 1989. The UN imposed sanctions in April, 1992 cutting Libya off from new aircraft supplies until these were lifted in 1999 following the arrest of the suspects in the Lockerbie air disaster in which a bomb planted aboard a Pan Am Boeing 747 caused it to explode in mid-air. Libya has since discussed having its MiG-23s, MiG-25s and Su-24s upgraded, and purchasing MiG-29 and MiG-31 interceptors, as well as long-range SAMs. Meanwhile, serviceability is believed to be very low, with many aircraft in storage.

Throughout its post-revolutionary history, the LARAF has maintained a very high ratio of aircraft to personnel. It uses 'advisers', effectively mercenary pilots, from North Korea, Pakistan and Syria.

The LARAF has 23,000 personnel. Flying hours are average, limited by a shortage of spares. A bomber squadron operates the six surviving Tupolev Tu-22A bombers, plus two Tu-22U conversion trainers. There are nine fighter squadrons, which between them operate 50 MiG-21, 75 MiG-23, 60

Libya

MiG-25, with three MiG-25U trainers, and 15 Mirage F1ED. Another 13 squadrons are in the FGA role, with a total of 40 MiG-23BN, 15 MiG-23BU, 30 Mirage 5D/DE, 14 Mirage 5DD and 14 Mirage F1AD, 12 surviving Su-24MK and 45 Su-20/22. Two reconnaissance squadrons have four Mirage 5DR and seven MiG-25R. There are 22 Mi-24 attack helicopters. Transport aircraft include ten Lockheed C-130H/L-100-20/30 Hercules and ten C-130 tankers, which pre-date sanctions, 18 Ilyushin Il-76, loaned to Libyan Arab Airlines, 16 Alenia G222, eight Antonov An-26, 15 Let-410OUVP and a Boeing 707, plus a VIP Falcon 50. Transport helicopters include four CH-47 Chinook, four SA321 Super Frelon, two Agusta-Bell 212 and seven Mi-8. Training aircraft comprise 150 Aero L-39ZO Albatros, also available for light attack duties, as are 190 SIAI SF260WL Warrior. There are 20 Mi-2 helicopter trainers. French and Russian AAM and ASM weapons are deployed.

Army aircraft include 40 SA342L Gazelle for anti-tank and liaison duties, with the latter role also falling to five Agusta-Bell 205 and five AB206, and ten Cessna O-1E Bird Dog, while ten CH-47 Chinook provide a heavy-lift capability.

Naval aircraft include seven SA321 Super Frelon ASW and SAR helicopters, and 12 Mi-14PL ASW helicopters.

Lithuania

POPULATION: 3.7 million

LAND AREA: 25,170 square miles, 65,201 sq km

GDP: $11.2bn (£7.8bn), per capita $6,000 (£4,195)

DEFENCE EXPENDITURE: $199m (£139.2m)

SERVICE PERSONNEL: 12,190 (32% conscript) active, plus 336,000 reserves.

LITHUANIAN AIR FORCE

Re-formed: 1992

Lithuania's history of military aviation dates from 1919, as the country fought for its independence in the wake of the Russian Revolution and the civil war that followed. A school of military aviation was established that year, while in 1920 the fledgling Lithuanian Air Force found itself fighting briefly against Polish forces. A number of Lithuanian aircraft were built during the 1920s and 1930s, initially with three trainer designs, the ANBO-1, II and III. Next came the ANBO-IV reconnaissance aircraft of 1932, also used as a light bomber and which made a round-Europe tour in 1934. It was the most numerous Lithuanian aircraft, with 14 built, plus 20 of its ANBO-41 development. The Lithuanian Air Force was too small to counter the massive might of the Soviet Union when it invaded in 1940, annexing the country and making it part of the USSR. Lithuanian personnel were absorbed into the Soviet armed forces.

After the USSR broke up, Lithuania seized her independence. In early 1992, the Aviation Service, Aviacijos Tarnyba, was formed, but adopted its current title in 1993. Most Soviet aircraft were removed before independence, but four L-39 Albatros were obtained from the training school force in Krygyzstan, and in 1999 these were joined by two new aircraft bought direct from the manufacturer. The

ABOVE: *Lithuania maintained a small air arm with its own nationally-designed aircraft before World War II, when the country was annexed by the USSR. These are ANBO-IV reconnaissance aircraft and light bombers in 1934. (Lithuanian Air Force)*

Lithuania

LAF has a requirement for up to 12 fighters, but this has been affected by funding.

The LAF has 800 personnel. It now has eight Aero L-39ZA/C Albatros for training and light attack duties, while 17 Mi-8 undertake transport and SAR. Other transport aircraft include six Antonov An-2 Colt and three An-26, as well as four Mi-2 helicopters. There is also a separate National Guard, with ten An-2 and 4 PZL Wilga.

RIGHT: *Lithuanian Mi-8MT. (Mader-Acaes Collection)*

Macedonia

POPULATION: 2 million

LAND AREA: 10,229 square miles, 27,436 sq km

GDP: $3.6bn (£2.5bn), per capita $4,827 (£3,375)

DEFENCE EXPENDITURE: $77.1m (£53.9m)

SERVICE PERSONNEL: 16,000 (50% conscript) active, plus 60,000 reserves.

MACEDONIAN ARMY

Formed: 1998

A small air arm was formed within the Macedonian Army as the country broke away from the former Yugoslavia during the late 1990s, initially operating four Zlin-242 trainers and an Antonov An-2, as well as four Mi-17 helicopters. This force is now being augmented by a handful of Su25K/UB after up to 20 ex-Turkish Northrop F-5A/Bs were rejected in favour of the more familiar Russian equipment. In 2000 Germany donated two MBB BO105 and two Bell UH-1H helicopters, and Ukraine has also provided four surplus Mi-8/Mi-17 transport and four Mi-24 attack helicopters. An Mi-8 was shot down by Albanian-rebels in 2001. Personnel strength of around 600-700 can be expected to increase rapidly.

Madagascar

POPULATION: 16.4 million

LAND AREA: 228,600 square miles, 590,002 sq km

GDP: $5.6bn (£3.9bn), per capita $700 (£489)

DEFENCE EXPENDITURE: $42m (£29.4m)

SERVICE PERSONNEL: 13,500 active.

MALAGASY AIR FORCE/ARMEE DE L'AIR MALGACHE

Formed: 1960

The island of Madagascar became independent from France in 1960, and received the standard French military aid package of three Broussards (Bush Rangers), a C-47 and an Alouette III helicopter. This was augmented with additional aircraft so that it grew to include three Douglas C-47s, six Max Holste 1521M Broussards, two Dassault MD315 Flamant light transports, with two helicopters, a Bell 47G Sioux and the Alouette III.

A combat capability was created later using Soviet military aid, with the arrival of first MiG-17s and then MiG-21s. These aircraft now equip a single fighter and ground-attack squadron with four MiG-17s and eight MiG-21FLs, plus another two MiG-21U for conversion training. These aircraft are maintained by North Korean advisers, presumably not included in the AdlAM's 500 personnel. Transport is provided by five Antonov An-26, with only believed to be airworthy, a C-47, a BAe748, two C-212 and two VIP Yakovlev Yak-40. A helicopter squadron has six Mi-8. A Piper Aztec and three Cessna 172 are employed on liaison duties, with four Cessna 172 for training. New equipment is needed, but is unlikely to be afforded in the near future.

Malawi

POPULATION: 11.2 million

LAND AREA: 36,686 square miles, 117,614 sq km

GDP: $1.5bn (£1bn), per capita $900 (£629)

DEFENCE EXPENDITURE: $26m (£18.2m)

SERVICE PERSONNEL: 5,300 active.

MALAWI ARMY AIR WING

Formed: c. 1992

Malawi, the former British colony of Nyasaland, became independent in 1963, with a small army, which eventually acquired an air wing. A Police Air Wing also operated aircraft at first, but this was disbanded. Currently, just 80 personnel are involved in aviation, which is confined to transport and communications using two Basler Turbo 67 (turboprop C-47), and a VIP BAe125-800, with two Eurocopter SA330F Puma and two SA332 Super Puma, an AS350B Ecureuil and an AS365N Dauphin. Four Dornier 228 might still be in service, but some reports suggest these may have been withdrawn.

Malaysia

POPULATION: 22.1 million

LAND AREA: 128,693 square miles, 334,110 sq km

GDP: $88bn (£61.5bn), per capita $12,900 (£9,020)

DEFENCE EXPENDITURE: $2.8bn (£2bn)

SERVICE PERSONNEL: 100,500 active, plus 42,800 reserves.

ABOVE: *C-130H Hercules of the Royal Malaysian Air Force (Jeremy Flack/ Aviation Photographs International)*

ROYAL MALAYSIAN AIR FORCE

Formed: 1958

Malaysia came into existence in 1963 as a federation of the former British colonies of Malaya, Sabah (North Borneo), Sarawak and Singapore, although the latter withdrew in 1965. Malaysian military aviation started with the creation of the Straits Settlements Volunteer Air Force in 1936, equipped with Hawker Audax aircraft. In 1940, it became the Malayan Volunteer Air Force, using civil aircraft, mainly de Havilland Tiger and Leopard Moths and Rapides that had been requisitioned on the outbreak of World War II in Europe in 1939. This force escaped from the Japanese invasion of Malaya, but it ceased to exist during the invasion of Sumatra.

In the immediate aftermath of the war, military aviation in Malaya was left to the RAF until 1950, when a Malayan Auxiliary Air Force was formed. At first, this was a training organization using de Havilland Tiger Moths, which were joined by North American T-6 Harvard trainers. The first combat aircraft were Supermarine Spitfire F21 fighters. The Tiger Moths were replaced by DHC-1 Chipmunks in 1956.

Malaya (but not the other territories) became independent in 1958, and the MAuxAF became the Royal Malayan Air Force, flying Scottish Aviation Twin Pioneer transports and the Chipmunk trainers. With the formation of the Federation of Malaysia in 1963, the present title was adopted. New aircraft included Handley Page Herald, DHC-4 Caribou, de Havilland Devon (military Dove) and Heron transports, and Canadair CL-41 Tutor armed jet trainers, known locally as the Tebaun (Wasp). The new Federation was born at a time of crisis, threatened by the territorial ambitions of neighbouring Indonesia, and the RMAF fought alongside the RAF and Fleet Air Arm against Indonesian infiltration. Afterwards, although Singapore left the Federation, the RMAF also had the support of the RAF, RAAF and RNZAF, all of whom maintained forces in Malaysia for some years afterwards. Co-operation with Singapore continued on defence.

Australian-built Commonwealth CA-27 Avon-Sabres, were delivered in 1969, and these were joined by ten Dassault Mirage Vs during the early 1970s, by which time the RMAF had grown to 4,500 personnel and was operating 20 Alouette III and ten Sikorsky S-61A helicopters. Training by this

Malaysia

ABOVE: *MiG-29N (Jane's)*

time was using 18 BAC Provost T-51 and 15 Scottish Aviation Bulldog trainers.

Malaysia has accorded a high priority to defence since the formation of the federation. Lockheed C-130 Hercules were obtained to enhance the transport capability, followed by Northrop F-5E/F Tiger II interceptors and reconnaissance aircraft, and later by F/A-18D Hornets. Trainers included the Pilatus PC-7, with Aermacchi MB339A armed-trainers and BAe Hawk light fighters in the ground-attack role. Growing variety in procurement sources was underlined in 1995 when the RMAF acquired MiG-29N air superiority fighters. It has supplemented the surviving PC-7s with the MkII version, and the C-130 force has also been upgraded, including stretching some of the aircraft.

Defence expenditure was checked during the late 1990s by the Asian economic crisis. CASA CN235 transports were delivered three years late in 1999, while a decision on new helicopters has been postponed, but is likely to involve the Mi-17. Additional MiG-29s may be purchased, but the opriginal aircraft are receiving an MLU. The RMAF is also believed to be seeking an AEW capability, which could be filled by ex-USN E-2C Hawkeyes if any become available.

The RMAF has 8,000 personnel. Its fighter capability is provided by three squadrons, two of which operate 15 MiG-29N and three MiG-29U, while the third is a fighter lead-in force with four F-5E, three F-5F and two RF-5E. Strike duties are handled by four squadrons, three of which operate a total of eight BAe Hawk 108, 17 Hawk 208 and nine MB339A. Four Beech King Air 200 operate limited MR. There are four transport squadrons: one tactical squadron with 11 DHC-4 Caribou and six IPTN CN235; one with five Lockheed C-130H Hercules, while another operates seven C-130H-30/-MP and two KC-130H tankers, as well as nine Cessna 402B, of which two are used for aerial survey work. The fourth squadron mainly handles Royal Flight VIP duties with a Bombardier Global Express, a Fokker F-28-1000 Fellowship and a Dassault Falcon 900, as well as two Sikorsky AS-61N and two S-70A, and an Agusta A109C helicopter. Other helicopters include 30 S-61A for transport and SAR duties and 12 SA-316A/B Alouette III, as well as two fire-fighting Mi-17s. It is likely that the now elderly S-61A force could be replaced by up to 40 additional Mi-17s. Trainers include 38 Pilatus PC-7 and nine PC-7 MkII, and nine MD3-160, as well as use of the MB339A and Cessna 402B aircraft.

ROYAL MALAYSIAN NAVY

Formed: 1988

The naval air arm was formed in 1988 with 12 Westland Wasp helicopters obtained from the Royal Navy for shipboard ASW duties. This small force grew to 17 Wasps, with 160 personnel. The Wasps are being replaced by six AS555N Fennecs and six Westland Super Lynx, while up to 30 additional helicopters, possibly Super Lynx may be ordered to operate from 27 German-built corvettes.

ROYAL MALAYSIAN ARMY

Formed: 1977

A small air arm was formed by the Royal Malaysian Army in 1977 with ten SA316B Alouette III helicopters for observation, liaison and communications duties. Ambitious plans exist for the force to develop into a fully capable strike and transport force with attack and transport helicopters. The CSH-2 Roivalk was selected during the mid-1990s, but not ordered due to the economic crisis. There are also plans to buy CH-47 heavy lift helicopters, as well as light observation helicopters in the A109 or MD900 Explorer class.

Maldives

POPULATION: 250,000

LAND AREA: 115 square miles, 298 sq km

MALDIVES DEFENCE FORCE

The Maldives Defence Force charters aircraft from Air Maldives, the national airline, as required, while an Indian Air Force Mi-8 helicopter is also on loan.

Mali

POPULATION: 11.5 million

LAND AREA: 464,875 square miles, 1,204,350 sq km

GDP: $3.1bn (£2.2bn), per capita $697 (£487)

DEFENCE EXPENDITURE: $30m (£21m)

SERVICE PERSONNEL: 7,350 (includes conscripts) active.

MALIAN REPUBLIC AIR FORCE/FORCE AERIENNE DE LA REPUBIC DU MALI

Formed: 1960

The former French colony of Mali became independent in 1960, and received the standard French post-colonial package of a Douglas C-47 and two Max Holste 1521 Broussard (Bushranger), to which the USA added two C-47s and the USSR six obsolete MiG-15 fighter-bombers. Initially known as the Mali Air Force, the present title was adopted later. Further aid came from the USSR during the late 1960s, with the delivery of MiG-17s, followed by MiG-21s in 1986, following 1985's brief six day war with neighbouring Burkina Faso. Transport aircraft were also supplied. This intermittent flow of obsolete aircraft ceased with the collapse of the USSR, since when the only development has been the conversion of a C-47 as a Basler Turbo 67.

Today, the FARM has 400 personnel. It is believed that the 11 MiG-21 are non-operational, as are six L-29 Delfin, two Yakovlev Yak-18 and two Yak-11 training aircraft. It has become a transport and communications force, with two Antonov An-2 Colt, an An-24 and an An-26, with two helicopters, a Mi-8 and an A350B Ecureuil.

Malta

POPULATION: 393,000

LAND AREA: 122 square miles, 316 sq km

GDP: $3.6bn (£2.5bn), per capita $9,300 (£6,503)

DEFENCE EXPENDITURE: $27m (£18.9m)

SERVICE PERSONNEL: 2,140 active.

ARMED FORCES OF MALTA AIR SQUADRON

Malta became independent of the UK in 1964, basing its armed forces on former Maltese units within the British Army and locally raised personnel of the Royal Navy. An aviation element was not formed immediately. Malta has an integrated defence force, with the aviation element part of No2 Composite Regiment. Currently, 76 personnel are allocated to aviation. There are no combat or transport aircraft. Two BN-2B Islanders provide patrol and SAR duties, while Cessna O-1 Bird Dogs were replaced in the training role in 2000 by four ex-RAF BAe Bulldog, with a fifth added in 2001. There are five SA316B/3160 Alouette helicopters for communications and SAR duties, with two Hughes NH-369M and two Agusta-Bell 47G Sioux for training and communications. Two SAR Italian AB212 helicopters are manned by Maltese and Italian AF personnel.

Mauritania

POPULATION: 2.8 million

LAND AREA: 398,000 square miles, 1,085,210 sq km

GDP: $0.8bn (£559m), per capita $1,900 (£1,328)

DEFENCE EXPENDITURE: $23.6m (£16.5m)

SERVICE PERSONNEL: 15,650 (includes conscripts) active.

MAURITANIAN ISLAMIC AIR FORCE/FORCE AERIENNE ISLAMIQUE DE MAURITANIE

Formed: 1960

The former French colony of Mauritania became independent in 1960, and received the standard parting gift of a Douglas C-47 and two Max Holste 1521M Broussards. Uncertainty over sovereignty of the region bordering Morocco, disputed with Polisario guerrillas, led to acquisition of combat aircraft, with five Britten-Norman BN-2A Defender aircraft were acquired for reconnaissance and COIN in the mid-1970s, with four Cessna FTB337F for the same roles. Plans to acquire six IA-58 Pucara from Argentina were abandoned after a coup in 1978. Despite withdrawal from the region bordering Morocco, patrols have continued to avoid Mauritania being drawn into the conflict between the Polisario and Morocco.

The FAIM has 150 personnel. It has seven BN-2A Defender and four Cessna FTB337F for COIN and patrol duties, with two Piper Cheyenne II for offshore patrol. A DHC-5 Buffalo and Xian Y-7 replaced the C-47, while two Harbin Y-12 transports were added in 1995. Five SIAI SF260E trainers were acquired in 2000.

Mauritius

POPULATION: 1.2 million

LAND AREA: 720 square miles, 2,038 sq km

GDP: $5.1bn (£3.6bn), per capita $19,000 (£13,287)

DEFENCE EXPENDITURE: $89m (£62m)

SERVICE PERSONNEL: 1,600 paramilitary active.

MAURITIUS COAST GUARD

Mauritius maintains two paramilitary organisations, the so-called Special Mobile Force and the Coast Guard, with 500 personnel and an aviation unit. Currently, it operates a Dornier Do228 and a Britten-Norman BN-2T Defender on EEZ patrols. A police air wing operates two SA316B Alouette helicopters.

Mexico

POPULATION: 100.6 million

LAND AREA: 760,373 square miles, 1,972,360 sq km

GDP: $554bn (£387bn), per capita $8,800 (£6,153)

DEFENCE EXPENDITURE: $5.3bn (£3.7bn)

SERVICE PERSONNEL: 192,770 (includes 30% conscripts) active, plus 300,000 reserves.

MEXICAN AIR FORCE/FUERZA AEREA MEXICANA

Formed: 1924

Mexico was amongst the first Latin American countries involved with military aviation, as revolutionary and counter-revolutionary groups used aircraft flown by mercenary pilots in 1910. During World War I, an aircraft industry was founded with government backing to overcome the shortage of aircraft. In the Fuerza Aerea Mexicana was formed with Bristol general-purpose biplanes, American-built DH4B bombers and Douglas O-2 AOP aircraft, for policing and COIN. In 1930, Avro 504K trainers were obtained, followed by the Mexican designed Azcarate-E. Vought O2U Corsair biplanes were obtained during the early 1930s, but little further development occurred until the end of the decade when Grumman G-23 fighters, Waco D-6 general-purpose aircraft, Consolidated Fleet 21 and Ryan S-T trainers entered service.

Although neutral initially, Mexico joined the Allies in 1942, placing bases at their disposal. The USA supplied Douglas A-24 Dauntless ASW aircraft, Vought-Sikorsky OS2U AOP and North American NA-16 training aircraft. Mexico's aircraft industry provided the Tezuitlan trainer. In 1945, a FAM Republic F-47D Thunderbolt fighter-bomber squadron was due to join the Allied forces in the Pacific, but Japanese surrender came before its arrival. Post-war, Douglas C-47, Lockheed Lodestar and Beech C-45 transports arrived, with Fleet, Fairchild and Vultee trainers. US military

Mexico

ABOVE: *Mexican Vought V-99M, the export variant of the O3U Corsair, awaiting delivery in 1937. (Vought)*

aid followed Mexico's membership of the Organisation of American States.

Jet aircraft did not arrive until the late 1950s, de Havilland Vampire F3 fighters and T55 trainers, joined in 1961 by Lockheed T-33A armed-trainers. By this time, Bell 47G Sioux helicopters were in service, followed later by Alouette II helicopters. Beech T-11 Kansan and T-34 Mentor aircraft entered service. During the 1970s, Northrop F-5E/F Tiger II fighters were obtained, while the transport fleet was upgraded with Lockheed C-130A Hercules. Substantial numbers of Pilatus PC-7 armed-trainers were stationed at strategic locations.

Mexico has no obvious regional military threat, with a powerful and friendly northern neighbour and the nearest threat, Cuba, with economic problems seriously undermining military capability. COIN remains important due to the existence of five significant rebel groups. A new role is that of operations against drug traffickers while EEZ operations have also grown in importance, with two maritime patrol Embraer EMB-145 and a third for AEW due in 2004. Transport and liaison have always been important.

The FAM has almost doubled its personnel to 11,770 over the past quarter century. It is unusual in maintaining its own paratroops, a battalion of 2,000 men. It has eight F-5E Tiger II and two F-5F in a fighter squadron. Seven close-support COIN squadrons are positioned in strategic locations with a total of 75 PC-7, while another two squadrons have 30 AT-33A Shooting Stars. An armed helicopter squadron operates a Bell 205, 15 Bell 206B JetRanger, five Bell 206L LongRanger, and up to 15 212s, while there have been reports of an order for Mi-24s. A reconnaissance squadron operates ten Commander 500S, four Fairchild C-26 and two Schweizer SA237 helicopters. Five transport squadrons, operate a total of nine Lockheed C-130A Hercules, 12 Douglas C-47 and three Boeing 727-100, while a Presidential Flight operates three 737, a 757-200, two Gulfstream III and a Lockheed Jetstar, as well as two Sikorsky S-70A and three AS332L Super Puma helicopters. Communications aircraft include an assortment of older or smaller aircraft, including Merlins, a Convair 580, nine T-39 Sabreliners, Commander 500s and Turbo Commanders, Beech King Airs and four Pilatus PC-6 Turbo Porter utility aircraft. There are some 60 Bell UH-1H Iroquois; nine Sikorsky UH-60L Black Hawk; two SA330F Puma and two AS355F Ecureuil; and 17 MD530/F transport helicopters. The COIN armed-trainers also undertake the training role complementing 35 Beech Bonanza, 20 Maule MX-7-80 and 71 new Cessna 182S.

MEXICAN NAVAL AVIATION/AVIACION DE LA ARMADA DE MEXICO

Mexican naval aviation was established after World War II as a land-based MR and SAR force. For many years it operated a number of Consolidated Catalina PBY-5A amphibians, later adding Alouette II and Bell 47J Sioux helicopters. Today, it has 1,100 out of the Mexican Navy's 37,000 personnel and its role has expanded to include COIN aircraft, while helicopters are operated from destroyers, corvettes and frigates. There is little standardization, with many types available in ones and twos. There are ten L-90 Redigo armed-trainers with a COIN role. Offshore patrol duties fall to eight CASA C212-200 Aviocar, eight MD900 Explorer helicopters, with a further three armed Explorers introduced in 2002, Piper Aztec and Navajo, Cessna 182, 206, 210, 337G and 402, 404 and 421 aircraft, and the Mexican Tonatiuh MX-1. SAR is provided by three Alouette III and 11 BO105C/CB (usually shipboard) helicopters. Other recent acquisitions have included a DHC-8 Dash 8Q200 transport, with a second likely to follow, joining 20 Mi-8, two UH-1H Iroquois, and three AS550 Fennec helicopters, a DHC-5D Buffalo, four Antonov An-32, a Beech King Air 90, and a Fairchild-Hiller FH227. A Learjet 24D and an AS365 Dauphin are used on VIP duties. Training aircraft include 11 Maule MX-7-180, ten Beech F-33C Bonanza and ten B55 Baron, six Cessna 152, four MD500E and a Rotorway Exec 162F, while two Robinson R22 and an R44 replaced five UHJ-12E Raven in 2001.

Moldova

POPULATION: 4.4 million

LAND AREA: 13,000 square miles, 34,188 sq km

GDP: $1.3bn (£881m), per capita $3,343 (£2,337)

DEFENCE EXPENDITURE: $21.1m (£14.7m)

SERVICE PERSONNEL: 8,220 (63 per cent conscript) active, plus 66,000 reserves.

MOLDOVAN AIR FORCE

Formed: 1991

In common with many other former 'republics' within the USSR, Moldova acquired the Soviet aircraft based within its territory, in this case the most significant being 34 MiG-29 of a naval fighter wing. These aircraft were sold to Eritrea and the Yemen, while the USA bought 21 to prevent them being sold to Iran, at a price reported to be around US $40 million in 1997. This has left the Moldovan Air Force as a transport and communications unit, with ten Antonov An-2, three An-72s and an An-24, a Tupolev Tu-134 and an Ilyushin Il-18, and eight Mi-8 helicopters. Combat helicopters may be purchased in the future. The MAF has 800 personnel.

Mongolia

BELOW: *USPU-gun mont and 9-A-624 machine gun on Mi-24V Hind (Jeremy Flack/Aviation Photographs International)*

POPULATION: 2.7 million

LAND AREA: 604,095 square miles, 1,564,360 sq km

GDP: $1bn (£699m), per capita $2,200 (£1,538)

DEFENCE EXPENDITURE: $19.6m (£13.7m)

SERVICE PERSONNEL: 9,100 (40% conscript) active, plus 137,000 reserves.

MONGOLIAN AIR FORCE

Part of the Mongolian Army, the Mongolian Air Force dates from 1926, when four Russian-built aircraft were supplied. The following year, a flying school was established. There is little factual history about the early years, although by 1933, around 100 aircraft of Soviet origin were in service, and Soviet aid continued to counter the growing Japanese threat throughout the rest of the 1930s. The peak strength of 450 aircraft was reached by 1938, and these included Polikarpov I-15 and I-16 fighters, Tupolev TB-3 bombers and R-5 AOP aircraft. There was an encounter with Japanese forces in 1939. During World War II, a Lavochkin La-5 fighter unit saw action against the Germans inside the USSR.

Post-war, aircraft supplied by the USSR included Polikarpov Po-2s and Antonov An-2s, followed during the 1950s by Ilyushin Il-14 and Lisunov Li-2 (C-47) transports, as well as Yakovlev Yak-11 and Yak-18 trainers, while MiG-15 jet fighters entered service toward the end of the decade, when they were obsolete. Mil Mi-4 helicopters were also introduced.

The collapse of the USSR cut the Mongolian Air Force, for many years known as the Air Force of the Mongolian People's Republic, off from its main source of arms. Despite its geographical position, Communist China has not been an alternative source other than for a small number of transports. Its current strength is 800 personnel. Ten MiG-21s were in a fighter squadron, but have been abandoned in the open and are not operational. There are 11 Mi-24 attack helicopters, believed to be stored. The remaining aircraft are transports, including ten Antonov An-2, four An-24 and three An-26, four HAMC Y-12 and 12 Mi-8 helicopters.

Morocco

POPULATION: 28.5 million

LAND AREA: 171,388 square miles, 466,200 sq km

GDP: $33bn (£23.1bn), per capita $4,200 (£2,937)

DEFENCE EXPENDITURE: $1.7bn (£1,2bn)

SERVICE PERSONNEL: 198,500 (50% conscript) active, plus 150,000 reserves.

ROYAL MOROCCAN AIR FORCE/FORCE AERIENNE ROYAL MOROCAINE

Formed: 1956

Morocco became independent of both France and Spain in 1956. An air force was formed, Aviation Royal Cherifienne, using this title for the next twenty years. Initially, personnel were trained in France and Spain, with Armee de l'Air personnel seconded to Morocco. The primary role was army support, using six Morane-Saulnier MS500 Criquet AOP aircraft. In 1957, a de Havilland Heron, two Beech Twin Bonanza, three Max Holste 1521M Broussard and a Bell 47 Sioux helicopter were delivered for communications work.

Morocco's important strategic position with Mediterranean and Atlantic coastlines, soon led to military aid from both the USSR and the USA. The USSR provided MiG-17 fighter-bombers, Ilyushin Il-28 jet bombers and MiG-15UTI trainers in 1961. In 1966, the USA provided Northrop F-5A fighter-bombers, Fairchild C-119G Packet and Douglas C-47 transports, Sikorsky H-34 and Kaman HH-43 Huskie helicopters, and North American T-6 Texan trainers. Aircraft were acquired from other sources, including Agusta-Bell 205 helicopters and Potez Magister jet trainers. Relations with the USSR cooled during the late 1960s and early 1970s, with the USSR favouring Algeria instead. By the early 1970s, a shortage of spares saw the MiG-17s withdrawn. Relations with Spain also cooled because of Moroccan claims for the return of the two Spanish enclaves of Melilla and Ceuta.

Algeria supported an uprising by Polisario guerrillas in the Western Sahara, with the FARM supporting ground forces from 1975 until a UN brokered cease-fire was agreed in 1988. Despite peace negotiations in 1989, the cease-fire has been breached on many occasions. Morocco has built a barrier several hundred kilometres long to keep the Polisario out, and the FARM patrols this barrier.

During the 1980s, F-5As were joined by F-5Es and Dassault Mirage F1 interceptors, and by specialised COIN aircraft including the Rockwell OV-10A Bronco and Franco-German Alphajet armed-trainers. A substantial force of Lockheed C-130 Hercules transports entered service, including tankers. Helicopters acquired at this time included Boeing CH-47C Chinook and Aerospatiale Puma, with Gazelle helicopters having an anti-tank and a COIN capability. An offer of surplus F-16 aircraft by the USA in 1991 was allowed to lapse. The Mirages received an extensive upgrade in 1996-97, but an attempt to acquire additional Alphajets surplus to Luftwaffe requirements to cover attrition fell foul of strict export rules.

Over the past quarter century, the FARM has more than trebled in size to 13,500 personnel. Flying hours are relatively high for a developing country, at 100 plus. There are 19 Mirage F-1CH interceptors and another 15 F-1EH strike aircraft, eight F-5A and 22 F-5E, with three F-5B and four F-5F trainers. Two RF-5As provide a reconnaissance capability. Other strike aircraft include 20

BELOW: *The Royal Moroccan Air Force includes F-5E strike aircraft as its main offensive capability. (Jane's).*

Morocco

Alphajet armed-trainers, 20 Magister armed-trainers, 24 SA342 Gazelle, 12 with armed with HOT anti-tank missiles and 12 with cannon, and six OV-10A Bronco, which also operate in the forward air control role with ten AS202A Bravo. Given the long coastline, MR is surprisingly limited, with just two Dornier Do28D Skyservant aircraft, although 11 Defenders are operated by the fisheries ministry on EEZ patrols. ELINT is undertaken by two Falcon 20. Transport aircraft include 13 Lockheed C-130H Hercules, two KC-130H tankers and two RC-130 surveillance aircraft, seven CASA CN235M, and Boeing 707. There are now just nine CH-47C Chinook, but there are 30 SA330C Puma, 38 AB205 and five AB212 helicopters. An assortment of aircraft operates in the communications role, including a Falcon 50, Gulfstream II and III, two Cessna Citation V, eleven Beech King Air 100/200/300 and 20 AB206 JetRanger. Apart from the Alphajets and Magisters, training also takes place on 14 Cessna T-37B and ten Beech T-34C Turbo Mentor aircraft, with up to 20 K-8s reported on order. An aerobatic team has nine CAP10B/231. Missiles are mainly AIM-9B/D/J Sidewinder, R-530 and R-550 Magic AAM, with HOT anti-tank missiles and, for the F-5Es, AGM-65B.

The paramilitary Gendarmerie Royale provides a wide range of functions, including a coast guard service, and has an 'Escadron Aerien'. This operates two Sikorsky S-70A Black Hawk helicopters, as well as two Eurocopter SA315B Lama, seven SA330C Puma, five SA342K Gazelle and two AS365N Dauphin.

Mozambique

POPULATION: 20.1 million

LAND AREA: 302,250 square miles, 771,820 sq km

GDP: $2.4bn (£1.7bn), per capita $1,500 (£1,048)

DEFENCE EXPENDITURE: $87m (£61m)

SERVICE PERSONNEL: c.11,000 (includes some conscripts) active.

ABOVE: *Mi-24 'Hind E' similiar to that operated by FAM (Jeremy Flack/Aviation Photographs International)*

MOZAMBIQUE AIR FORCE/FORCA AEREA MOCAMBIQUE

Formed: 1975

Mozambique became independent from Portugal in 1975 and a civil war started that lasted until 1995. An air force was established almost immediately, the Forca Popular Aerea de Libertacao de Mocambique, with equipment provided by the USSR and operated by Cuban 'advisers'. The FPALM received more than 30 MiG-17F fighter-bombers and MiG-15UTI and seven Zlin 326 trainers, as well as nine Antonov An-26 transports, several Mi-8 and a number of Mi-24 helicopters. These were in addition to Cessna and Piper training and liaison aircraft, and four Alouette III helicopters. At one time, some 50 MiG-21s were stationed in the country, flown by Cuban and East German pilots, and these were incorporated into the FPALM in 1992 after an attempt to sell the aircraft failed, replacing the MiG-17s.

The collapse of the USSR brought major problems for the FPALM, still engaged in fighting a civil war. Spares were abruptly cut off, serviceability levels dropped, and the Cuban and East Germans left. The civil war continued, and the attrition rate of aircraft remained high, despite support from Zimbabwe which deployed forces in support of the government. The FPALM quickly deteriorated into a transport and communications force.

The present title was adopted in 1998. The FAM has 1,000 personnel. It has a nominal combat force of four Mi-24 helicopters and is believed to be non-operational, while the six Mi-8 transport helicopters were conspicuous by their absence during the severe floods of 1999. Transport aircraft are eight An-26, and again availability is in doubt. Liaison duties are covered by a Cessna 172 and two 152s, and four Piper Cherokee Six light aircraft. There are still seven Zlin 326 left for training.

Myanmar

POPULATION: 45.4 million

LAND AREA: 261,789 square miles, 676,580 sq km

GDP: $37bn (£25.9bn), per capita, $1,400 (£979)

DEFENCE EXPENDITURE: $2.1bn (£1.5bn)

SERVICE PERSONNEL: 444,000 active.

BURMA ARMED FORCES

Formed: 1955

Although Burma became independent from the UK in 1948, it was not until 1955 that an air force, the Union of Burma Air Force, was established with British assistance. The UK provided a gift of ex-RAF Supermarine Spitfire Mk18 fighters, de Havilland Mosquito Mk6 fighter-bombers, and Airspeed Oxford and de Havilland Tiger Moth trainers. This initial force was soon expanded with Spitfire Mk9s, and later by Hawker Sea Fury FB11 fighter-bombers, Bristol 170 Mk31M Freighters, Douglas C-47s and Beech D18S. Cessna 180s were introduced for liaison duties. All of these aircraft were provided in small numbers. During the late 1950s, 30 Hunting Provost T53 trainers were introduced, with some de Havilland Vampire T55 jet trainers. In 1960, the UBAF received DHC-3 Otters, and Japan supplied Kawasaki-Bell 47 Sioux helicopters. North American F-86F Sabre fighters and Lockheed T-33A armed trainers, Alouette III and Kaman Huskie helicopters were introduced in the late 1960s, with DHC-1 Chipmunk basic trainers.

In 1989, the country changed its name to Myanmar, and the present title was adopted. Three Mi-4 helicopters supplied by the USSR were followed in 1990 by Chinese aircraft, including F-6s (MiG-19) and F-7Ms (MiG-21), as well as the Polish WZL W-3 Sokol light helicopter.

Burma is a military dictatorship spurned by the West. Despite defence being given a high priority to quell internal dissent, the country's poor economic condition has meant few modern aircraft. Aircraft are bought by barter, as with the trading of teak to Yugoslavia in payment for 20 Super Galeb armed-trainers delivered in 1991. In 1998-99, 12 Sino-Pakistani K-8 trainers were introduced.

As part of the integrated defence force, the air arm has 9,000 personnel, up 50 per cent over the past 30 years. Russia is providing 12 MiG-29C/UB for an interceptor squadron. There are three fighter

ABOVE: *The Mi-17 is the mainstay of the Myanmar's helicopter force. (Mader - Acaes Collection)*

squadrons with a total of 34 F-7M and FT-7s that may be upgraded by Israel. Ground-attack is provided by two squadrons with 22 A-5M. The dual roles of COIN and training are covered by ten remaining Super Galeb G4 and 12 Pilatus PC-7 Turbo Trainers. Transport is provided by two Fokker F27 and four FH227 Friendship, five PC-6B Turbo Porter and four Shaanxi Y-8D (An-12). Although there are 12 Bell 205A helicopters, the main force is now 12 Mi-17. Communications aircraft include six Cessna 180 and six Bell 206 JetRanger, as well as ten W-3 Sokol, also used for SAR, and 15 Mi-2 helicopters. There is a Cessna Citation II for VIP duties. Training uses armed trainers, 12 K-8 Karakorum and nine PC-9s.

Namibia

POPULATION: 1.7 million

LAND AREA: 318,261 square miles, 824,296 sq km

GDP: $2.9bn (£2bn), per capita $5,537 (£3,872)

DEFENCE EXPENDITURE: $105m (£73.4m)

SERVICE PERSONNEL: 9,000 active.

NAMIBIA DEFENCE FORCE AIR WING

Formed: 1994

Formerly South-West Africa, Namibia was mandated to South Africa following World War I by the League of Nations and then by the United Nations. It became independent in 1991, creating a small defence force, with an air wing forming in 1994. The first aircraft were two Hindustan SA315 Cheetahs (Alouette II), lost in a mid-air collision in 1999, and two SA316B Cheetaks (Alouette III). During the intervening period, six ex-USAF Cessna O-2A Super Skymasters and a Cessna F406 Caravan II were acquired for surveillance. Two HAMC Y-12 transports were purchased later. VIP transport is provided by a Falcon 900 and a Learjet 31. Four K-8 trainers entered service in 2001, with two Mi-8s leased from Moldava. There have been reports of small numbers of Mi-24s and MiG-23s being supplied, possibly on loan and doubtless with mercenary pilots.

The Fisheries Ministry is responsible for the Coast Guard service, which has a Cessna F406 Caravan II and a SAR Sikorsky S-61L helicopter.

NATO

NATO AIRBORNE EARLY WARNING FORCE

Formed: 1980

The high cost of sophisticated airborne early warning and control aircraft led NATO to create a force of AEW aircraft which started to enter service in 1980. The force is theoretically based in Luxembourg, and unusually the aircraft all have that country's civil registrations, although operationally the aircraft are based in Germany. There are Boeing E-3A Sentries in the force, operating in three squadrons and a training unit, with crews drawn from Belgium, Canada, Denmark, Germany, Greece, Italy, Luxembourg, the Netherlands, Norway, Portugal, Turkey and the USA. NATO AEWF also can call upon the RAF's AEW force when necessary, although the latter also has a national as well as a NATO role. The USAF and the Armee de l'Air also have their own AEW, or AWACS, forces. The former has a NATO and a national role, while the latter is strictly national.

Luxembourg also has an Airbus A400M on order that will be operated within a Belgian transport squadron.

BELOW: *This NATO Boeing E-3A carries a flamboyant colour scheme to celebrate NATO's 50th anniversary and its enlargement with the admission of Poland, Hungary and the Czech Republic. (NATO)*

Nepal

POPULATION: 24.4 million

LAND AREA: 54,606 square miles, 141,414 sq km

GDP: $5.4bn(£3.8bn), per capita $2,300(£1,608)

DEFENCE EXPENDITURE: $50m(£35m)

SERVICE PERSONNEL: 48,000 active.

ROYAL NEPALESE ARMY AIR SERVICE

Formed: 1971

Nepalese military aviation started in 1971, when the Royal Nepalese Army received its first aircraft, a Short Skyvan 3M, although the Royal Family had an Alouette III helicopter, which was joined in 1970 by a VIP version of the Short Skyvan. The new Royal Nepalese Army Air Service took over the royal family's aircraft. It has evolved as a communications and transport operation, but a combat capability may be acquired in the future to counter Communist insurgents. The mountainous terrain makes surface communications difficult and the construction of major airfields costly. The RNAAS has 215 personnel. Currently, the aircraft are split between the Royal Flight, with its two AS332L Super Puma and two Bell 206L LongRanger helicopters, and the air battalion, operating the rest of the aircraft, one BAe748-2A, two Short Skyvan 3M, two SA330C/G Puma and two Hindustan SA316B Chetak (Alouette III). Two Mi-17s are reported to be on order.

Netherlands

POPULATION: 15.9 million

LAND AREA: 13,959 square miles, 36,175 sq km

GDP: $347bn (£243bn), per capita $25,171 (£17,602)

DEFENCE EXPENDITURE: $6.5bn (£4.5bn)

SERVICE PERSONNEL: 50,430 active, plus 32,200 reserves.

ROYAL NETHERLANDS AIR FORCE/KONINKLIJKE LUCHTMACHT

Formed: 1953

Dutch military interest in air matters started with an artillery observation balloon operated for a brief period in 1886, but sustained involvement did not occur until 1911, when aircraft and balloons were acquired for manoeuvres. It was another two years before the Royal Netherlands Army formed an Aviation Division with two van Meel and three Farman F-22 aircraft. In 1914, the Royal Netherlands Navy obtained two Martin TT seaplanes, buying another 12 in 1917. Meanwhile, the Royal Netherlands Army had obtained additional Farman F-22s, bringing its total of this type to 20 by 1915. In 1917, 20 Nieuport 17C-1 Bebe and ten Fokker DIII fighters were obtained, followed by additional Nieuports and some Caudron GIII reconnaissance aircraft in 1918.

ABOVE: *Fokker DXXI fighters were no match for the Luftwaffe in 1940 – note the fixed undercarriage! (Royal Netherlands Air Force's History Unit)*

The Netherlands remained neutral throughout World War I, and unlike the air forces of the combatant nations which contracted once peace returned, continued to grow, albeit slowly, in the post-war years, acquiring former German aircraft, 40 Rumpler CVs and 36 Trumpenburg Spijke trainers. Ambitious post-war re-equipment plans were pruned as the economic situation deteriorated. New equipment included 16 Thalin K and 20 Fokker DVII fighters, and 56 Fokker CI reconnaissance aircraft. Throughout the 1920s, new aircraft consisted of 15 Fokker DXVI and 10 DXVII fighters, 32 CIV, 100 CVI, four CVIII and five CIX reconnaissance aircraft, three FVIIA Trimotor transports, and 30 SIV trainers - with many of these lasting well into the following decade. The Royal Netherlands Indies Army received more modern equipment. Its initial post-war equipment included 25 DH9 bombers and 25 Avro 504K trainers. During the 1920s, ten Vickers Viking amphibians, six Fokker DVII and four DCI fighters, and 20 CIV reconnaissance aircraft entered service. In 1930, nine Curtiss P-6E Hawk fighters were delivered, followed during the late 1930s by more than 100 Martin 139-W bombers.

BELOW: *The first Dutch jets were Gloster Meteor fighters, many manufactured under licence by Fokker. (Royal Netherlands Air Force's History Unit)*

Neglect of Dutch air defences was founded on a policy of neutrality. No plans were made for co-operation with Belgium, the UK or France, the most likely allies.

A last minute attempt to re-arm came in 1938. Reorganization turned the Aviation Division into the Army Air Service. Aircraft were ordered from Fokker and US manufacturers, but there was insufficient time. On 10 May, 1940, German forces rushed into the Netherlands, using paratroops and air-landed troops, but the AAS had too few aircraft. It had 30 Fokker DXXI, some DXVII and 20 G1 fighters, 16 Fokker TV bombers, 11 Douglas DB-8A attack aircraft, 40 Fokker CV and CX reconnaissance aircraft, and few Koolhaven trainers on AOP duties. Overwhelmed by the Luftwaffe, this small force fought valiantly for five days, after which the survivors and the flying school escaped to England.

The Royal Netherlands Indies Army received many of the aircraft ordered for the AAS. In the eighteen months before Japan entered the war, the RNIA received 24 Curtiss Hawk 75-A, 24 Curtiss-Wright CW-22 Falcon, and 72 Brewster 339 and 439 fighters; a few Douglas DB-7 bombers, 36 Consolidated PBY-5 Catalina flying boats, 48 Sikorsky S-43 amphibians, and 20 Lockheed Lodestar transports. Once the Netherlands government-in-exile declared war on Japan in December, 1941, this force fought the advancing Japanese, but by March, 1942, the remnants were in Australia, with Japan in control of the Dutch East Indies.

In Europe, Dutch squadrons operated alongside the RAF and the Royal Navy's Fleet Air Arm, and formed the nucleus of post-war Dutch military aviation following the liberation of the Netherlands. Aircraft included Supermarine Spitfire fighters, North American B-25 Mitchell bombers, Douglas C-47 transports and Auster AOP3s. Post-war re-equipment started in 1947, with priority given to training aircraft, including North American T-6 Harvard, de Havilland Dominie (Rapide) and Tiger Moth, Percival Proctor, Avro Anson and Airspeed Oxford trainers; a total of 350 aircraft. Lockheed 12 and 14 transports were also supplied. In the Far East, Japanese surrender saw the RNIA dealing with Indonesian rebels as it tried to regain control over the East Indies, using North American F-51D Mustang and Curtiss P-40N Warhawk fighters, B-25 Mitchell bombers and C-47 transports.

The first jets were delivered in 1948, 200 Gloster Meteor F4 and F8 fighters, many built under licence by Fokker, which also supplied 40 of its S-11 trainers in the same year.

The Netherlands became a member of the North Atlantic Treaty Organisation in 1952.

In 1953, the AAS became the Royal Netherlands Air Force, Koninklijke Luchtmacht, an autonomous air arm. That year, it received the first of 200 Republic F-84E Thunderjet fighter-bombers, later replaced by F-84F Thunderstreaks, Hawker Hunters and North American F-86F Sabres. These were accompanied by Piper L-18 Super Cubs for liaison duties, DHC-2 Beaver light transports, Hiller H-23 helicopters and 20 Fokker S14 jet trainers, while Lockheed T-33A trainers were also delivered. The first of 120 licence-built Lockheed F-104G Starfighter fighter-bombers entered service in 1963, followed later by 105 Canadian-built Northrop F-5A/Bs for Thunderstreak replacement. Surface-to-air missiles were deployed for the first time in the 1960s, with Nike-Ajax and Nike-Hercules. The Starfighters were augmented by RF-104G reconnaissance aircraft for a single squadron. A squadron of 12 Fokker F-27M Troopships provided most of the transport, replacing earlier types while the Beavers operated under army control.

The Netherlands was amongst the European NATO members participating in the Lockheed F-16 fighter-bomber pro-

ABOVE: *The Royal Dutch Navy operated Westland Wasps from its* Van Speijk*-class frigates, ships based on the British Leander-class. (GKN Westland)*

BELOW: *Mainstay of the Klu-Royal Netherlands Air Force combat squadrons is the Lockheed Martin F-16A(R), which have been up-graded under the European Mid-Life Update Programme. This aircraft has AMRAAM missile and reconnaissance pod. These aircraft may be replaced by the US JSF, or by the Eurofighter Typhoon or Saab Grippen. (Klu)*

Netherlands

ABOVE: *Possibly the last of a long line of Dutch-designed and built aircraft, a Klu Fokker F-50 transport. (Klu)*

ABOVE: *This KC-10 Extender tanker/transport is one of two operated by the Klu. (Klu)*

ABOVE: *The Klu has a limited transport capability, including just two Lockheed C-130H-30 Hercules. (Klu)*

gramme, replacing the Starfighters and F-5A/Bs, with the first of more than 200 entering service in early 1981. An unusual role for the KLu during the 1970s through to the mid-1990s was the operation of two Fokker F-27 Maritimes in the Netherlands Antilles, a role normally handled by the Navy. The F-27M Troopships were eventually replaced by a mixture of aircraft, including second-hand converted KDC-10A Extender tanker/transports, Lockheed Hercules and Fokker 60s. The provision of aircraft for the army has continued, with Apache anti-tank helicopters and Chinooks for heavy lift, as well as smaller transport and liaison machines.

In common with most air forces, the KLu suffered from repeated defence cuts, more than halving in personnel over the past thirty years from 23,000 to 10,000 today. There were major cuts in 1991, 1993 and 1999; the last of these cutting the F-16 force from seven squadrons to six. The aircraft have benefited from the KLu's participation in the F-16 European Mid-Life Update programme. Flying hours remain reasonably good, at an average of 180. Today, the KLu has six squadrons with five operating 108 F-16AM, equipped AIM-9 Sidewinder and AIM-120B, AMRAAM and AGM-65G Maverick, and the sixth as a conversion unit with 21 F-16B. There are 30 AGM-114K Hellfire-equipped Boeing NAH-64D Apache helicopters in two squadrons. Other helicopters include 17 Eurocopter AS352U2 Cougar and nine SA3160 Alouette in a transport squadron, a squadron with 27 BO105CB/DB on liaison duties, and a squadron with 13 Boeing CH-47D Chinook heavy-lift helicopters, all operated for the army. Three Bell 412SP provide SAR. Transport is provided by two KDC-10A Extender tanker transports, two Lockheed C-130H-30 Hercules, four Fokker 60, two Fokker 50, and a VIP Gulfstream IV, in a single composite squadron. The main trainer is the Pilatus PC-7 Turbo Trainer, owned by the KLu, but operated by a contractor. Fast jet training takes place in the USA, as does training for the Apache, but other training is in the Netherlands. SAM includes Hawk, Patriot and Stinger.

Netherlands

ROYAL NETHERLANDS NAVAL AIR SERVICE/MARINE LUCHTVAARTDIENST

ABOVE: *Support for ground forces is provided by the Klu, including these Boeing AH-64D Apache attack helicopters which entered service in 2000. (Klu)*

Re-formed: 1944

The Royal Netherlands Navy's air arm dates from 1917, when it was formed with six Martin seaplanes and three Farman F-22 aircraft. The country was neutral during World War I, but the many islands and vast territorial spread of the Netherlands East Indies gave naval aviation the chance to prove its worth. During the 1920s, the MLD operated some 40 Hansa-Brandenburg W-12 seaplanes in the Netherlands East Indies, later augmenting these with Dornier Wal flying boats which remained in service up to the Japanese invasion of 1942. In the immediate pre-war period, Dornier Do24K flying boats were also acquired for operations in the East Indies, but the rapid advance of Japanese forces after the invasion saw the MLD's personnel escaping to Ceylon (Sri Lanka) and Australia. In the Netherlands, the MLD operated Fokker TVIIIW seaplanes, many of which were flown to France as the Netherlands fell, and from there to England before the French surrender in June, 1940.

For the rest of the war, Dutch naval air personnel in the UK flew with the Royal Navy's Fleet Air Arm. They flew Fairey Swordfish on anti-submarine duties from MAC-ships, merchant aircraft carriers, which were grain ships and tankers modified with a wooden flight deck and a side island for convoy escort duties in the North Atlantic, and able to carry three or four Swordfish. Once the MAC-ships were withdrawn, the Swordfish were replaced by Barracudas, in turn replaced by 30 Fairey Firefly fighter-bombers. In 1946, the Royal Navy escort carrier HMS *Nairana* was loaned to the MLD, re-named *Karel Doorman*, until the Royal Netherlands

BELOW: *Heavy lift helicopters in the Klu, primarily to support ground forces, are 13 Boeing CH-47D Chinook. (Klu)*

Netherlands

ABOVE: *The Royal Netherlands Navy is responsible for long-range maritime-reconnaissance, with 13 Lockheed P-3C-II up-graded Orions. (MLV)*

Navy's own light fleet carrier, also named *Karel Doorman*, formerly HMS *Venerable*, could enter service in 1948. Both ships were named in memory of the Dutch admiral lost during the Battle of the Java Sea in 1942. At first, the new carrier operated later versions of the Firefly, replaced by Hawker Sea Furies, themselves replaced in due course by Armstrong-Whitworth Sea Hawk jet fighter bombers. The Fireflies and Sea Furies were accompanied aboard by ASW Grumman TBM-3W and TBM-3S Avengers, until these were replaced by Grumman S-2 Trackers, of which there were eventually three squadrons.

Shore-based MR remained with the MLD in the post-war years. A squadron of Convair PBY-5A Catalina amphibians was replaced in 1951 by Lockheed PV-2 Harpoon landplanes, themselves replaced in turn by 15 P2V-5 Neptunes in 1953. Nine Breguet Br1150 Atlantic MR aircraft replaced the Neptunes in 1970. By this time, *Karel Doorman* had been withdrawn and sold to the Argentine Navy in 1969, and while the Sea Hawks were withdrawn, the three squadrons of 36 Trackers became shore-based. Naval aviation continued at sea, however, for by this time eight guided-missile frigates of the Van Speijk-class (based on the British Leander-class) had entered service with Westland Wasp helicopters, of which the MLD operated twelve. Eight Sikorsky SH-34J and six Bell UH-1 Iroquois helicopters were also in service by this time.

The Wasps were eventually replaced by Westland Lynx helicopters, initially aboard the Kortenaer-class frigates designed jointly with Germany. The Lynxes were modernized by 1989, and will be replaced by 20 European NH90 helicopters from 2007 onwards. The Atlantics and Trackers were replaced by Lockheed P-3C-II Orion aircraft in the late 1980s, some of which are undergoing a 'capability upkeep programme', while the others will be sold or stored. A Beech Super King Air 200 was leased in 1989 for pilot training.

Today, the MLD has 950 of the RNthN's 12,340 personnel. It operates from four remaining Kortenaer-class and eight Karel Doorman-class frigates, from a destroyer and from the LPD *Rotterdam*. It has 13 P-3C-11 Orions, reducing to ten upgraded aircraft, and 21 out of the original 22 Lynx. It retains the leased Super King Air 200, which is civil registered.

LEFT: *Clearly, far from home, is this Marine Luchtvaartdienst Westland SH-14D Lynx, up-graded to Super Lynx standard, aboard a Dutch frigate. (MLV)*

New Zealand

POPULATION: 3.9 million

LAND AREA: 103,736 square miles, 268,676 sq km

GDP: $53bn (£37.1bn), per capita $19,100 (£13,357)

DEFENCE EXPENDITURE: $804m (£562m)

SERVICE PERSONNEL: 9,230 active.

ABOVE: *New Zealand's decision to scrap its combat aircraft with the withdrawal of the A-4K Skyhawks at the end of 2001 has put the future of its Aermacchi MB399CB armed-trainers in doubt. (RNZAF)*

ROYAL NEW ZEALAND AIR FORCE

Formed: 1937

New Zealanders played an active part in World War I, flying with the RFC and RNAS, and ultimately, the RAF. New Zealand did not have any military aircraft of her own until 1923, when the New Zealand Permanent Air Force was formed within the New Zealand Army. This move was prompted by the 1920 British offer of surplus RAF aircraft to the dominions, with New Zealand's share of the Imperial Gift meant to be 100 aircraft. Indecision meant that there was little choice left by the time plans were finalized, and less than half the aircraft offered were accepted. Some aircraft were leased to commercial operators, leaving the NZPAF with ten Bristol F2B fighters, ten DH4 and nine DH9 bombers, and four Avro 504K trainers. The aircraft were shared with the New Zealand Air Force, as the reserves were known!

Economic difficulties limited new aircraft purchase during the 1920s to five Gloster Glebe and a small number of de Havilland Puss Moth and Moth trainers, and a few additional 504K trainers. Limited expansion started during the 1930s, when a Saunders-Roe Cutty Sark flying boat and ten Fairy IIIFs were followed in 1935 by 12 Hawker Vildebeest torpedo-bombers and some Avro 626 trainers. In 1936, the NZPAF became the Royal New Zealand Air Force, but remained part of the army until 1937, when it became a separate service on the recommendation of a seconded RAF officer. A major expansion programme was implemented in the short time left before the outbreak of World War II in Europe. Reserve squadrons gained ex-RAF Blackburn Baffins, with new aircraft for the regular squadrons. The priority was for trainers, Airspeed Oxfords, Fairey Gordons, Hawker Harts and Vickers Vincents. An order for 30 Vickers Wellington bombers was due for delivery on the eve of war, but these were transferred to the RAF with their crews to form a New Zealand element. The RNZAF's main wartime role was the training of pilots and other aircrew for the Empire Air Training Scheme, but it did field 27 combat squadrons, including twelve fighter squadrons, initially with Curtiss Kittyhawks and then with Chance-Vought FG1 Corsairs. Six bomber squadrons operated Lockheed Hudsons, followed by Venturas. Two flying boat squadrons operated Short Singapores, then replaced these with Consolidated PBY-5A Catalina amphibians and Short Sunderlands. There were two Vickers Vincent general reconnaissance squadrons, a Douglas Dauntless dive-bomber squadron and a Grumman Avenger squadron. Other squadrons operated transport aircraft, including Douglas C-47s and Lockheed Lodestars. The RNZAF mostly fought in the Pacific, and, post-war, formed part of the British Commonwealth Occupation Force in Japan.

The inevitable peacetime reduction in strength occurred, to five regular and four reserve squadrons, the latter equipped with North American F-51D Mustang fighters. During the 1950s, one regular squadron had de Havilland Vampire FB9 jet fighter-bombers, two had de Havilland Mosquito bombers until these were replaced with a single squadron of English Electric Canberra B1 bombers, while a fourth operated MR Short Sunderlands until replaced by Lockheed P-3B Orions in 1967. The fifth squadron operated transport aircraft, a mixture of Douglas C-47s, Bristol 170M freighters and Handley Page Hastings C3 transports. The reserve squadrons were reduced to Harvard armed-trainers before being disbanded in 1957. A squadron of de Havilland Venom fighter-bombers was leased from the RAF for a few years. The Venoms were replaced by ten McDonnell Douglas A-4K Skyhawk fighter-bombers, with additional aircraft delivered later. Lockheed C-130H Hercules updated the transport squadron while Bell 47G Sioux and UH-1H Iroquois helicopters were introduced. BAC 167 Strikemaster and Victa Airtourer trainers were used for basic and advanced training, with TA-4Ks for Skyhawk conversion.

New Zealand was for many years a member of the South East Asia Treaty

New Zealand

ABOVE: *The RNZAF's Lockheed P-3K Orion maritime-reconnaissance aircraft are likely to be up-graded and down-graded, being refurbished, but losing their ASW capability. (RNZAF)*

Organization, with the UK, USA, France, Pakistan, the Philippines and Australia. A squadron was based in Singapore during the 1960s and 1970s. SEATO, intended as a Far Eastern NATO, never fully realized its potential and lacked a formal command structure. The alliance was replaced by a series of bilateral arrangements with Australia and the United States.

The Skyhawks were updated during the late 1980s, when they provided two strike squadrons. The close support Strikemasters replaced during the early 1990s by Aermacchi MB339C armed-trainers. Ex-RAF Andover transports replaced the Bristol Freighters. The RNZAF was pulled out of Singapore, but stationed part of a squadron of Skyhawks at Nowra, near Sydney, to provide maritime air defence training for the Royal Australian Navy.

In 1999, it was decided to replace the Skyhawks with 28 leased ex-USAF F-16 Hornets, but a change of government led to this plan being abandoned, before deciding to scrap the fast jet combat squadrons altogether in 2001. Plans to upgrade the Orion force were rejected, instead downgrading the aircraft to a surveillance role.

Today, the RNZAF provides SAR, EEZ patrols and transport, plus Kaman SH-2G Seasprite helicopters operated on behalf of the RNZN from three frigates. It has 2,500 personnel. The future of the 17 Aermacchi MB339CB armed-trainers is doubtful following withdrawal of the Skyhawks. Commercial contractors are being considered for EEZ, while FRU duties for the RNZN are being met by hired preserved aircraft! Five Lockheed C-130H Hercules are being replaced one-for-one by Lockheed C-130J Hercules II purchased as part of a joint RNZAF/RAAF order. There are two Boeing 727-22QC transports capable of freight and VIP roles. Five SH-2G helicopters are operated for the RNZN, while 14 Bell UH-1H Iroquois helicopters operate on SAR and also support the army. Five Bell 47G Sioux

BELOW: *The RNZAF Lockheed C-130H Hercules are being replaced with C-130J Hercules II in a joint programme with the RAAF (RNZAF)*

New Zealand

training helicopters are due to be replaced, and there are also three Beech King Air B200 aircraft in the training role.

ABOVE: *The RNZAF provides all air power in New Zealand, including Bell UH-1H Iroquois operated for the Army and on SAR. (RNZAF)*

ROYAL NEW ZEALAND NAVY

During World War II, many New Zealanders served with the Royal Navy's Fleet Air Arm, but naval aviation was not introduced in New Zealand until the arrival of two Westland Wasp helicopters in the 1960s to operate from a new frigate. Although the frigate force peaked at four, with seven Wasp helicopters, it was decided that this was too small for an independent air arm. Five Kaman SH-2G helicopters are now operated by the RNZAF from three frigates.

Nicaragua

POPULATION: 5.2 million

LAND AREA: 57,143 square miles, 148,006 sq km

GDP: $3.1bn (£2.2bn), per capita $2,200 (£1,538)

DEFENCE EXPENDITURE: $26m (£18.2m)

SERVICE PERSONNEL: 16,000 active.

FUERZA AEREA NICARAGUA

Re-formed: 1996

Nicaragua became involved with military aviation shortly after the end of World War I, when the National Guard received four Curtiss JN-4 'Jenny' trainers and some war surplus DH4 bombers from the USA in return for assistance in the Panama Canal Zone. Although the US Army provided some pilots, the small force dwindled so that by 1927, it had fewer than 20 personnel and just three Swallow biplanes. A second attempt came in 1938, with US assistance in establishing the National Guard Air Force, or Fuerza Aerea de la Guardia Nacional. Aircraft provided included Grumman G-23 fighters and Waco D biplanes, used as light bombers. A flying school was supplied with Boeing-Stearman PT-13A Kaydet, Fairchild PT-19 and North American AT-6 trainers, with USAAF instructors.

During World War II personnel fell to less than 70. The US provided war surplus aircraft, including a number of Lockheed P-38 Lightning twin-engined fighters and some trainers. Nicaragua became a member of the Organization of American States in 1948, and the title of the Nicaraguan Air Force, or Fuerza Aerea de Nicaragua was adopted. More aircraft were supplied, including one squadron each of North American F-51 Mustang and Republic F-47D Thunderbolt fighter-bombers, and Douglas C-47 transports. The FAN grew steadily to some 1,500 personnel over the next 25 years. Obsolete aircraft continued to predominate, including Douglas B-26 Invader bombers acquired during the 1950s, and COIN Lockheed T-33A armed-trainers. Beech C-45 transports and AT-11 trainers and North American T-28 Trojan trainers followed, as well as a Hughes 269 and four OH-6A helicopters for army co-operation. A civil war started during the 1960s, ending in 1970, before a more prolonged civil war from 1982 to 1998. Left-wing Sandinista guerrillas formed a government, and the FAN became the Sandinista Revolutionary Air Force, Fuerza Aerea Sandinista, operating against US-supported guerrillas. The new regime built and improved airfields, while pilots were trained in Bulgaria to fly the MiG-21, most of which deployed in Cuba and never arrived in Nicaragua, which did receive many Mil helicopters and Antonov transports.

Elections led to a change of government in 1990. Fighting continued, but the new regime immediately reduced the FAS, selling Mi-24 attack helicopters to Peru. The title Fuerza Aerea Nicaragua was re-adopted in 1996.

The FAN has 1,200 personnel. Its sole combat capability is 15 Mi-17 armed transport helicopters. The main role is transport, with two Antonov An-2 and four An-26s. There is a VIP Cessna 404 Titan. Other aircraft undertake training and communications work, including three Piper PA-23 Cherokee, five Cessna 172, a 180, two U-17 and ten Piper Super Cubs.

Niger

POPULATION: 11.1 million

LAND AREA: 458,596 square miles, 1,253,560 sq km

GDP: $1.8bn (£1.3bn), per capita $844 (£591)

DEFENCE EXPENDITURE: $27m (£19m)

SERVICE PERSONNEL: 5,300 active.

NIGER NATIONAL AIR SQUADRON/ESCADRILLE NATIONALE DE NIGER

Formed: 1960

A former French colony, Niger became independent in 1960, with the standard French package of aircraft for communications and transport duties - a Douglas C-47 Dakota transport and three single-engined Max Holste 1521M Broussards. From the outset, it has been part of the army. Three ex-Luftwaffe Noratlas transports eventually replaced the C-47, while the Broussards were replaced by two Cessna F337s and two Dornier Do28s. A Douglas C-54 Skymaster was also operated.

The ENN has not developed a combat capability. In 1991, two Lockheed C-130H Hercules replaced the Noratlases, but just one remains, badly damaged after taxiing off a runway in 2000. Two Dornier Do228s were also acquired for light transport duties, although again just one of these survives, with a single Do28, a VIP Boeing 737-200. There are 100 personnel. In addition to the aircraft mentioned, there is one An-26, aquired from Libya.

Nigeria

POPULATION: 113 million

LAND AREA: 356,669 square miles, 923,773 sq km

GDP: $53bn (£37.1bn), per capita $1,359 (£950)

DEFENCE EXPENDITURE: $2.4bn (£1.7bn)

SERVICE PERSONNEL: 78,500 active.

NIGERIAN AIR FORCE

Formed: 1964

The Federation of Nigeria dates from independence from the UK in 1960. The Federal Nigerian Air Force was formed in 1964 with assistance from India and Germany. Initial equipment included ten Nord Noratlas transports, 30 Dornier Do27 liaison aircraft and 26 Piaggio P149D trainers. A civil war erupted during the late 1960s as Biafra attempted to break away from the Federation. Although air power played a relatively small role during the conflict, Soviet military aid started, including MiG-15 and MiG-17 fighter-bombers, and a handful of Ilyushin Il-28 bombers, with Egyptian pilots.

ABOVE: *Nigerian C-130H Hercules (Jeremy Flack/Aviation Photographs International)*

Successive regimes in Nigeria have given defence a high priority, especially after a military coup in late 1983. Nevertheless, competing demands for finance and oil price fluctuations have meant that contracts for new equipment have frequently had to be renegotiated and options have lapsed. Doubts persist over whether some aircraft have been paid for, notably 15 Jaguar strike aircraft delivered during the late 1970s. There have also been problems with serviceability due to the climate and the scarcity of skilled labour. Attempts to sell aircraft to raise funds for new equipment have come to nothing. At one time, the Jaguar force was due to return to the UK for refurbishment and transfers to the RAF to cover attrition losses. In 1999 it was announced that the Jaguars, G222s and MiG-21s would be sold, but in 2000 plans were announced to keep the MiG-21s and upgrade these in a programme similar to that conducted in India. In 2000, US aid enabled eight C-130H Hercules and 12 Alphajets to return to operational use. Equipment purchases have continued, and have included Mi-35 Hind attack helicopters and Mi-34 training helicopters delivered by Russia during 2000. In 2001 20 CN235s were ordered on a barter deal for

Nigeria

crude oil. AIEP Air Beetle trainers have been assembled from kits using local labour.

Nigeria has considerable regional ambitions, and since the late 1990s has deployed substantial forces in Sierra Leone to assist internal security. Defence continues to receive a high priority, but often funding lags behind planned purchases.

Now known simply as the Nigerian Air Force, it has 9,500 personnel. Equipment serviceability is estimated to be around 50 per cent, with a high attrition rate. Three fighter squadrons, include one with 18 Alphajet, one with 12 MiG-21MF, with AA-2 AAM, and three MiG-21U trainers, and a third with 12 Jaguar SN and three Jaguar BN. Many of the 18 Aero L-39MS Albatros trainers are armed. There are 18 BO105D armed helicopters in a single squadron, as well as seven Mi-35P. Two transport squadrons have five C-130H and three C-130H-30 Hercules, five G222, eight Dornier 228 (including VIP), with three Eurocopter SA330 Puma and two AS332 Super Puma. There are 17 Dornier Do28D and 15 Do128 in the utility role. A presidential flight has a Boeing 727-200Adv, a BAe 125-1000, a Citation II, two Falcon 900s, a Gulfstream II and a IV. Training uses 12 MB339AN, 58 AIEP Air Beetle T18 (Vans RV-6A), 14 Hughes 300C and three Mi-34 helicopters.

NIGERIAN NAVY

Formed: 1988

The Nigerian Navy has formed a small air arm, initially with three Westland Lynx ASW helicopters to operate from a German-built frigate. One of the aircraft crashed in 1989. Two A109s were acquired in 2001.

Norway

POPULATION: 4.5 million

LAND AREA: 125,379 square miles, 322,600 sq km

GDP: $162bn (£113bn), per capita $26,400 (£18,461)

DEFENCE EXPENDITURE: $2.9bn (£2bn)

SERVICE PERSONNEL: 26,700 (55% conscript) active, plus 222,000 reserves.

ABOVE: *Norwegian frigates operate Westland Lynx helicopters, and the search is on for a replacement, complicated by the fact that the Nordic Standard helicopter is likely to be too big. (GKN Westland)*

ROYAL NORWEGIAN AIR FORCE/KONGELIGE NORSKE LUFTVORSVARET

Formed: 1944

Norwegian military aviation dates from 1912, when the Royal Norwegian Navy was given a Taube and the Royal Norwegian Army was given a Maurice Farman. Official support followed in 1915, when the Naval Air Service, Marinens Flyvevaesen, and Army Air Force, Haerens Flyvapen, were formed. Both services had their own aircraft factories so that neutral Norway survived the wartime aircraft famine for non-combatant nations. The factories built aircraft under licence, including Maurice Farman types, Bristol F2B fighters and Hansa Brandenburg W33 seaplane fighters.

Post-war, the peacetime strength of the two services was established at 36 fighters and 36 bombers for the HF, with 20 fighters, 20 torpedo-bombers and 24 reconnaissance aircraft for the MF. Norwegian aircraft were provided for both services, including MF9 fighters, MF11 reconnaissance seaplanes and the MF8 and MF10 seaplane trainers. The MF also received Douglas DT-2B torpedo-biplanes and Heinkel He115 seaplanes. The HF received 30 Curtiss Hawk 75A and Gloster Gladiator fighters, Caproni Ca310 and Ca312 bombers, Douglas DB-8A attack aircraft, and Fokker CV and CVD reconnaissance-bombers. The small size of the two services and their elderly aircraft, meant that the spirited resistance mounted against the German invasion of April, 1940, was to no avail, despite the support of the RAF and the Fleet Air Arm. Aircraft and personnel escaped to the UK. HF personnel in the UK were formed into two fighter squadrons, initially flying Hawker Hurricanes, but these were later replaced by Supermarine Spitfires, with one of the squadrons becoming the highest scoring Allied unit during the war, and also having the lowest accident rate. MF personnel went to Canada to train on Northrop N-3PB seaplanes, ordered before the invasion, and afterwards operated these from British bases under the control of RAF Coastal Command.

Norway

ABOVE: *The Royal Norwegian Air Force has four squadrons of F-16AMs, recently up-graded. (Jane's)*

LEFT: *Maritime reconnaissance along Norway's long coastline is provided by four Lockheed P-3C Orion, plus another two P-3N operated for the coastguard. (Michael J. Gething)*

The two air arms were merged in 1944 to form the Royal Norwegian Air Force, Kongelige Norske Luftforsvaret. On its creation, the new KNL had three fighter squadrons, two bomber squadrons, a reconnaissance squadron and a transport squadron. Aircraft included Spitfire IXs, de Havilland Mosquito VIs, Consolidated PBY-5 Catalinas, Airspeed Oxfords, Avro Ansons, ex-BOAC and RAF Lockheed Lodestars, Fairchild PT-26 and North American T-6 Harvards. The first jets, de Havilland Vampire Mk3s, were introduced in 1948, and a further 25 were ordered the following year. The initial composition was changed by a Royal Commission in 1949 that proposed a strength of eight interceptor squadrons, two photo-reconnaissance squadrons, a bomber squadron and a transport squadron, each with eight aircraft. Norwegian membership of NATO ensured military aid from the United States, starting with 200 Republic F-84E Thunderjets for eight KNL fighter-bomber squadrons. In 1956, one squadron converted to RF-84F Thunderflash reconnaissance-fighters. North American F-86F and F-86K Sabres started to replace the F-84s in 1957. Transport aircraft were not neglected during this period, with Douglas C-47 and Fairchild C-119F Packet transports introduced, with CCF Norseman and de Havilland Canada DHC-3 Otters for communications duties. The first helicopters, Bell 47D/G Sioux, arrived, and eight Grumman HU-16 Albatross amphibians replaced the Catalinas. Lockheed F-104G Starfighters were introduced in 1963, with 20, and during the late 1960s these were joined by Northrop F-5A fighter-bombers and RF-5A reconnaissance-fighters. By this time, squadron strengths were 20 aircraft for the single F-104 squadron, and 16 for the four fighter-bomber and one reconnaissance-fighter squadrons. The 1960s saw six MR Lockheed P-3B Orion enter service, while six C-130H Hercules were also introduced, as were SAAB-91 Safir and Lockheed T-33A jet trainers. The helicopter force expanded, with 32 Bell UH-1H Iroquois for army co-operation duties, a number of Sikorsky UH-19s and, in 1971, ten Westland Sea King Mk43 (S-61) ASW helicopters. Cessna O-1E Bird Dog and Piper L-21 AOP aircraft were also provided for duties with the army.

Norway has taken defence more seriously than many European nations, or indeed many western countries with a small population. Lockheed F-16s were introduced to replace the F-104s and many of the F-5s during the 1980s, and upgraded during the 1990s, along with the surviving F-5s. The Orions were replaced by the P-3C variant in 1989, with the older P-3Bs sold to Spain with the exception of two modified as P-3Ns for coastguard duties. The UH-1Hs were replaced by Bell 412SPs for army support, but it was not until 1992 that the Cessna O-1 Bird Dogs were replaced by helicopters. Despite this, post-Cold War defence cuts also affected Norway, reducing F-16 numbers from 60 to 48. Although a decision on a new fighter was due to be taken in 2000, this has been deferred, and a replacement will not enter service until 2012. A C-130H replacement was also shelved, although the existing force will either have to be extensively refurbished or replaced with C-130Js. Personnel numbers have fallen over the past 25 years from 9,000 to 5,000.

The KLD has a reasonable 180 average flying hours per year. It has four squadrons with 48 upgraded F-16AM, with AIM-9L/N Sidewinder and AIM-120 AMRAAM, and 10 F-16B, with ten aircraft in reserve. It withdrew an F-5 squadron that had operated primarily in the training role for some years. Four MR P-3C Orion are augmented by two P-3N operated for the coastguard. There are six Lockheed C-130 Hercules in the transport force, augmented by three DHC-6 Twin Otter, also used on SAR duties with 12 Sea King Mk43B ASW, upgraded between 1989 and 1995. There are eighteen utility Bell 412SP to support, while six Westland Lynx Mk86 are used on coastal patrol and also operated from three coastguard vessels. Norway participated in the Nordic Standard Helicopter Programme, which will see 14 ASW NH90s replace the Sea Kings, and the ten aircraft on option might replace the Bell 412s. The NH90 will be too heavy to be a Lynx replacement. ASM includes CRV-7 and Penguin Mk-3.

Oman

POPULATION: 2.7 million

LAND AREA: 82,000 square miles, 212,380 sq km

GDP: $17.7bn (£12.4bn), per capita $8,200 (£5,734)

DEFENCE EXPENDITURE: $1.7bn (£1.2bn)

SERVICE PERSONNEL: 43,400 active.

ROYAL AIR FORCE OF OMAN

Formed: 1959

Known as Muscat and Oman until 1970, military aviation started in 1959 with the formation of the Sultan of Oman's Air Force for police duties. Initially, it was largely manned by seconded RAF personnel and by ex-RAF personnel. The original equipment included Percival Provost T52 armed trainers and DHC-2 Beaver transports, later augmented by BAC 167 Strikemaster armed jet trainers and Short Skyvan 3M transports. During the 1970s, Hawker Hunter fighters entered service, followed by Sepecat Jaguar S/B strike aircraft, BAe One-Eleven and Lockheed C-130H Hercules transports, Agusta-Bell 205, 206 and Bell 214 helicopters. Boeing 747SP and McDonnell Douglas DC-8-73 airliners, and two Grumman Gulfstream II corporate jets were acquired for the Royal Flight. Plans to acquire Panavia Tornado strike aircraft were abandoned in favour of BAe Hawk 200s and 100s.

Now known as the Royal Air Force of Oman, it has 4,100 personnel. Twelve F-16C/D Bloc 50-Plus are in course of delivery for an interceptor squadron, joining two fighter/ground-attack squadrons each of which has eight Jaguar S(O) Mk1, upgraded during the late 1990s to Jaguar 97 standard and expected to remain in service until at least 2005. A ground-attack squadron has 12 BAe Hawk 203s, while a close support squadron has 12 Pilatus PC-9 and four Hawk 103 armed trainers. Three transport squadrons include two with a total of three C-130H Hercules and ten Skyvan 3M, and one with three BAC One-Eleven. The Hercules may be replaced by C-130Js. Seven of the Skyvans also undertake MR. Helicopters include 20 Super Lynx 300s in course of delivery for transport and SAR, which may replace some of the 20 Agusta-Bell AB205A. Three AB212 undertake VIP duties. There are also three Bell 206B JetRanger and five 214B. The Royal Flight includes two Boeing 747SP, two Gulfstream IV, three AS330J Puma and three AS332C/L1 Super Puma. In addition to the Hawks and PC-9s, training uses three PAC Mushak, two Super Falke and four AS202 Bravo.

BELOW: *The Royal Air Force of Oman at one time had 12 BAC Strikemaster armed-trainers. (BAE SYSTEMS)*

Pakistan

P A K I S T A N

POPULATION: 161.8 million

LAND AREA: 310,403 square miles, 803,944 sq km

GDP: $62.8bn(£43.9bn), per capita $2,400(£1,678)

DEFENCE EXPENDITURE: $3.65bn (£2.55bn)

SERVICE PERSONNEL: 620,000 active, plus 513,000 reserves.

ABOVE: *Shenyang F-6 after overhaul at the PAC rebuild factory (Jane's)*

PAKISTAN AIR FORCE

Formed: 1947

Indian independence in 1947 divided the country into two states, India and Pakistan, with the latter divided into two, West Pakistan and East Pakistan (now Bangladesh), separated by 1,000 miles of Indian territory. The pre-independence Indian armed forces were also split. The Royal Pakistan Air Force inherited two former Royal Indian Air Force squadrons, one with Hawker Tempest fighters, the other with Douglas C-47 transports, supported by de Havilland Tiger Moth and North American T-6 Harvard trainers. Assistance was provided by the RAF, which seconded personnel. The RPAF built up to a planned strength of three Tempest fighter squadrons; a Handley Page Halifax bomber squadron; a C-47 transport squadron; and a communications squadron with Auster AOP5s, a C-47, Harvard, a Vickers Viking and two de Havilland Doves. In 1950, the PAF received Hawker Fury fighters and the first of an order for 62 Bristol 170M Freighter transports ordered specifically for the operation of the air link between the two parts of Pakistan. The first jets, 36 Supermarine Attacker FB2 fighter-bombers, arrived during the early 1950s.

In 1954, Pakistan became a member of the South East Asia Treaty Organization, SEATO, and US military aid started. Later, Pakistan joined the other regional defensive alliance, the Baghdad Pact (which later became the Central Treaty Organisation after a revolution in Iraq). Pakistan became a republic in 1956, with the RPAF dropping the 'Royal' prefix. North American F-86F Sabre jet fighters were delivered in 1956, and followed in 1958 by Martin B-57B (Canberra) jet bombers and a VIP Vickers Viscount. Lockheed T-33A jet trainers and tactical reconnaissance aircraft were also delivered, as well as Sikorsky UH-19 and Bell 47G Sioux helicopters, while Lockheed C-130B Hercules started to supplement and then replace the C-47s and Bristol 170Ms. The US supplied Lockheed F-104A Starfighter interceptors in 1962. US military aid was suspended in 1965 as a result of territorial disputes between Pakistan and India over Kashmir and the Rann of Kutch, which had seen aerial combat in which the PAF's Sabres fared badly against the lighter and more agile Indian Gnats. The IAF's superior performance also resulted from using cannon rather than air-to-air guided missiles, as heat-seeking versions proved unreliable in tropical conditions. Pakistan turned to Communist China for armaments, starting during the late 1960s with deliveries of Shenyang F6s (MiG-19s) and Ilyushin Il-28 bombers, and Mi-6 helicopters. These were joined by15 Dassault Mirage IIIE and three IIIR fighter-bombers, while Alouette III helicopters were also obtained.

Growing tension between India and Pakistan was compounded by unrest in East Pakistan, smaller but more populous than West Pakistan. Indian support for East Pakistan was resented by Pakistan. On 3 December, 1971, the PAF struck at ten IAF bases, including two in Kashmir, timing the attacks at dusk on a Friday, the Moslem Sabbath, when the Indians would be least likely to expect an attack. The attack was largely unsuccessful - the IAF claimed to have lost just three aircraft. The IAF had not deployed many aircraft forward close to the frontier, and those that were had been stored in concrete shelters invulnerable to anything other than a direct hit. The IAF also had the advantage of Soviet-supplied Tupolev Tu-126 AWACs aircraft. After the initial attacks, the PAF deployed relatively few of its 300 combat aircraft, and did little to support the land campaign, in a classic example of an air force without the support of an indigenous aircraft industry struggling to conserve its equipment. The war lasted until 16 December, when East Pakistan declared independence as the new state of Bangladesh.

The Mirage IIIs were followed by 30 Mirage V fighter-bombers, replacing the surviving Sabres during the early 1970s. Additional F-6s and A-5s, also a MiG-19 derivative, were obtained from China. US arms supplies resumed after the USSR invaded neighbouring Afghanistan, projecting Pakistan into the position of a Cold War front line state. During the 1980s, Lockheed F-16 Hornets were delivered for both the interceptor and strike roles. When Soviet forces withdrew from Afghanistan, Pakistan supported Mujahideen rebels in their fight against the Afghan government. While additional F-16s were delivered in 1990, further aircraft were embargoed by the US. Despite objections from India, 50 ex-RAAF Mirage IIIs were obtained, mainly as spares for the existing Mirages. The PAF managed to grow throughout this period, maintaining its links with China and introducing the Chengdu F-7M Airguard (MiG-21) and its F-7P Skybolt derivative, while collaborating with the Chinese aircraft industry in the development of the K-8 jet trainer, which first flew in late 1990. Pakistan had already built a basic trainer, the MFI-17B Mushshak, under licence from Saab.

ABOVE: *MFI-17B Mushanks. (Peacock/Aviation Photographs International)*

The 'on-off' nature of US arms supplies continued throughout the 1990s, but came to a complete stop in 1998 in protest against Pakistan testing nuclear weapons. The situation deteriorated further with Pakistani incursions into Indian Kashmir and the overthrow of the civilian government by the military in 1999. Collaboration with China continues to include a 50:50 joint project for the FC-1/Super Seven fighter, on which work started in 1999. K-8 deliveries have been slow, partly because the Chinese Air Force will not accept the aircraft because of its western engine.

Today, the Pakistan Air Force has some 40,000 personnel, having risen from 15,000 in the last 30 years. Flying hours are report-

BELOW: *Mirage IIIEP. (Peacock/Aviation Photographs International)*

ed to be in the region of 210 - implying a high state of operational readiness. There are 12 fighter squadrons: Three, including an OCU, operate up to 50 F-6 and 30 FT-6 trainers; two, including an OCU, operate 22 F-16A and 10 F-16B; one squadron has 44 Mirage IIIO/OD; and six, including an OCU, operate 143 Chengdu F-7MP and 13 FT-7, with another 40 in course of delivery. In the ground-attack role, three squadrons, again including an OCU, operate 50 Mirage 5PA and two 5DPA; one operates 16 Mirage IIIEP and three IIIDP trainers; and two operate 42 Nanchang A5 Fantan. There are relatively few transport aircraft. Mainstay of the transport element remains eight Lockheed C-130B Hercules, with four C-130E and one L-100-20. There are also two Harbin Y-12 transports, while four CN235Ms are entering service. Two Boeing 707-320 and a 737-300 are used on VIP duties, with two Fokker F27-200 Friendship and a Dassault Falcon 20E, another two of these aircraft are modified for electronic warfare. The communications role uses some of the 43 MFI-17B Mushshak also used for training, and being upgraded to Super Mushshak status, while there are four Cessna 172N, a Citation V and a Piper Senaca on communications duties. Training uses 30 Shenyang FT-5 (MiG-17), the dozen or so K-8 Karakorum that have been delivered, and the Mushshaks. There are 15 SA316B/SA319 Alouette III helicopters for liaison and SAR duties. The PAF operates two Atlantic and two P-3C Orion aircraft on behalf of the Pakistan Navy. Missiles include ASM AM-39 Exocet, AGM-65 Maverick, AS30 and AGM-84 Harpoon, all Western, in the ASM role, with AAM provided by AIM-7 Sparrow, AIM-9L/P Sidewinder and R-530 Magic, as well as AGM-88 Harm anti-radiation missiles, with six SAM batteries operating Crotale and one with CSA-1 (SA-2). The future shape of the PAF will be interesting, as US arms supplies might resume given the country's support for US action in Afghanistan, but at present, spares for the F-16s have assumed a higher priority than new aircraft, while Pakistan might hesitate to become dependent on a single source for arms.

ABOVE: *Mirage IIIDP. (Peacock/Aviation Photographs International)*

BELOW: *L-100-20 Hercules (Peacock/Aviation Photographs International)*

PAKISTAN NAVAL AVIATION

Pakistan naval aviation initially used a handful of Sikorsky H-19 helicopters acquired during the late 1950s for transport and SAR, later augmented by Alouette IIIs for liaison and ASW duties, as well as two Fokker F27-200 Friendships for transport and maritime-reconnaissance training. Plans to acquire Kaman Seasprite helicopters during the early 1990s for operation from escort vessels were abandoned, and instead Westland Lynxes obtained. Three Lockheed P-3C Orion and four Breguet Atlantic MR aircraft have been operated by the PAF, but one of each was lost in 1999 - the Atlantic being shot down by the Indian Air Force close to the border.

Today, PNA has two P-3C Orion and

ABOVE: *SA.330J Puma (Peacock/Aviation Photographs International)*

three Atlantic for MR, with two Britten-Norman BN-2T Defender for EEZ patrols. Three Fokker F27-200 and two F27-300s provide transport. There are six Sea King 45 and three Westland Lynx HAS3 for ASW and SAR, with the latter operating from warships, six Tariq-class (Amazon-class) and two Shamsher-class (Leander-class) frigates acquired from the Royal Navy, with eight Alouette III.

PAKISTAN ARMY AVIATION CORPS

The Pakistan Army moved into aviation with Beech L-23s and Cessna O-1E Bird Dog aircraft for AOP duties, later adding 90 Bell 47G Sioux helicopters. In contrast with most armies, Pakistan has continued to use fixed-wing aircraft in the AOP and liaison role, including more than 100 of the licence-built MFI-17 Mushshak, also used for training. While Mi-8 helicopters were added, the mainstay of the transport element has been the Aerospatiale Puma, of which 35 were delivered. The PAAC received 20 Bell AH-1S Cobra anti-tank helicopters in 1984 and 1985. In recent years, Mi-17 helicopters have augmented the earlier Mi-8s, while the Bell UH-1H force has been much reduced.

The arms embargo has meant that the current strength of the PAAC consists of just 18 AH-1F Cobra attack helicopters, with the remaining force consisting of unarmed helicopters for transport, communications and training. There are 115 MFI-17 Mushshak in the AOP and training role with 30 remaining Cessna O-1E Bird Dogs. Fixed-wing transport aircraft include two Harbin Y-12(II) delivered in 1997, while a Commander 690 and an 840, and a Cessna 421, are used on communications and survey work. Transport helicopters include 25 SA330J Puma, ten Mi-8, some of which may be retired shortly and up to 33 Mi-17, five Bell 205A and five UH-1H, and 28 206B JetRanger, some of which are also used on training. Communications uses 20 SA316B Alouette III and 15 SA315B Lama helicopters, with 12 Bell 47G used on liaison and training, with the latter role also filled by ten Schweizer 300.

Palestine

POPULATION: 3 million

Reliable official statistics are not available

SERVICE PERSONNEL: 35,000 paramilitary active.

PALESTINE AIR FORCE

Palestine covers the former Gaza Strip and West Bank area, formerly part of Jordan and now known as Jericho. There are a number of paramilitary organisations, and the paramilitary Palestine Air Force provides transport. It currently operates a Lockheed Jetstar VIP transport, two Mi-8 and two Mi-17 helicopters.

Panama

POPULATION: 2.8 million

LAND AREA: 28,575 square miles, 74,009 sq km

GDP: $10.2bn (£7.1bn), per capita $7,100 (£4,965)

SECURITY EXPENDITURE: $135m (£94m)

SERVICE PERSONNEL: 11,800 paramilitary.

PANAMA NATIONAL AIR SERVICE/PANAMA SERVICIO AEREO NACIONAL

Panama has only paramilitary organisations, providing internal security and policing the approaches to the Panama Canal, as well as transport and liaison. The former Panama Air Force, Fuerza Aerea Panama, attempted to obtain Cessna A-37B Dragonfly armed jet trainers for COIN duties in 1983, but was disbanded after the US invasion of Panama in 1989. The PSAN has 400 personnel. Aircraft include a CASA CN235M for patrol and transport, two C212-200s, three C212-300s, and a Britten Norman BN-2A Islander. A Piper Seneca handles communications duties. VIP aircraft include a Boeing 727 and an S-76. There are 13 Bell 205 and UH-1H helicopters, of which five were purchased from Taiwan in 1997, and four 212s. Training uses six Chilean-built Enaer T-35D Pillan.

Papua New Guinea

POPULATION: 4.9 million

LAND AREA: 183,540 square miles, 461,693 sq km

GDP: $4.5bn (£3.1bn), per capita $2,800 (£1,958)

DEFENCE EXPENDITURE: $56m (£39.2m)

SERVICE PERSONNEL: 4,400 active.

ABOVE: *IAI Arava. (Mader-Acaes Collection)*

PAPUA NEW GUINEA DEFENCE FORCE

Formed: 1975

Papua New Guinea was administered by Australia until independence in 1975. A Papua New Guinea Defence Force was established after independence to maintain internal security, with an air element for coastal and border patrols. Four GAF Nomad Mission Masters, fitted with search radar, and five ex-RAAF Douglas C-47s for patrols along the border with the Indonesian territory of Irian Jaya were soon operational. Three IAI-201 Aravas were later acquired and four surplus Bell UH-1H were provided by Australia in 1989. The PNGDF joined Australian forces in rescue missions after Papua New Guinea was hit by a tidal wave in July, 1998.

Today, just 200 of the PNGDF's 4,400 personnel are involved in aviation. There are four GAF N22B/N22SB Nomad Mission Masters with Searchwater radar for coastal patrols, with three IAI-201 Arava and two CASA-IPTN CN235M transports, and four Bell UH-1H Iroquois for transport and SAR. Australia donated an NBO105 in 1999.

Paraguay

ABOVE: *The Chilean civil registered Enaer T-35 Pillan demostrator (Jeremy Flack/Aviation Photographs International)*

POPULATION: 5.6 million

LAND AREA: 157,047 square miles, 406,630 sq km

GDP: $9.5bn (£6.6bn), per capita $3,800 (£2,657)

DEFENCE EXPENDITURE: $123m (£86m)

SERVICE PERSONNEL: 18,600 (65% conscript) active, plus 164,500 reserves.

PARAGUAYAN AIR FORCE/FUERZA AEREA PARAGUAY

Paraguayan military aviation originated with Army Potez XXV reconnaissance-bombers during the late 1920s as tension arose with neighbouring Bolivia over a boundary dispute in the Gran Chaco. Aided by an Italian military aviation mission and mercenary pilots, the army's air corps, or Fuerzas Aereas Nacionales, acquired Fiat CR30 fighters, Caproni Ca101 bombers and Breda Ba25 trainers during the 1930s as war broke out with Bolivia. Fiat CR32 fighters were obtained later. Paraguay won the war. In 1938, Breda Ba65 ground-attack aircraft were obtained, and, in 1939, Muniz M-9 trainers from Brazil. The late 1940s saw Beech C-45 and Douglas C-47 transports, and Vultee Valiant, Fairchild M-62, North American NA-16 and Boeing-Stearman PT-17 Kaydet trainers in service.

In 1948, Paraguay was a founder member of the Organization of American States, by which FAN had become the Fuerza Aerea Paraguay, although still under army control.

During the 1950s and 1960s, a variety of transport types augmented the earlier aircraft, including Douglas C-54s and a Convair 240, while some North American T-6 Texan trainers were armed for COIN duties. The first helicopters were introduced, Bell 47G Sioux and Hiller UH-12E. Although land-locked, Paraguay sits across the Paraguay River and much of the terrain is swampy, so a Grumman JRF Goose amphibian was operated for many years. Combat aircraft were acquired during the 1980s, Brazilian-built Embraer EMB326B Xavantes (Aermacchi MB326), augmenting a number of AT-33As. Enaer T-35D Pillan trainers were acquired from Chile in 1991 and 1992. The most significant development in recent years has been the arrival in 1998 of Northrop F-5E Tigers donated by Taiwan, which also provided six Bell UH-1H helicopters in 2001.

The FAP has 1,700 personnel, although this may fall as total personnel numbers in the armed forces are being reduced. It has ten Northrop F-5E and two F-5F Tiger fighter-bombers, six Lockheed AT-33, which are in need of major overhaul to return to airworthiness, six Embraer EMB-326 Xavante and four T-27 Tucano armed-trainers. Three Douglas C-47 and four CASA C212-200 Aviocar, as well as five Cessna 206/210, two 402, a 185 and a T-41D Mescalero, a DHC-6 Twin Otter and two Piper Aztec, a Beech Baron and a King Air, as well as two PZL104 Wilga are used on transport and communications. There is a VIP Boeing 707-320 and a Cessna Citation II. Eleven Bell UH-1B/H Iroquois helicopters are used for transport, with two HB350 Esquilo (Ecureuil), an Agusta A109HO and a Hughes 300. Training is on two North American AT-6 Harvard, 12 surviving Enaer T-35A/B Pillan and three A122 Uirapuru.

PARAGUAYAN NAVAL AVIATION/AVIACION DE LA ARMADA PARAGUAYA

Paraguay's navy handles river patrols and SAR, with 100 out of its 2,000 personnel engaged in naval aviation. Aircraft are used for communications and patrol duties, including a Cessna 210, two 310 and a 401, with two HB350 Esquilo and an UH-12E Raven helicopters. Training uses two Cessna 150 and an OH-13H Sioux.

Peru

POPULATION: 26.1 million

LAND AREA: 496,093 square miles, 1,249,048 sq km

GDP: $66bn (£46.2bn), per capita $4,700 (£3,286)

DEFENCE EXPENDITURE: $878m (£613m)

SERVICE PERSONNEL: 100,000 (64% conscript) active, plus 188,000 reserves.

PERUVIAN AIR FORCE/FUERZA AEREA PERUANA

Formed: 1950

Peru established army and naval air arms in 1920. The government acquired 12 Avro 504K trainers for the army, as well as a Curtiss JN4, four Morane-Saulnier Parasols, two Spad S-7C and three Bristol F2B fighters, and a Blackburn Kangaroo bomber. The USA supplied two Curtiss Seagull flying boats for the new Peruvian Naval Air Service, which later received DH9 bombers and Vought UO-Q Corsair observation aircraft. The two air arms merged in 1929, forming the Peruvian Aeronautical Corps, Cuerpo de Aeronautica del Peru. Initially, the CAP operated Nieuport 121C and Curtiss Hawk fighters, Douglas M-4 seaplane bombers, Vought UO-Q and O2U-1E Corsair and Potez 39 observation aircraft, Boeing 40B transports, and Avro 504R Gosport, Stearman C-3R, Hanriot 240 and Morane-Saulnier MS110 trainers. New aircraft during the early 1930s included Fairey Fox II bombers and Fairey Gordon general-purpose aircraft. In 1933, the CAP fought in a border dispute with neighbouring Colombia.

An Italian air mission visited Peru in 1935, leading to the purchase of Caproni Ca114 fighters, Ca135 bombers, Ca111 transports, Ca310 general-purpose aircraft and Ca100 trainers. An attempt was made to establish a Caproni factory in Peru, for local assembly of the aircraft. The late 1930s saw Curtiss Hawk 75-A and North American NA-50A fighters, Douglas DB-8A bombers, Vultee 54 and Curtiss-Wright CW-22 trainers introduced, with some Faucett F-19 transports. During World War II, there were few new aircraft as Peru remained neutral. The CAP did see action in 1941, against Ecuadorian forces in a border dispute.

Peru became a member of the Organisation of American States in 1948. US military aid, included 20 Republic F-47D Thunderbolt fighter-bombers, 20 North American B-25J Mitchell bombers, and 12 Lockheed PV-2 Harpoon MR aircraft. There were also Consolidated PBY-5A Catalina amphibians, Curtiss C-46 and Douglas C-47 transports, Beech T-11 Kansan, North American T-6 Texan, Boeing-Stearman PT-17 Kaydet and Fairchild PT-26 trainers. These were followed in 1949 by four DHC-2 Beaver transports purchased from Canada. Some pre-war aircraft remained operational until the mid-1950s.

BELOW: *Canberra B.56 bomber. (Jeremy Flack/Aviation Photographs International)*

The CAP became the Peruvian Air Force, Fuerza Aerea Peruana, in 1950. No further aircraft were introduced until 1955, when the FAP received its first jets, North American F-86F Sabre fighters and Lockheed T-33A trainers. These were followed in 1956 by 16 Hawker Hunter F52 fighter-bombers and eight English Electric Canberra B8 bombers, which joined eight ex-USAF Douglas B-26 Invader bombers introduced in 1955. Additional Canberras were purchased later, as well as Lockheed F-80C Shooting Star fighters. Peru has had border disputes with its neighbours, Bolivia, Chile, Colombia and Ecuador, and anxious to avoid an arms race, the USA requested the UK to embago BAC Lightning interceptors, but 14 Dassault Mirage 5 fighters were ordered instead.

Peru became the first South American country to buy Soviet equipment, introducing Sukhoi Su-22 fighter-bombers, Antonov An-26 transports, and Mil Mi-6 and Mi-8 helicopters in 1976. Despite the Su-22s' poor serviceability, a second batch followed in 1980. The An-26s were replaced with An-32s in 1987. A substantial increase in transport helicopters occurred during 1989-90, with 32 Mi-17s. Six Harbin Y-12 transports were purchased from China in 1991. Soviet equipment was both available and cheaper than Western aircraft. The USSR's collapse brought both opportunities and problems in procurement. The opportunities included buying MiG-29s and Su-25s from Belarus in 1996, at an attractive price. The problems lay in the manufacturer refusing to sell spares, as the aircraft had not been purchased from them! The issue was resolved by a further Peruvian purchase of three MiG-29SEs from the manufacturer in 1998, on condition that spares be supplied for the original aircraft, which replaced some Canberras and Su-22s. Mi-24 attack heli-

Peru

ABOVE: *Shipborne Sea King (Jane's)*

copters were purchased from El Salvador in 1993. Peru also purchased Dassault Mirage 2000 interceptors from France, and Embraer EMB312 Tucano trainers from Brazil. At this time, US assistance resumed to help Peru fight the Maoist Sendero Luminso, 'Shining Path', guerrillas and drug traffickers.

Today, the FAP has 15,000 personnel, an increase of 6,000 over the past 30 years. It operates a wide variety of aircraft, giving logistics and maintenance problems. There are three fighter squadrons, with one operating ten Mirage 2000P, with two 2000DP; one operating twelve Mirage 5P and three 5DP; and one operating 19 MiG-29 and two MiG-29U, which are receiving major overhauls. Two ground-attack groups have a total of six squadrons; three of which have a total of 24 Su-22 and four Su-22Us; and three have a total of 23 Cessna A-37B Dragonfly armed-trainers. There are still ten Canberra bombers. Ten Mi-24 provide an attack helicopter squadron, but may be replaced soon. There are three Boeing 707-320 tanker transports, with two McDonnell Douglas DC-8-62CF, 18 An-32, three An-74, three Lockheed C-130A and five L-200-20 Hercules, six DHC-6 Twin Otter, eight Y-12(II) and two Pilatus PC-6 Porter transports. The Presidential Flight has a Boeing 737-500, a Fokker F28-1000 Fellowship and a Falcon 20F. There are five Mi-6, three Mi-8 and 35 Mi-17, ten Bell UH-1H Iroquois, ten 212 and five 214ST, and two 412, and five BO105C helicopters. A Beech Super King Air and four Metro IIIs are employed on communications duties. Two Learjet 36A are used on surveillance and calibration duties. Training uses 15 T-41A/D Mescalero, 19 EMB312 Tucano, some of which may still be for sale after six were sold to Angola, six Il-103, 18 Zlin 242L, 13 MB339AP and three AS350B Ecureuil.

PERUVIAN NAVAL AIR SERVICE/SERVICIO AERONAVALE DE LA MARINA PERUANA

The Peruvian Navy initially used eight Bell 47G Sioux helicopters, but the introduction of ex-Dutch guided missile cruisers in 1973 and 1976, followed by Carvajal-class (Lupo-class) frigates during 1978-81, led to shipboard helicopters. Only one of the guided missile cruisers remains in service, but this can carry three small helicopters or a single medium helicopter. The Aeronavale is also responsible for MR, but lacks sophisticated long-range aircraft. Currently, aviation accounts for 800 out of the navy's 25,000 personnel. MR is handled by five Beech Super King Air B200Ts, which can also be used in the light transport role, three Embraer EMB-111 Bandeirante and three Fokker F27 Friendships. A DHC-6 Twin Otter is operated on behalf of the coast guard. Five Agusta-Bell AB212 helicopters can be deployed aboard the four frigates on ASW and ASV duties, while there are six Agusta-built AS-61D Sea Kings for a similar role aboard the guided-missile cruiser. Two An-32 and four Mi-8 helicopters are used for transport, with a Cessna 206 and an EMB-120 Brasilia for communications. Training uses five Beech T-34C Turbo Mentor and four Bell 206 helicopters.

PERUVIAN ARMY AVIATION/AVIACION DEL EJERCITO PERUANA

Peruvian Army Aviation is primarily for transport, although its ten Agusta A109K helicopters can be armed. It has a small number of fixed-wing aircraft, including four Antonov An-32, a Cessna 150 and a 172, and seven Cessna 206 and two 303. There are three Mi-26 heavy lift helicopters, which may have replaced two earlier Mi-6, 26 Mi-8 and 13 Mi-17, as well as five SA315B Lama, two Bell 412 and eight Enstrom F28, the latter being used mainly for training.

Philippines

PHILIPPINES

POPULATION: 77.3 million

LAND AREA: 115,707 square miles, 299,681 sq km

GDP: $82.4bn (£57.6bn), per capita $3,400 (£2,378)

DEFENCE EXPENDITURE: $1.5bn (£1bn)

SERVICE PERSONNEL: 107,000 active, plus 131,000 reserves.

ABOVE: *Northrop F-5A (Mader-Acaes Collection)*

PHILIPPINE AIR FORCE

Formed: 1947

In 1935 the Philippine Constabulary formed an aviation branch using Stearman Model 76 AOP aircraft and Curtiss JN-4 trainers to assist ground forces in detecting bandit camps. This grew into the Philippine Army Air Force with US assistance in 1940, equipped with 12 Boeing P-26 fighters, most of which were destroyed on the ground when Japan invaded on 10 December, 1941. Those Philippine personnel that could escape were absorbed into the US forces for the duration of the war.

The Philippines were liberated in 1944, and became a republic the following year. In 1947, the Philippine Air Force was established as a separate service, with North American F-51D Mustang fighter-bombers and Douglas C-47 Dakota transports, used primarily on COIN. A US defence treaty in 1951 was followed in 1955 by the Philippines becoming a member of the South East Asia Treaty Organization, and obtaining further US aid. The Mustangs were replaced by the PAF's first jets, North American F-86F Sabres, in 1957. Japanese-built Beech T-34 Mentor trainers arrived in 1959. Other aircraft included Sikorsky H-19A and H-34, and Hiller FH-1100 helicopters, Lockheed T-33A jet trainers, Grumman HU-16A Albatross amphibians, and North American T-6G Texan and T-28 Trojan trainers. Northrop F-5A/Bs replaced Sabres during the late 1960s, when Fokker F27 Troopship and NAMC YS-11 transports also arrived, and Mitsubishi-built Sikorsky S-62A helicopters were introduced for SAR.

Defence was accorded a high priority throughout the 1970s and 1980s by the unpopular Marcos regime, and when this was overthrown in 1986, severe spending constraints were introduced. The F-5A/Bs remained a viable force through the provision of ex-Taiwanese aircraft through the US Military Assistance Programme in 1989. Economy measures saw the withdrawal of many older aircraft, and a number of SIAI SF260M/W armed-trainers. Funds were made available for 18 SIAI S211 trainers assembled in the Philippines. Plans to replace the F-5A/B force were postponed through the turn of the century while the force shrunk to just ten aircraft from the original 20. The decision to increase the defence budget with a 15-year plan starting during the mid-1990s, in response to a reduction in US forces in the region and growing Chinese interest in the Spratly Islands, were hampered by the Asian economic crisis of the late 1990s.

Currently, the PAF has 16,000 personnel, 7,000 higher than the figure in the late 1960s. It has a considerable 'wish list', replacing F-5A/Bs with either A-4 Skyhawks from Kuwait or New Zealand, or surplus USAF F-16s, as well as maritime patrol aircraft, with the ATR42MP favoured, light attack aircraft, such as the BAe Hawk or the AMX, and up to eight SAR helicopters. Eight F-5As and two F-5Bs form a single fighter squadron. There are 20 Fairchild OV-10 Bronco and four Lockheed RT-33A, many of which have been refurbished and returned to service after becoming unserviceable, for COIN, on which the 18 S211 can also be used. There are 15 Cessna T-41D Mescalero (172) also available for light reconnaissance as well as training, as are 33 SF260s that have been converted to turboprop power. The importance of COIN is emphasised by the presence of three squadrons of armed helicopters operating a total of 59 Bell UH-1H Iroquois, six Sikorsky AUH-76 (S-76 gunship) and 33 MD520. There is a single MR F27-200MPA, with another six F27-200 transports, four MR GAF N22SL Nomad, with another eight N22B transports. Transport is also provided by eight Lockheed C-130B Hercules, two C-130H, two L100 and four ex-RAF C-130K, and two Britten-Norman BN2 Islanders. VIP transport includes two SA330L Super Puma and six Bell 412 helicopters, as well as an F27 and an F28 Fellowship.

PHILIPPINE NAVAL AVIATION

The Philippines Navy has a small force of aircraft and although no helicopters are embarked, two offshore patrol craft and two LST have helicopter decks. Current aircraft include eight Britten-Norman BN-2 Islanders, a Cessna 177 Cardinal and a Cessna 152, with the latter used for training. Other aircraft and seven BO105 helicopters are used for SAR, communications and utility duties. A 15-year defence plan envisages a surface fleet with ASW/ASuW helicopters.

Poland

POPULATION: 38.8 million

LAND AREA: 120,733 square miles, 311,700 sq km

GDP: $160bn (£111.9bn), per capita $8,422 (£5,889)

DEFENCE EXPENDITURE: $3.3bn (£2.3bn)

SERVICE PERSONNEL: 206,045 (44% conscripts) active, plus 406,000 reserves.

POLISH AIR DEFENCE AND AVIATION FORCE

Re-formed: 1944

Poland did not become an independent nation until 1918, during the chaos of the Russian Civil War. Polish personnel had served as aviators in the Russian, German and Austrian armed forces. In 1919, a seven squadron air corps was formed as part of the army, with almost 100 aircraft including Spad S7C fighters, Breguet Br14A reconnaissance and Br14B2 bomber aircraft, and Salmson SAL2-A2 reconnaissance-bombers. Aircraft and personnel of volunteer flying units were absorbed, giving a wide variety of military and civil aircraft. Bristol F2B fighters, Sopwith Dolphins and Martinsyde F4s were bought from the UK, which donated war surplus DH9 bombers and Sopwith Camel fighters. Former German and Italian aircraft were also obtained, as the new state prepared to fight the USSR for its independence. The conflict started in late 1919 and lasted for a year, with air power saving Poland from defeat.

Between the two world wars, a determined attempt was made to establish a Polish aircraft industry, with both private and state-owned concerns producing their own designs and building foreign designs, including Fokker FVII transports and Avia BH33 trainers, under licence. Significant Polish aircraft included the PZL P-1, P-6, P-7 and P-11 fighters that appeared during the 1930s, as well as P-23B reconnaissance-bombers, and immediately before the outbreak of World War II, the PZL P-37 bomber. Other aircraft included the RWD-8, RWD-14 and Lublin R-XIII AOP aircraft and PWS-2 trainers. In 1938, the Polish Air Force, Polskie Lotnictwo Wojskowe, became a separate air service following army reorganization. It had 55 fighters and 76 bombers, as well as transport, liaison and training aircraft. More than 250 aircraft were attached to the army for AOP and liaison duties. Ambitious plans for expansion of the PAF recognized that many of its aircraft were obsolete. Yet, of the many modern aircraft planned, only the PZL P-37 bomber was in service when German forces invaded on 1 September.

ABOVE: *On the outbreak of World War II in Europe in 1939, there were few modern aircraft in the Polish Air Force. One of the few modern types was the PZL-37 bomber, seen here nearest the camera, but the elderly fighters in the background were more typical. (IWM)*

The overwhelming strength of the Luftwaffe, with its modern aircraft, gave the Germans aerial superiority, aided by the rapid advance of the Panzer divisions and invasion by Soviet forces on 17 September, although Poland did not finally surrender until 27 September. Polish personnel took the few surviving aircraft to Rumania, where they were abandoned, and continued to France, arriving in time to provide a pool of experienced pilots for the Armee de l'Air. As German forces advanced westwards, Polish personnel flew Morane-Saulnier MS40C and Caudron C714 fighters before the fall of France, when they fled to the UK. Polish squadrons were formed in the RAF, initially being equipped with Hawker Hurricane fighters and Fairey Battle bombers, later replaced by Supermarine Spitfires and Vickers Wellingtons respectively. One squadron flew Boulton-Paul Defiant night-fighters, replacing these with Bristol Beaufighters and then with de Havilland Mosquitoes. By the end of the war, Polish

Poland

ABOVE: *MiG-29 Fulcram (Jane's)*

pilots had also flown North American F-51D Mustang fighters and B-25 Mitchell bombers, Handley Page Halifax and Consolidated B-24 Liberator bombers. Meanwhile, other Polish airmen had flown with the Soviet forces once Germany invaded Russia. Poland was liberated by advancing Soviet forces in 1944, but few Polish personnel serving with the RAF returned home, and of those that did, few were allowed to stay in the reformed PAF before being sent to detention camps.

The post-war Polish Air Force continued to use its wartime Soviet aircraft at first, including Polikarpov Po-2, Yakovlev Yak-1, Yak-3 and Yak-9 fighters; Ilyushin Il-2 bombers, and Yakovlev and Petlyakov Pe-2 trainers. The fighter units standardised on the Yak-9, which were joined by Ilyushin Il-10 ground-attack aircraft and Tupolev Tu-2 bombers. Later, Lisunov Li-2 (C-47) transports and Yak-11 and Yak-18 trainers were introduced. The first jets arrived in 1950, Yak-23 fighters. Over the next few years, these were joined by MiG-15 fighters, many of which were built in Poland as the LIM-2. The Polish-designed Junak-3 trainer was introduced, with Yak-17 jet trainers, Ilyushin Il-28 jet bombers and Mil Mi-1 helicopters, produced in Poland as the SM-1. In 1957, the Yak-23s and MiG-15s started to be replaced by MiG-17 jet fighters, and in turn these gave way to first the MiG-19 and then the MiG-21, while ground-attack duties passed to the Sukhoi Su-7. These aircraft were accompanied by the inevitable conversion trainers, with the MiG-15UTI filling the role of advanced jet trainer. A substantial transport force of almost 50 aircraft developed, including Antonov An-2 and An-12, Ilyushin Il-14 and Il-18, as well as the Li-2. Mil Mi-4 and the heavy lift Mi-6 helicopters joined the Mi-1s and SM-1s. The PAF became one of the largest and best-equipped air forces of the Warsaw Pact, second only to that of the USSR itself, due to the country's size and population, but also because dissent came much later to Polish political life than elsewhere in Eastern Europe. MiG-23 and Su-20 and Su-22 aircraft updated the combat squadrons, as well as Mi-8 and Mi-17 transport helicopters, and Mi-24 attack helicopters.

The changing political situation at the end of the 1980s forced a reappraisal of the strengths of the Polish armed forces. The last new aircraft to be delivered by the USSR were 12 MiG-29s in 1989, and further orders for this type were cancelled. The Soviet 37th Air Army moved its 350 aircraft from Polish territory in 1990, and on 1 April, 1991, the Warsaw Pact ceased to exist. The PAF was merged in 1990 with the Air Defence Force, Wojska Obrony Powietrznej Kraju, taking its present title of Polish Air Defence and Aviation Force/Polska Wojska Lotnice I Obrony Powietrznej, PWLiOP. The new service had almost 400 combat aircraft and another 250 or more transport and training aircraft. Initially, Soviet structures continued, with two squadrons of 12-16 aircraft in an air defence regiment and three squadrons of 9-12 aircraft within other air regiments. There were six air defence fighter regiments, three fighter and three fighter-bomber regiments, a reconnaissance-bomber and a tactical reconnaissance regiment. In 1995, many of the helicopter units used in support of the army were transferred to army control, creating an Army Air Force.

Plans to acquire Western equipment were put on hold due to the economic situation, but Poland became a member of NATO in April, 1999. In September, 1999, the MiG-23s were withdrawn. A replacement aircraft has still to be chosen, but the PWLiOP would like up to 60 aircraft, probably Lockheed Martin F-16s, but the F/A-18, Mirage 2000, Saab Gripen and Eurofighter Typhoon have all been considered. As an interim solution, 23 ex-Luftwaffe MiG-29s were introduced during 2002. An advanced trainer to replace the ill-starred I-22/M-93 Iryda, which had to be grounded, is another priority. Meanwhile, the MiG-29s are being upgraded in Germany to meet NATO standards and communications requirements, as are the Su-22s in Israel.

The PWLiOP has 43,735 personnel, to fall to 36,000 by 2006. Annual flying hours are around 60-120 hours, too low for full combat effectiveness. The PWLiOP is divided into two air defence corps, North and South. The fighter force consists of a squadron of 18 MiG-29U, backed by 4 MiG-29UB, although this force is slightly reduced as aircraft are sent to Germany to be upgraded. The ground-attack fighter role is filled by nine squadrons, of which four have 74 Sukhoi Su-22M4 and 18 Su-22UM3K, and five have 114 MiG-21PFM/M/MF/bis/UM/US/R. A reconnaissance squadron has seven Su-22M4. Transport is provided by two regiments operating 25 Antonov An-2, ten An-26 and two An-28s, as well as the nine Yak-40s, some of which join two Tupolev Tu-154M on VIP duties. Helicopters include eight Mi-8, 65 Mi-2, many of which are also used for training, and a Bell 412HP. Training uses the TS-11 Iska, of which there are 137, as well as 18 W-3 Sokol and 35 PZL-130 Turbo-Orlik. The I-22/M-93 Iryda replacement is likely to be either the BAe Hawk or the Aero L-159, either of which would be built under licence. Missiles include AA-2, AA-3, AA-8 and AA-11 AAM; AS-7 ASM; and S-200 WEGA, SA-2, SA-3, SA-4 and SA-5.

POLISH NAVAL AIR FORCE

Polish naval aviation has traditionally been land-based, with a single air regiment operating helicopters and fixed-wing aircraft, but this is changing with the acquisition from the USN of two FFG-7 (Perry-class) frigates that will operate Kaman SH-2G Seasprite helicopters. Currently, naval aviation accounts for 2,500 out of the navy's 16,860 personnel. There are two squadrons with a total of 22 MiG-21bis and five MiG-21UMs, which may be replaced by ex-air force Su-22s. MR, EEZ and pollution control uses two PZL-Mielec M-28E Bryza, with another two M-28TD on transport duties, also provided by five Antonov An-2 and five An-28, with another two for MR duties, and two Mi-17. SAR and ASW duties are handled by PZL-Swidnik W-3RM helicopter, 13 Mil Mi-14PL/PS, with the ten ASW machines upgraded during 1995-97. The FFG-7 frigates operate four ex-USN Kaman S-2 Seasprites. There are six Mi-2RM for communications duties, with two W-3 Sokol, and another four for SAR. Training uses 12 TS-11 Iskra, with another six on reconnaissance duties.

POLISH ARMY AIR FORCE

Formed: 1995

Polish army aviation today dates from the decision to transfer tactical helicopters from the PWLiOP, starting in 1995. The initial force of Mi-24 attack helicopters was augmented by an additional 18 machines formerly belonging to East Germany and handed over after German reunification. The plan was to bolster this force with 100 PZL W-3 armed helicopters, but this plan was scrapped in 1998 because of difficulties integrating the Rafael NT-D anti-tank missile and Elbit sensors, leading the PLWL to consider buying 10-15 secondhand combat helicopters, possibly the AH-1 or A129s. The Mi-8 and Mi-17 force is a combination of former PWLiOP machines and those operated by the Ministry of the Interior. Today, the PLWL has 40 Mil Mi-24 attack helicopters, 35 Mi-8 and three Mi-17 transport helicopters. While 32 Mi-2 are used for transport, another 23 operate in the attack role, and 18 are used in combat support. There are just 20 W-3W Sokol helicopters in service, and these may be withdrawn in the near future.

ABOVE: *S-22 (NATO)*

Portugal

POPULATION: 9.9 million

LAND AREA: 34,831 square miles, 91,945 sq km

GDP: $104bn (£72.7bn), per capita $16,370 (£11,447)

DEFENCE EXPENDITURE: $2.3bn (£1.6bn)

SERVICE PERSONNEL: 43,600 (19% conscript) active, plus 210,930 reserves.

PORTUGUESE AIR FORCE/FORCA AEREA PORTUGUESA

Formed: 1957

By the time that Portugal established both army and navy air arms in 1917, a flying school had been in existence for some years and officers from both services had received flying training in the UK and France. The Arma de Aeronautica used Spad S7C fighters and Breguet Br14A2 bombers, while the Aviacao Maritima introduced Fairey Campania seaplanes and Short Felixstowe F3 flying boats, and later acquired Fairey III seaplanes. By 1924 the AdA had 25 aircraft, including Martinsyde F4 Buzzard fighters, Caudron GIII and Avro 504K trainers, and its future establishment had been set at three squadrons, one each with fighters, bombers and reconnaissance aircraft. In 1927, the AM had three Fokker TIV reconnaissance-seaplanes, three HS2L and seven CAMS37 flying boats, and five Hanriot H41 training seaplanes. Many of the older aircraft, such as the AdA's Br14A2s, soldiered on while more up-to-date equipment entered service, including 16 licence-built Potez XXV bombers and 20 Vickers Valparaiso reconnaissance aircraft. In the late 1930s, the AdA received Hawker Hinds and 30 Gloster Gladiator fighters, Breda Ba65 ground-attack aircraft and Junkers Ju86K bombers, finally replacing the wartime vintage aircraft. For training, the AdA had 22 de Havilland Tiger Moths, but both services received licence-built Morane-Saulnier MS233 and Avro 626 trainers.

Portugal remained neutral throughout World War II, although retaining cordial relations with the Allies, even after they occupied the Azores, with the RAF using Santa Maria as a maritime-reconnaissance base. The AdA had received a small number of Hawker Hurricane and Supermarine Spitfire I fighters, and to these were added Bell Airacobra and Curtiss Hawk 75A fighters, Bristol Blenheim and Consolidated B-24 Liberator bombers, Miles Master, Magister, Martinet and Airspeed Oxford trainers. The AM received Bristol Beaufort and Blenheim bombers, Short Sunderland flying boats, Grumman G-21 amphibians, and Lockheed Hudson bombers. Post-war, Portugal became member of NATO, but re-equipment was slow, with only 20 Republic F-47D Thunderbolt fighter-bombers being delivered during the late 1940s, with a few Douglas C-47 and C-54 transports.

In 1952, the two air arms were amalgamated to form the autonomous Portuguese Air Force, Forca Aerea Portuguesa. The first jets arrived in 1953, Republic F-84G Thunderjets, finally replacing the Hurricanes (used for training, making Portugal the last country to operate this aircraft) and Spitfires, with Lockheed T-33A jet trainers and DHC-1 Chipmunk basic trainers. The Royal Navy provided 15 North American T-6 Harvard trainers as a gift. The FAP also introduced Boeing SB-17G Fortress and Douglas B-26 Invader bombers and Lockheed PV-2 Harpoon MR aircraft, Grumman SA-16A Albatross SAR amphibians, and Alouette II and Sikorsky H-19A Chickasaw helicopters. North American F-86F Sabre fighters later joined the F-84Gs, while Lockheed P-2E Neptunes were introduced for MR. Other acquisitions were 36 Fiat G91R-4 fighter-bombers and 25 Dornier Do27 liaison aircraft.

Despite its continued membership of NATO, the 1960s and 1970s found Portugal facing difficulty in defence procurement because of international hostility to her African colonies, where the FAP provided support for ground forces. The international community failed to act when India occupied Goa and Indonesia annexed East Timor. The situation eased considerably after both Angola and Mozambique were granted independence in 1975.

The loss of the colonial commitments and the general reduction in defence budgets in the 1990s has meant that the FAP has seen its personnel numbers fall from 17,500 to 7,400 over the past three decades. Nevertheless, re-equipment and modernisa-

BELOW: *LTV A-7 Corsairs provided the mainstay of the Portuguese Air Force's strike capability until the mid-1990s. (Vought)*

ABOVE: *The Portuguese Navy's helicopter flight was the first European operator of the Westland Super Lynx, with five Mk95s for operations from Vasco da Gama-class frigates. More are likely to be ordered. (GKN Westland)*

tion has taken place, helped by the grant of a renewed lease of the Azores base to the United States in 1989. This resulted in the sale to Portugal of Lockheed F-16s for the interceptor and attack role, joining six refurbished ex-USN Lockheed P-3P Orions for MR duties delivered in 1988, and LTV A-7 Corsair IIs used in the strike role. The arrival of the Orions released Lockheed C-130H Hercules from MR for their transport role. The Luftwaffe transferred surplus Alphajets during the early 1990s to replace the earlier Lockheed T-33A and de Havilland Vampire jet trainers. The first F-16s arrived in 1994, and a second batch was delivered in 2001.

Today, the FAP has been upgrading its F-16s under the European Mid-Life Upgrade Programme, and is replacing its armed Alphajets with the second squadron of F-16s, making a total of 33 F-16As and seven F-16Bs. The six Lockheed P-3P Orions are also enjoying refurbishment and systems upgrades to match those of other NATO operators. An army air arm is being created, and this will take over many, if not all, of the 12 Cessna FTB337G used for army co-operation, and possibly some of the 17 SE3160 Alouette III and ten SA330C Puma helicopters. Transport uses three Lockheed C-130H and three C-130-30 Hercules, which also undertake some SAR work with the Pumas and Orions, as well as the 24 CASA C212-100/300 Aviocar, also used for transport, SAR, ECM and EEZ duties. A Dassault Falcon 20 is used for calibration duties, with three Falcon 50s for VIP use. Training still uses seven DHC-1 Chipmunks, but much training is conducted in the United States. Fourteen EH101s have been ordered, with two more on option, for SAR and EEZ work.

PORTUGUESE NAVAL AVIATION/MARINHA

Formed: 1993

Portugal re-established a naval air arm in 1993 with the delivery of five Westland Lynx Mk95 helicopters to form the Naval Helicopter Squadron, Esquadrilha de Helicopteros de Marinha, to operate from three Vasco da Gama-class (MEKO 200) frigates. Up to eight additional helicopters are likely to be ordered.

PORTUGUESE ARMY AVIATION

Formed: 2001

Portuguese army aviation returned in 2001 with the delivery of nine Eurocopter EC635s, and both the French and German armies are providing assistance in training and integration of the helicopters into the army. It may take over the FAP's FTB337Gs. Ten NH90TTHs are on order.

Qatar

POPULATION: 610,000

LAND AREA: 4,000 square miles, 10,360 sq km

GDP: $12.4bn (£8.7bn), per capita $23,800 (£16,643)

DEFENCE EXPENDITURE: $1.4bn (£979m)

SERVICE PERSONNEL: 12,330 active.

QATAR EMIRI AIR FORCE

Qatar has a small but potent air force, initially built around Alphajet armed-trainers and Dassault Mirage F1s, with Westland Commando (S-61) and Eurocopter Puma transports. The Qatar Emiri Air Force was active during the Gulf War in 1991 for the liberation of Kuwait, in addition to providing base facilities for other Coalition air forces. The Mirage F1s were sold to Spain in 1997 following the delivery of Mirage 2000-5s multi-role fighters. The QEAF may buy BAe Hawk trainers, or follow Bahrain in buying a flying academy service from the UK.

Current strength is 2,100 personnel. One squadron operates nine Mirage 2000-5EDA and three 2000-5DDA, with another operating six Alphajets. An attack helicopter squadron operates ten Eurocopter SA342L Gazelles equipped with HOT anti-tank missiles and another operates in the anti-surface vessel role with eight Westland Commando Mk3 (S-61) equipped with Exocet ASuW missiles. Three Commando helicopters provide transport, with a fourth for VIP duties alongside an Airbus A340-200, a Boeing 707-320, a 727-200 and a 747SP, and a Dassault Falcon 900.

Romania

POPULATION: 22.2 million

LAND AREA: 91,671 square miles, 237,428 sq km

GDP: $38.4bn (£26.9bn), per capita $4,583 (£3,204)

DEFENCE EXPENDITURE: $995m (£695.8m)

SERVICE PERSONNEL: 103,000 (33% conscript) active, plus 470,000 reserves.

ROMANIAN AIR FORCE

Formed: 1945

The Romanian Army formed a Flying Corps in 1910. By late 1911, it was operating four Bleriots and four Henri Farmans, to which a small number of Bristols and Morane Type Fs were added the following year, as well as some Nieuports before the outbreak of World War I. The Central Powers soon defeated the Romanian Flying Corps.

Post-war, a Directorate of Army Aviation was established, with a planned strength of three groups, each of three squadrons. The fighter group operated Spad S7C1s, the bomber group de Havilland DH9s and Breguet Br14Bs, with Breguet Br14A2 and Brandenburgs in the reconnaissance group. The three groups had a total of 72 aircraft. During the late 1920s, 70 Armstrong-Whitworth Siskin III and 60 Spad S61C1 fighters, 120 Potez XXV and 30 XXVII reconnaissance aircraft, as well as Breguet Br19B2 bombers and Savoia S59 flying boats, with Morane-Saulnier MS35 trainers, were obtained. Many of the Potez and Morane-Saulnier aircraft were assembled in Romania. The early 1930s saw the Romanian-designed SET XV fighter, SET VIIK reconnaissance aircraft; SET VII and SET X trainers enter service with 50 PZL P-11b fighters and 20 Consolidated Fleet Model 10G trainers. These were followed in 1936 by Miles Hawk and Nighthawk trainers, as well as the Romanian IAR 37, 38 and 39 light bombers. Hawker Hurricane fighters and Bristol Blenheim bombers were bought from the UK in 1939.

Romania supported the Axis Powers during World War II. Germany provided Messerschmitt Bf109E and Heinkel He112B fighters, Heinkel He111H bombers and Junkers Ju87D Stuka dive-bombers, as well as Heinkel He114 reconnaissance seaplanes and Fieseler Fi156C liaison aircraft. Savoia-Marchetti SM79 bombers were built under licence and the Romanian aircraft industry provided the IAP80 fighter. Romanian forces joined Operation Barbarossa, the invasion of the Soviet Union. Later in the war, Germany provided Junkers Ju88A bombers and Henschel Hs129A-O ground-attack aircraft, but by this time the strategic situation had swung against the Axis, and in late August, 1944, Romania was overrun by Soviet forces.

Post-war, the new Romanian Air Force continued to operate its wartime equipment, until the USSR supplied Yakovlev Yak-9 fighters. A 1947 peace treaty restricted Romania to 150 military aircraft and 8,000 air force personnel, but this was swept aside when she joined the Warsaw Pact and started to receive Soviet military aid. The first jets, MiG-15 fighters, arrived in 1953, accompanied by Ilyushin Il-10 ground-attack aircraft, Lisunov Li-2 (C-47) transports, and Yak-11 and Yak-18 trainers. Within a couple of years, these aircraft were joined by MiG-17 fighters and Ilyushin Il-28 jet bombers, as well as the first helicopters, Mil Mi-4s. During the 1960s, MiG-19 and MiG-21 fighters and Antonov An-2 transports also entered service. Later aircraft included Ilyushin Il-12 and Il-14 transports, and Aero L-29 Delfin trainers. By the late 1960s, the FAR had kept to its personnel limit of 8,000, but combat aircraft numbers had reached 250. Continual modernisation

Romania

ABOVE: *The mainstay of the Romanian Air Force's transport force are four elderly Lockheed Martin C-130Bs. (NATO photo)*

of the FAR continued into the late 1980s, with the delivery of MiG-23s and then MiG-29s, An-24, An-26 and An-30m transports, and the inevitable Mi-8 helicopters. Licence-built Britten-Norman BN-2A Islanders were also introduced in the utility role, while the Alouette III helicopter was also built under licence as the IAR-316. A number of Chinese aircraft were also put into service, including the H-5R (Il-28) light bomber.

The FAR was seriously affected by the collapse of the USSR and the end of the Warsaw Pact in April, 1991. By this time, it had 12 air defence regiments flying MiG-21s, MiG-23s and MiG-29s, while two ground-attack regiments operated the Yugoslav and Romanian joint venture IAR-93 Arao. The Romanian aircraft industry provided the IAR-28M training aircraft, and for advanced training, the IAR-99 Soim to replace the L-29 Delfin. The MiG-23s were retired, with the MiG-21s and MiG-29s upgraded. Transport aircraft delivered in recent years have included ex-USAF Lockheed C-130B Hercules, and additional aircraft will be obtained, once the economy allows, to replace the An-24 and An-26.

The FAR has 18,900 personnel, and although it maintains a substantial number of aircraft on paper, with an annual average flying hours per pilot of just 40, these spend little time in the air! Fighter aircraft consist of a single air regiment with 15 MiG-29A and three MiG-29UB, while in the ground-attack role there are four regiments with 73 IAR93B, five with 182 MiG-21, of which 110 have been upgraded to MiG-21 Lancer A/C standard, including 10 Lancer B two-seat conversion trainers. In the reconnaissance role, there are still a number of Il-28 and H-5SC. There up to 100 IAR316 Alouette III on attack and utility duties, although the fomer probably uses just 16 helicopters. Transport is provided by four elderly Lockheed C-130B Hercules, 17 An-2 light transports and three An-30. There are 80 Romanian-built IAR330H/L Puma helicopters in the tactical transport role, along with 24 Mi-8 and two Mi-17. Three SA365N Dauphin provide VIP transport. Training uses 45 Aero L-29 and 21 L-39, up to 40 IAR-99 Soim, many of which have been upgraded, 23 Yak-52, and MiG conversion trainers.

ROMANIAN NAVAL AVIATION

The Romanian Navy has a small number of helicopters, mainly shore-based but also capable of operating from a destroyer. The aircraft consist of six IAR330 Puma ASW helicopters and five IAR316 Alouette III, with the latter capable of shipboard use.

Russia

POPULATION: 146.7 million

LAND AREA: 6,501,500 square miles, 16,838,885 sq km

GDP: $1,200bn (£839bn), per capita $7,600 (£5,314)

DEFENCE EXPENDITURE: $60bn (£42bn)

SERVICE PERSONNEL: 977,100 (33% conscript) active, plus 20,000,000 reserves.

ABOVE: *The Yakovlev Yak-1 was one of the aircraft used against invading German forces during World War II. (Yakovlev)*

RUSSIAN MILITARY AVIATION

Formed: 1924

The structure of Russian military aviation remains unique, a continuation of the structure used in the former Soviet Union, with separate air forces for long-range aviation, frontal aviation (as opposed to the army's own air elements), air defence and military air transport. These are in addition to the distinct elements of naval aviation, army aviation and border guards aviation.

The Imperial Russian Army had been using balloons for observation purposes since the turn of the century when a Central Aviation School was established in 1910. The Imperial Russian Navy also established a flying school later that same year. In 1912, the Imperial Government bought small quantities of British and French aircraft for the army, and Curtiss flying boats for the navy. Licences were obtained for production of these aircraft so that on the outbreak of war in 1914, the Imperial Russian Flying Corps had in service 244 aircraft of Bristol, Farman, Morane and Nieuport design. The Imperial Russian Navy also had a substantial number of aircraft. The Russian aircraft industry mainly comprised of firms with other interests, such as the Russo-Baltic Wagon Factory, and was incapable of meeting wartime demand. Some indigenous designs were put into production in small batches during 1915, notably Igor Sikorsky's RBVZS-16, RBVZS-17 and RBVZS-20 single and twin-seat fighters. Later, other types entered production, including the Lebed-7 and Lebed-10 fighters, and Lebed-212, *Anasal* and *Anade* reconnaissance aircraft for the IRFC, while Dmitrii Grigorovich produced the M-5 and M-9 flying boats for the navy. Most ambitious was Sikorsky's *Ilya Mourametz*, or 'Giant', a four-engined heavy bomber originally developed as an 'airbus' from his earlier twin-engined *Bolschoi*, or 'Grand', of 1913. By late 1914, sufficient of these aircraft were available to form the 'Squadron of Aerial Ships', *Eskadrilya Vozdushnykh Korablei*, in Poland, and in February, 1915, one of the squadron's aircraft dropped a 600-lb bomb on German forces near Plotsk. A total of 72 Giants were built, and were able to slow the German advance.

While Russian aircraft production reached almost 1,800 aircraft and some 660 engines during 1916, foreign designs predominated, including Morane MB and Parasol, Nieuport 11, 17 and 21, Spad S7, Sopwith 1 1/2-Strutter, BE2E and RE8 fighters, Voisin, DH4 and DH9 bombers, and Farman trainers. Most of the aircraft in IRFC service were French, but in 1916, Britain supplied more than 250 aircraft. War on the Eastern Front differed from that on the Western Front in having relatively little fighter activity, and the IRFC found itself mainly engaged in reconnaissance and bombing, with some close-support strafing.

Probably the most loyal of the Russian armed forces, the IRFC continued to function throughout the initial period of transition from the Tsarist regime to the Provisional Government. The rapid deterioration in the situation after the revolution in November, 1917, meant that many aircraft fell into the hands of revolutionaries. In December, 1917, the revolutionaries started to establish their own aircraft and balloon units, leading to the formation of the Workers' and Peasants' Red Air Fleet in May, 1918, retaining the title of Red Air Fleet until 1924, although officially becoming part of the Army, Aviadarm, in the meantime. By the end of the revolution and the Russian Civil War, the Red Air Fleet had some 300 aircraft, and over the next few years substantial numbers of foreign aircraft were ordered, including Fokker DXIII fighters and CIV reconnaissance aircraft, and a number of Ansaldo types. DH9A bombers were copied and produced in Russia as the R1. Several German designs were produced as the Treaty of Versailles had banned military aircraft production in Germany.

The Soviet Military Aviation Forces, Sovietskaya Voenno-Vozdushnye Sily, were formed in 1924, absorbing the Red Air Fleet. The first post-revolution Soviet designs included the Polikarpov I-1 and I-2 and Grigorovich I-1 and I-2 fighters. The confusing designations arising because the 'I' prefix denoted a fighter design. It became Soviet practice to assign designers' names rather than those of the manufacturer to aircraft. Other aircraft included the Tupolev TB-1 bomber, R-3 and R-6 reconnaissance aircraft, and G-1 transport. These were followed by the Polikarpov I-3 and Tupolev I-4, Polikarpov-Grigorovich I-5, Grigorovich I-6 and DI-3 fighters; Polikarpov TB-2 bomber, R-5 reconnaissance aircraft and V-2 training aircraft; and Tupolev G-2 transport. By 1930, the SV-VS had 1,000 aircraft in 20 air regiments, and had experienced combat during border clashes with China in 1929.

ABOVE: *A wartime Soviet bomber was the Tupolev SB-2, seen here on a typical Russian airfield. (Novosti)*

The 1930s saw large orders for heavy bombers and the development of mass drops by paratroops, using Tupolev's G-2 transport. Heavy bombers included the Tupolev TB-2 and the Kalinin K-7, while Tupolev also designed the SCh-1 and SCh-2 ground-attack aircraft, the SB-2 light bomber and the I-14 fighter. Ilyushin designed the TB-3 bomber, and Polikarpov the I-15, I-15B and I-16 fighters.

Despite the emphasis on heavy bombers, Soviet military doctrine was similar to that of the Germans, with air power used in close support of ground forces rather than strategically, as with British and American forces.

In 1936, SV-VS personnel and aircraft were sent to Spain to fight on the Republican side during the Spanish Civil War, with some 1,400 of the latest Soviet fighter and bomber aircraft deployed in Spain, many of them flown by Spanish pilots. Their performance has since been claimed to have been as good as that of the German and Italian opposition flying with the Nationalist side, although the Republicans lost the war! The explanation could be that the first generation of German aircraft deployed in Spain were soon shown to be inferior, but the second generation proved to be far more effective. This period was also marked by the Soviet dictator Stalin's purges, with Andrei Tupolev imprisoned and his ANT-42 heavy bomber design, which entered service as the TB-7, later passing to Petlyakov and becoming the Pe-8, while Tupolev's light bomber design, the SB-2, became the Pe-2.

Warfare flared up in 1938 and 1939, fighting Japanese forces and eventually pushing them out of Mongolia with heavy Soviet casualties. This conflict used the same aircraft that had been sent to Spain. In addition, from 1937 onwards, large numbers of Soviet aircraft were provided for Chinese Communist forces, often with 'volunteer' pilots and ground crew. The SV-VS had by this time grown to 6,000 aircraft, but its efficiency was affected, in line with the rest of the Soviet forces, by the regular purging of senior officers on political grounds, with subsequent imprisonment and often execution. In common with the rest of the Soviet forces, during the Communist period each unit had a political commissar as well as the more usual commanding officer. Examples of US and German aircraft were obtained during this period to gain an insight into rapidly changing aviation technology, and this was aided further in 1938 by a Russo-German agreement which provided a number of Messerchmitt, Dornier and Heinkel aircraft. The Polikarpov I-17 fighter was being introduced in limited numbers when war started in 1939, and was accompanied by ground-attack and dive-bombing aircraft, including the Polikarpov VIT-1 and VIT-2, the Ilyushin Il-2, the Archangelski Ar-2 and the Tupolev R-10 reconnaissance-bomber.

A Russo-German pact allowed Soviet forces to invade Eastern Poland in September, 1939, after German forces had all but crushed Polish resistance. Later that year, Soviet forces invaded Finland, which had seized its independence during the revolution and civil war. Determined Finnish resistance and poor Soviet planning meant that the war ended in 1940 with Finland making territorial concessions but retaining her independence, despite the fact that 2,000 Soviet aircraft had been deployed against Finland's numerically inferior forces. New aircraft types continued to enter service, including Lavochkin I-22 and Yakovlev I-26 (which became the Yak-1 shortly afterwards when designations were changed to reflect the design bureau rather than the aircraft's role), Yatsenko I-28, Mikoyan I-61 (MiG-1) and I-200 (MiG-3) fighters; Sukhoi Su-1 strike aircraft; Ilyushin DB-3 (Il-4), Petlyakov Pe-2 and Pe-8, and Yermolaev Yer-2 and Yer-6 bombers; Antonov A-7 transport and SS-2 liaison aircraft.

The number of aircraft actually entering

service remained small. By 1941, the Soviet armed forces had some 18,000 aircraft, but just a fifth of these could be regarded as modern. Attempts to boost production were hampered both by the Soviet system and by the very necessary movement of the key factories eastwards, out of the range of German medium bombers, in anticipation of a possible war with Germany. In June, 1941, Germany launched the invasion of the Soviet Union, Operation Barbarossa, using the well-tried *blitzkrieg* strategy of combining air power with fast moving armoured formations. Half of the USSR's aircraft were deployed in the west, in anticipation of a German attack, and against these 9,000 aircraft, the Luftwaffe had 1,945 aircraft, joined by another 1,000 aircraft from Germany's allies, mainly Italy, Hungary and Romania, as well as some from Finland and Croatia. The Luftwaffe caught most of the Soviet airfields by surprise, striking at the 66 airfields that accommodated 70 per cent of Soviet air power in the West, although as many as 50 per cent of the Soviet aircraft are believed to have been non-operational. One Soviet commander committed suicide after losing 600 aircraft without making any impact on the invading German forces. Individual pilots tried desperate measures, even ramming German bombers by inserting their fighter's propeller into the elevators.

ABOVE: *The Yak-1 was followed by the Yak-3. (Yakovlev)*

Only a determined strategic heavy bombing campaign backed by strong and well deployed fighter defences, plus tactical aircraft such as dive-bombers and tank-busting rocket-firing aircraft could have stopped the invasion. Instead, it was the severity of the Russian winter and by the over-stretched logistics of the Germans that proved to be the best defence. German technical superiority meant that it took massive American and British military aid to revive the SV-VS. Even here, inefficiency ruled, with fighters delivered at great cost in the infamous Arctic Convoys being left idle on airfields rather than deployed quickly to operational units. The USA supplied Bell P-39N Airacobra and P-63C Kingcobra, Curtiss P-40 and Republic F-47 Thunderbolt fighter-bombers; Douglas A-20 ground-attack aircraft; North American B-25 Mitchell bombers and AT-6 trainers; with Consolidated PBY-5A Catalina amphibians: and Douglas C-47 transports, which were also built in the USSR as the Lisunov Li-2. The UK provided Hawker Hurricane and Supermarine Spitfire fighters; de Havilland Mosquito fighter-bombers; and Armstrong-Whitworth Albermarle AOP aircraft. Between them, the two Allied powers supplied some 15,000 aircraft, helping the USSR to go on to the counter-attack and carry the offensive to Berlin in 1945.

It was not until late in the war that new and improved Soviet aircraft entered service. These included the Yakovlev Yak-3 and Yak-9, Lavochkin La-7 and La-9, and Mikoyan MiG-5 fighters; Tupolev Tu-2 bombers and Ilyushin Il-10 ground-attack aircraft. Despite heavy losses, the SV-VS had 20,000 aircraft when peace returned. Despite earlier conflict with Japan in Mongolia, the USSR did not declare war on Japan until August, 1945, and the SV-VS played no part in Japan's defeat.

BELOW: *Using captured German designs, Yakovlev produced the Yak-15 jet fighter. (Yakovlev)*

Captured German airframes and engines were shipped to the Soviet Union, and the Yakovlev Yak-15 and Mikoyan MiG-9 jet fighters designed in great haste, and followed by the Yakovlev Yak-17. It was not until a number of Rolls-Royce Nene and Derwent jet engines were purchased from the UK, and copied in the USSR, that the famous MiG-15 fighter could be designed, based on captured German airframe plans. Boeing B-29 Superfortresses, that had force-landed on Soviet territory in the mistaken belief that this was still friendly, were seized and copied, entering production and service as the Tupolev Tu-4. Other, less well known and less successful, aircraft of this period included the Yakovlev Yak-23 and Lavochkin La-15 jet fighters, and the Tupolev Tu-77 twin-jet attack aircraft, from which the Tu-12 and Tu-14, and eventually the Ilyushin Il-28, jet bombers were developed. The famous Antonov An-2 biplane utility transport first flew in 1947.

An opportunity arose for the Soviet Union to test its new aircraft with the outbreak of the Korean War in 1950. With covert Soviet and open Chinese support, the North Koreans had invaded South Korea. The USSR was boycotting the UN Security Council at the time, enabling the United Nations to muster an international force, but with the United States providing the largest contingent, to drive the North Korean forces back. The MiG-15 was matched against British Gloster Meteor and North American F-86 Sabre jets, as well as North American F-51D Mustangs, Fairey Fireflies and Hawker Sea Furies, with a Sea Fury accounting for a MiG-15. The MiG-15 was to survive in production as the MiG-15UTI, initially meant as a conversion trainer but in fact an advanced jet trainer, when its deficiencies led to its replacement in service by the MiG-17 from 1952 onwards. Deliveries of aircraft to satellite countries, those supposedly liberated by Soviet forces during the closing stages of the war, resumed, and enabled the Russian aircraft industry to mass produce even the most mundane designs. Some countries, notably Poland and China, were allowed to build Soviet types under licence. The less important client states, especially those in Africa, received retired SV-VS aircraft.

ABOVE: *The Yak-17 was clearly a development of the Yak-15. (Yakovlev)*

A Sukhoi experimental aircraft was the first Soviet aircraft to break the sound barrier. In 1955, the supersonic MiG-19 fighter was introduced. This was joined in the SV-VS by the Yakovlev Yak-25 night-fighter; the Tupolev Tu-16, the world's only long-range

BELOW: *MiG-15 'Fagot' preserved in Champlin Museum. (Jeremy Flack/Aviation Photographs International)*

ABOVE: *Mi-1 'Hare'. (Jeremy Flack/Aviation Photographs International)*

turboprop heavy bomber, and the Tu-20 jet bomber; Ilyushin Il-12 and Il-14 transports. Helicopters were also introduced at an early date, with the Mil Mi-1 helicopter of 1950 and the Mi-4 of 1951; the Yak-4 twin-rotor helicopter of 1954, the Mi-6 heavy lift helicopter of 1958, and later by the Mi-8 and Mi-10 helicopters. Heavy bombers included the Myasishchev Mya-4, introduced in 1955. Antonov produced the An-12 transport. The MiG-21 interceptor first entered service in 1958, at the same time as the Sukhoi Su-7 ground-attack aircraft, with both types rapidly becoming standard equipment for the air forces of the Warsaw Pact, while the Su-9 was an interceptor development of the Su-7.

The steady flow of new aircraft into service, often unannounced and with their true designations not always immediately apparent, led NATO to resurrect the system of code-names, first used during war to identify Japanese aircraft in the Pacific. Under this scheme, fighters and fighter-bombers were allocated code-names beginning with 'F'; bombers gained code-names beginning with 'B'; maritime-reconnaissance aircraft began with 'M'; transport code-names began with 'C'; and helicopters with 'H'. Sub-types were given a suffix after the code-name, usually indicating an improved version of an original aircraft. This meant that the MiG-15 was identified as 'Fagot'. Some improved and special versions of an aircraft retained the original designation, as with the Mi-8 and the Mi-17, and the Mi-24 and Mi-35.

Although there may be doubts about the serviceability of many Soviet aircraft, and especially the transports, the vast resources put into defence production at this time ensured that NATO forces were always outnumbered. In 1968, the Mach 3 MiG-23 interceptor started to enter service, at about the same time as the supersonic Tu-22 rear-engined bomber. The Soviet regime also ensured that the largest transport aircraft were provided and the helicopters with the heaviest lift, while the Mi-24 Hind attack helicopter set a standard for other nations to follow. There were a number of idiosyncrasies in both aircraft design and in operations. Many Soviet aircraft, including transport aircraft, retained the tail-gunner's position long after this disappeared from western bombers. There was the continued interest in the flying boat and amphibian, long after these were almost abandoned in the west. The Mi-24 differed from other attack helicopters through having space for passengers or freight, even though this made the machine heavier and larger, increasing its radar signature and reducing manoeuvrability. Operationally, under the Soviet regime the functions of the military air transport fleet and that of the state-owned airline monopoly Aeroflot were often overlapping, with the airline organized on military lines. The structure evolved differently from that used elsewhere, with Soviet air power divided between the air force, long-range aviation, *Dalnaya Aviatsiya,* frontal aviation, *Frontovaya Aviatsiya,* air defence aviation, and military transport aviation, *Voennaya Transportnaya Aviatsiya,* as well as the more usual separation of naval, army and paramilitary border guard aviation. These divisions survived the collapse of the Soviet Union, although in 1998 the anti-aircraft defence force, PVO, was merged into the air force, or VVS.

For the most part, the Soviets used air power sparingly in suppressing dissent in the European satellite territories and other than for transporting their troops, preferring to use tanks and infantry. Soviet intervention in the internal affairs of Czechoslovakia - the so-called 'Prague Spring' - in 1968 were a case in point, with air power including the air-landing of troops at Prague, having deceived the local air traffic controllers into believing that these were civilian transport aircraft. The Cold War period saw massive elements of Soviet air power stationed at forward bases in East Germany and Czechoslovakia, and in northern Russia, with long-range aircraft used to probe western air defences and to monitor NATO naval exercises. As the early post-war alliance with Communist China fell apart, protection of the border also fell to Soviet air power, while a presence was maintained in Siberia to balance US forces in Japan. The use of client states and support for guerrilla movements in Africa, Asia and Latin America, by-passed direct confrontation between the USSR and its satellites and the network of Western alliances, of which NATO was the most important, largest and most organized, and most successful.

The big exception to these policies arose

in Afghanistan in late 1979. Civil war in Afghanistan threatened to destablize neighbouring Soviet states with substantial Moslem communities. On 24 December, 1979, an advance party of the 105th Guards Airborne Division was flown into Kabul, followed by the main party over the next two days, with 280 missions by Antonov An-12 and An-22 and Ilyushin Il-76 transports to land some 5,000 troops. In the heavy fighting that followed, close air support was provided by MiG-21 and MiG-23 fighter-bombers operating from bases in the Soviet Union, while later the Mi-24 Hind helicopter came to prominence for the first time in frequent battles with Afghan guerrilla groups. Helicopters proved invaluable not only in providing heavy fire-power quickly whenever and wherever it was needed, but in reducing the time and the number of Soviet casualties taken in seizing high ground. Nevertheless, even with modern weapons, Afghanistan proved to be difficult and costly to control, and eventually the USSR was forced to withdraw its forces, speeded by the collapse of the Soviet state.

ABOVE: *The long range bomber Tu-160 'Blackjack (Jane's)*

Meanwhile, continued technical progress saw the Tu-95 bomber enter service, armed with the AS-15 cruise missile, and later these were joined by a very small number of Tu-160 Blackjacks. The Ilyushin Il-76 transport also boosted Soviet capability, and was joined by the giant An-124, capable of carrying loads of up to 150 tons and originally intended to transport ballistic missiles. The advent of the nuclear submarine capable of carrying ballistic or cruise missiles did not undermine the maintenance of strong

BELOW: *Russian Tu-95 'Bear'. (Jeremy Flack/Aviation Photographs International)*

bomber forces during the Soviet era. Throughout this period the USSR provided every form of offensive weaponry in contrast to almost all other states which were forced by economic conditions to choose between the means of strategic warfare.

The collapse of the Soviet Union in 1990 was a triple blow to the Soviet Air Force. It lost manpower and bases as states broke away and proclaimed independence. It lost aircraft, as many were abandoned in airfields in the newly independent countries. Finally, the economic problems that followed during the switch from a Soviet style command economy to a capitalist market economy saw funding cut. The sudden interruption to the hitherto steady supply of new and more sophisticated aircraft meant that the average age started to increase, a situation that was soon aggravated by difficulties in the supply of spare parts. Within ten years, only a fifth of the aircraft could truly be described as modern, while total flying hours fell by 90 per cent from around 2 million annually to around 200,000. Amongst the few bright spots in this situation was an agreement with the Ukraine that resulted in the return of eight Tu-160 bombers during 1999 and 2000, enabling the fleet to grow to just 15 aircraft, while the Ukraine also returned a number of Tu-95s.

Today, the Russian Air Forces, in common with the rest of the Russian armed forces, are at a low level of operational readiness. Russian intervention in Kosovo in 1999 was aided by the movement of troops by air, but once in the area, logistics support fell apart in circumstances where an air re-supply operation was needed. Combat aircraft, mainly Su-24s, are believed to have been used in operations against rebel forces in Chechnya during 1999-2001. The total number of personnel for all of Russia's Air Forces is around 184,600, excluding naval and army aviation. Average annual flying hours are extremely low at 20 hours for pilots in long-range aviation and air defence, and even the military transport units, or VTA, only manage 44 hours, a point at which safety is compromised.

The Long-Range Aviation Command is also known as the 37th Air Army, and consists of two operational divisions and a combat conversion unit. There are 15 Tu-160MS, and around 68 Tu-95s. There are 158 Tu-22M/MR, of which almost 100 are in store, and some Tu-22M3 may be upgraded to provide an interim sub-strategic weapons platform until a new bomber can be put into production at some time after 2005. This force is supported by 20 Il-78 tankers. Training uses ten Tu-22M-2 and up to 30 Tu-134.

Air Defence Aviation has recently been split from Tactical Aviation. It has almost 900 fighters and some 2,000 surface-to-air guided missile launchers. Many of the aircraft are in store, but there are around 300 MiG-31s and 200 Su-27s, equipped with AA-8, AA-10 and AA-11 AAM, while the MiG-25 force is believed to be entirely in storage. Upgrades are planned to the MiG-31s, but the date for this appears to be receding. The force also includes 16 Beriev A-50 AEW aircraft.

Frontal Aviation, the other component of Tactical Aviation is based on a series of

BELOW: *Russian Frontal Aviation is based around the MiG-29.*

tactical and air defence armies, with the usual allocation being one air army per military district, although the Volga and Ural military districts have just an air corps each. Retirement of MiG-23, some MiG-25 and MiG-27 aircraft as well as Su-17 and Su-22 fighter-bombers have affected the operational capability, which is now based around the MiG-29, of which some 260 are believed to be in service, and up to 120 Su-24MR reconnaissance aircraft and 225 Su-25 fighters. The remaining 90 or MiG-25s are divided equally between MiG-25R/RB high altitude-reconnaissance aircraft and Kh-58 anti-radar missile-armed MiG-25BM defence suppression aircraft. Small numbers of Su-27IB and Su-30 are being introduced slowly, and the first operational units are unlikely to be formed until 2005. Around 450 Su-24s and 180 Su-25s are also used in the fighter-bomber role.

Both Air Defence and Frontal Aviation should benefit from the selection of Sukhoi Su-49 two-seaters for pilot screening and the maintenance of combat skills.

ABOVE: *Many of Russia's MiG-25s are now being retired.*

Military Transport Aviation, the VTA or *Voennaya Transportnaya Aviatsiya*, is currently replacing its fleet of Antonov An-12 transports with the new An-70, although it is highly unlikely that funding will allow all 250 to be replaced on a one-for-one basis. The promised doubling of defence procurement expenditure in 2000 will undoubtedly help, but there are many projects competing even for this substantial increase in funds, up from 31 billion roubles to 62 billion, approximately US$2 billion. The VTA is organised in two divisions as the 61st Air Army, with each division having four regiments, while there is also an independent regiment. The An-12 fleet is accompanied by Ilyushin Il-76M/MD, of which there may be as many as 250, as well as 45 An-22 and 25 An-124. An assortment of other types is also in service, while the VTA can still be reinforced by commercial aircraft, with some 1,500 aircraft available.

BELOW: *The Sukhoi Su-27 is one of the two main aircraft types in Air Defence Aviation, the other being the MiG-31.*

Russia

Training uses around 1,150 aircraft, including L-29 and L-39, and conversion training versions of frontline types.

INDEPENDENT NAVAL AVIATION

The Imperial Russian Navy has a long history of involvement with aviation, having started in 1910. In 1912, Curtiss flying boats were purchased from the United States, and at the start of World War I there was a force of around 60 flying boats. Grigorovich M-5 and M-9 flying boats entered service during the war, but operations were hampered by units of the Imperial Russian Navy being amongst the first to join the Russian Revolution in 1917.

After the revolution, naval and army aviation were merged into the air force. During the early 1920s, Tupolev TB-1P seaplane bombers entered service, and these were followed by Tupolev MDR-2 long-range flying boats and Beriev MBR-2 short-range flying boats. Shortly before the outbreak of World War II, the six-engined Tupolev MK-1 flying boat entered service. An independent naval air fleet was created on the eve of the outbreak of war. Although no aircraft carriers were obtained, shore-based fighter and bomber units existed until 1960, when these units were transferred back to air force control. The Soviet Navy was the forgotten service for much of its history, and it was not until the late 1950s and early 1960s that it was given the resources to gain a world role. Two Moskva-class helicopter cruisers were introduced in 1968 and 1969, each capable of carrying up to 18 anti-submarine helicopters. These were followed by the first of three Kiev-class aircraft carriers in 1976, each capable of carrying up to 35 helicopters and V/STOL aircraft, the new Yakovlev Yak-25 'Forger A' strike aircraft. A modified Kiev-class carrier, *Admiral Goshkov*,

ABOVE: *Pilots run to their Yakovlev Yak-38s, known as 'Forger' to NATO, aboard a Russian aircraft carrier. The collapse of the Soviet Union has seen these aircraft withdrawn. (Yakovlev)*

BELOW: *Tu-142M 'Bear'.(Jeremy Flack/Aviation Photographs International)*

followed, but the Yak-25s were withdrawn in 1992, leaving the carrier to operate only helicopters. The first aircraft carrier with an angled flight deck, *Admiral Kuznetsov*, has since joined the fleet and can operate up to 18 Sukhoi Su-33 fighters, Russia's most modern fighter, and a navalised version of the Su-27K as well as Su-25 ground-attack aircraft, and 15 Kamov Ka-27 ASW and Ka-31 AEW helicopters. Despite the angled flight deck, this ship still has a ski jump.

Today, naval aviation accounts for around 35,000 of the Russian Navy's estimated 171,500 personnel. Only one carrier remains operational, but cruisers and destroyers carry helicopters, while there are extensive shore-based aircraft to support the fleet, with a total of around 330 fixed-wing aircraft and 400 helicopters. Despite the carrier, the main strike capability remains shore-based, with around 160 Tu-22M3s, augmented by Su-24MPs for short-range strike duties. Long range maritime-reconnaissance used Tu-142s, which also undertake ELINT duties, along with Il-18s, Il-20s and An-12s. Additional ASW aircraft include Ilyushin Il-38 'May'. Beriev Be-12 flying boats provide ASW and SAR cover. While in addition to the Ka-31s, other helicopters include Mi-14s for mine-countermeasures and SAR.

ABOVE: *Russian naval Ilyushin Il-38 May (Jeremy Flack/Aviation Photographs International)*

RUSSIAN ARMY AVIATION

Russian Army Aviation developed after World War II with helicopters closely integrated with ground forces. The force has almost 3,000 helicopters available, but of these 600 are believed to be in store. Modernisation is also proving difficult, with the force relying heavily on the Mil Mi-24, with around 900 available in the attack role and another 140 for reconnaissance duties, while there are around 1,000 Mi-8 and Mi-17 transport helicopters, and a small number of heavy-lift Mi-26s. Only a handful of Kamov Ka-50 attack helicopters have been delivered so far, due to funding difficulties. As an interim measure, Mi-24s are being upgraded to enhance the type's night attack capability.

Independent of the army, the border guard has its own aviation element, with some 200 Ka-27, Mi-8, Mi-17 and Mi-24 helicopters, as well as SM92 surveillance aircraft and 70 transport aircraft, including An-24, An-26, An-72, Il-76, Tu-134 and Yak-40 types.

Rwanda

POPULATION: 8.8 million

LAND AREA: 10,169 square miles, 26,338 sq km

GDP: $2.4bn(£1.7bn), per capita $627 (£438)

DEFENCE EXPENDITURE: $125m (£87m)

SERVICE PERSONNEL: c.55-90, 000 - no accurate figures available.

RWANDA AIR FORCE/FORCE AERIENNE RWANDAISE

Formed: 1962

Rwanda was a German colony until 1918, then passed into Belgian administration until 1962. The Force Aerienne Rwandaise, was established on independence, mainly for transport and liaison duties, but later attempted to buy Aermacchi MB326 armed trainers in 1972, but had to cancel the order because of funding difficulties. An order for just three CM170 Magisters had to be cancelled three years later. Eventually two Rallye 235 Guerriers were fitted with machine guns for COIN operations. Civil war during the early 1990s reached a peak in 1994, when almost all of the FAR's aircraft were destroyed. Afterwards, two Mi-24 attack helicopters were acquired from Belarus in 1997, and in 1999 Russia provided four Mil Mi-17MD transport helicopters. There have been reports that Belarus might supply MiG-21s, but there is no training or support infrastructure. A Britten-Norman BN-2A Islander may still be operational. Personnel numbers are unlikely to exceed 1000.

Saudi Arabia

POPULATION: 22.2 million

LAND AREA: 927,000 square miles, 2,400,930 sq km

GDP: $185bn (£129.4bn), per capita $10,100 (£7,063)

DEFENCE EXPENDITURE: $18.7bn (£13.1bn)

SERVICE PERSONNEL: 201,500 (inc 75,000 National Guard) active.

ROYAL SAUDI AIR FORCE

Formed: 1950

Saudi Arabia came into existence with the union of the Nejd and the Hejaz in 1926. The first Saudi military aircraft were a gift from the British government to assist in suppressing dissident tribesmen. These aircraft were followed in 1931 by four ex-RAF Westland Wapiti general-purpose biplanes, operated by British aircrew. Italian military assistance was received in 1937, but World War II intervened and there was little further development for the next decade. The arrival of a British mission in 1950, saw fresh progress, organising the new Royal Saudi Air Force and providing Avro Anson light transports and de Havilland Tiger Moth trainers. Further development followed in 1952 when an American mission provided ten Temco TE-1A Buckaroo and a number of North American T-6 trainers, as well as Douglas C-47 transports. In 1953, 18 Saudi pilots were given further training in the UK by Airwork and Air Services Training, while a Saudi-Egyptian Defence Agreement saw a number of pilots trained in Egypt, which also supplied four de Havilland Vampire FB52 fighter-bombers before the agreement ended in 1957. The RSAF had meanwhile acquired nine Douglas B-26 Invader bombers and a number of DHC-1 Chipmunk basic trainers. During the late 1950s and early 1960s, 16 North American F-86F Sabre fighter-bombers were introduced, accompanied by Lockheed T-33 and Beech T-34 Mentor trainers, six Fairchild C-123 Provider and four Lockheed C-130E Hercules transports. A Westland Widgeon helicopter and a Vickers Varsity provided VIP transport.

ABOVE: *The Royal Saudi Air Force uses BAe Hawks for advanced training. (BAe)*

Growing tension in the Middle East, with Saudi Arabia taking a different line from Iraq, Egypt and Syria, resulted in the creation of strong Saudi forces during the 1960s. In 1966, 34 BAC Lightning F53 multi-role aircraft, were ordered, with six Lightning T55 and 24 BAC167 Strikemaster trainers. This marked the start of Saudi procurement of modern aircraft. The Lightnings were followed by Northrop F-5s for attack duties, while during the late 1980s and early 1990s, Saudi Arabia became one of the few export customers for the Anglo-German-Italian Panavia Tornado, buying 48 interdictor and 24 air defence versions. Mainstay of Saudi Arabia's air defences became the McDonnell Douglas F-15 Eagle, backed by five Boeing E-3A Sentry AEW aircraft and eight KE-3A tankers. Iraq's invasion of Kuwait in August, 1990, placed the RSAF on a high state of alert. The RSAF played an important role in the resulting war for the liberation of Kuwait, losing at least one Tornado IDS during strikes against Iraqi positions. The USAF accelerated delivery of F-15s to the RSAF, providing an extra 24 aircraft as well as stationing its own F-15s in the country.

Saudi Arabia's ability to maintain its relatively high level of defence expenditure has always been dependent on the price of crude oil, and a price fall during the late 1990s saw a number of modernization projects shelved. Higher oil prices in recent years mean that the RSAF is modernizing again, replacing its F-5E/F and RF-5E Tiger force with Lockheed Martin F-16s.

The RSAF has 20,000 personnel, having quadrupled in size over the past quarter century. It has nine fighter squadrons, of which one has 24 Tornado ADV, five operate 70 F-15C and 24 F-15D Eagles, and another three have 72 F-15S, with another 24 likely to be ordered. The strike role is handled by seven squadrons, with three operating 76 Tornado IDS, one with 15 Northrop F-5B and three with 53 F-5E, likely to be replaced by F-16s. A Tornado IDS squadron has ten Tornado GR1A in reconnaissance configuration, and these operate alongside a reconnaissance squadron with ten RF-5E Tigers. AEW is provided by a single squadron with five Boeing E-3A Sentry, while the tanker force now includes eight Boeing KE-3A and eight Lockheed KC-130H tanker transports. Additional transport capability lies in three squadrons with a total of seven Lockheed C-130E and 29 C-130H Hercules, as well as three L-100-30HS CASEVAC aircraft, with the introduction of C-130Js a strong possibility. There are also four CASA CN235 light transports. Two helicopter squadrons operate 22 Agusta-Bell AB-205, 13 AB-206A and 24 AB-212, with the latter also undertaking SAR as well as transport, and there are 12 AS532A2 Cougar for combat SAR. The Royal Flight operates two Lockheed TriStar and two McDonnell Douglas MD-11, a Boeing 737-200 and a 757-200, two 747SP and a 747-300, a 707-120 and two 707-320, and an Airbus A340-200. In addition, a VIP unit has four VC-130H Hercules, four BAe125-800, two Grumman Gulfstream III, two Lockheed JetStar and two AS61A. Training uses 45 BAe Hawk Mk65/A, 45 Pilatus PC-9, a BAe Jetstream 31 and 13 Cessna 172G/H/M, as well as a number of JetRangers and conversion trainers for the combat aircraft. There are separate air defence forces. The RSAF uses AIM-7F Sparrow and AIM-9J/L/P Sidewinder AAM, while AGM include AGM-65 Maverick, Sea Eagle, AGM-65 Maverick, AS-15 and AS-30.

ROYAL SAUDI NAVY

The Royal Saudi Navy became involved in aviation following the success enjoyed by Iraq and Iran in anti-shipping operations during the war between these two countries. Four frigates can each operate Eurocopter SA365SA Panthers, of which there are 24, mainly shore-based, while there are also 12 AS532AL Cougars, six of which can carry Exocet anti-shipping missiles. Four Panthers are used for SAR, but the other 20 have AS15TT anti-shipping missiles. It is likely that ASW helicopters will be purchased in the near future.

ROYAL SAUDI LAND FORCES ARMY AVIATION COMMAND

The Saudi Army, or Royal Saudi Land Forces, maintains a small but up-to-date aviation element, with 12 Boeing AH-64A Apache attack helicopters purchased in 1993, and which may be upgraded to AH-64D standard, supported by scout and utility helicopters, including 12 Bell 406CS Combat Scouts equipped with TOW anti-tank missiles, and 12 Sikorsky S-70A Desert Hawk.

BELOW: *The RSAF is one of the few export customers for the Anglo-German-Italian Panavia Tornado strike aircraft. (BAe)*

Senegal

POPULATION: 9.7 million

LAND AREA: 76,104 square miles, 197,109 sq km

GDP: $5.6bn (£3.9bn), per capita $2,118 (£1,481)

DEFENCE EXPENDITURE: $69m (£48.3m)

SERVICE PERSONNEL: c.9-10,000 (40% conscript) active.

ABOVE: *Obviously in retirement as a gate-guardian, this Potez CM170 Magister armed jet trainer is typical of the handful that remain in service with the Armee de l'Air Senegal. (AdlAS)*

SENEGAL AIR FORCE/ARMEE DE L'AIR SENEGAL

The AAS was established after independence from France in 1965, and is mainly a transport and communications operation with a very limited counter-insurgency capability. Its initial force was the standard French post-colonial package of Max Holste Broussards, an Alouette II helicopter and a Douglas C-47 transport. Despite an attempt at a pact with Gambia, which is surrounded on three sides by Senegal, in 1981, Senegal and Gambia have followed their own ways in defence, and Gambia has not established an air arm; it is not clear whether the AAS still operates over Gambia. Today, the AAS has 800 personnel, and operates five CM170 Magister armed jet trainers and three Rallye 235A Guerrier armed trainers in the training and COIN roles. Two DHC-6 Twin Otters are operated on EEZ duties. Transports include six Fokker F-27-400M Friendship, but there is also a Boeing 727-200 VIP transport, and a utility BN-2T Islander, two Alouette II helicopters and a SA341 Gazelle. Two Rallye 160 complete the training unit.

ABOVE: *The ageing Alouette III helicopter is still in service with many air forces, including that of Senegal. (AdlAS)*

Seychelles

POPULATION: 78,000

LAND AREA: 140 square miles, 404 sq km

GDP: $570m (£398), per capita $4,500 (£3,147)

DEFENCE EXPENDITURE: $10m (£7m)

SERVICE PERSONNEL: 450 active.

SEYCHELLES COAST GUARD AIR WING

The Seychelles are more than 100 small islands in the Indian Ocean, with a small army that includes a coast guard with 20 personnel in its air wing, briefly known as the Seychelles People's Air Force before the present title was adopted in the late 1990s. At one time it had a Merlin III for maritime patrols and replaced this with a Cessna Citation V, but now it has four aircraft, a Cessna F406 Caravan II and a 152, a Beech 1900D and a BN-2A Islander.

Sierra Leone

POPULATION: 4.8 million

LAND AREA: 27,925 square miles, 72,326 sq km

GDP: $770m (£538m), per capita $712 (£498)

DEFENCE EXPENDITURE: $9m (£6.3m)

SERVICE PERSONNEL: 6,000 active.

SIERRA LEONE DEFENCE FORCE

Formed: 1973

Sierra Leone in West Africa became independent of the UK in 1961. It was not until 1973 that a defence force was established with Swedish assistance, including four Saab MFI-17 Safaris, a Hughes 269C and two Hughes 369 helicopters. These aircraft were joined by a VIP BO105C helicopter in 1976. A Safari was soon lost, while most of the other aircraft were sold due to a shortage of money, and of spares. The BO105C remained, but was not joined by other aircraft until two AS355 Ecureuil helicopters were obtained in 1983. By this time, the country was embroiled in internal unrest, with the government fighting the so-called Revolutionary United Front with the assistance of a South African mercenary organisation, Executive Outcomes, which used two Mil-17 helicopters to support its operations. The government also used a Mi-8 and a Mi-24 flown by Belarussian pilots, and in 1999 acquired two more Mi-24 from the Ukraine. Successive peace agreements throughout the 1990s were broken, and both the United Nations and neighbouring African states, notably Nigeria, have been attempting to maintain the peace. British troops had to be deployed in 2000, and have since been rebuilding Sierra Leone's own defence forces with a planned 5,000-strong army. It is believed that just one Mi-24 remains operational, while the Mi-8s continue to be flown by contract pilots.

Singapore

POPULATION: 3.7 million

LAND AREA: 225 square miles, 580 sq km

GDP: $97bn (£69bn), per capita $26,000 (£18,182)

DEFENCE EXPENDITURE: $4.8bn (£3.4bn)

SERVICE PERSONNEL: 60,500 (66% conscript) active, plus 312,500 reserves.

ABOVE: *TA-4SU Skyhawk of the Singapore Air Force. (Michael J. Gething)*

REPUBLIC OF SINGAPORE AIR FORCE

Formed: 1965

Singapore withdrew from the Federation of Malaysia in 1965 and immediately established its own defence forces, while maintaining strong defence links with Malaysia. The embryonic air force was known initially as the Singapore Air Defence Command, and its early equipment included 20 Hawker Hunter FGA9 ground-attack aircraft and 16 BAC167 Strikemaster armed-trainers, as well as Alouette III helicopters and Cessna 172s for AOP and liaison. These were soon joined by Northrop F-5E/F Tiger II and, later, by ex-USN Douglas A-4S Skyhawk fighter-bombers with the first 47 Skyhawks refurbished and upgraded in the US, but a follow-on order for 86 aircraft was upgraded in Singapore. A limited MR capability was developed using Fokker 50 Enforcer 2s, with AEW using four Grumman E-2C Hawkeyes. Eight Lockheed F-16A/B Hornets were ordered in 1984, but now there are more than 60 F-16C/Ds. During 1986-89, Singapore upgraded its Skyhawks to 'Super Skyhawk' status. The F-5s were upgraded to F-5S standard in 1997.

An unusual feature of the RSiAF's training is that Singapore's limited territory has forced most of its training abroad, mainly in Australia and the United States. A Skyhawk squadron was based in France until 2001. The availability of training facilities does influence aircraft orders.

A substantial helicopter force has been developed, including Boeing CH-47D Chinooks and, currently being delivered, AH-64D Longbow Apache helicopters, augmenting a substantial Super Puma, Iroquois and Fennec operation. Tanker and transport aircraft also provide a longer reach than might normally be expected of a coun-

Singapore

try of this size. Initial pilot training is provided by BAe Flying Training Australia. In 2000 Singapore joined the NATO Flying Training in Canada, NFTC, programme.

The RSiAF has 6,000 personnel, half of whom are conscripts, with its relatively high aircraft to personnel ratio partly accounted for by 7,500 reservists and contracted out training. It is likely to remain a substantial and well-equipped air force given the political instability of much of the surrounding area. In the fighter and ground-attack role, there are six squadrons. Two FGA squadrons operate 50 A-4SU Skyhawks, while another two operate two F-16A, three F-16B, 22 F-16C and 20 F-16D, with another 24 F-16C/D based in the USA, leaving another two squadrons with 28 F-5S and nine F-5T trainers. A single reconnaissance squadron operates eight RF-5S. AEW is provided by four Grumman E-2C Hawkeye. There are five Fokker 50MPA Enforcer 2 aircraft for ASW and EEZ duties, with another four 50UTA transports, while there are six Lockheed C-130H and H-30 Hercules transports, and four KC-130B and four KC-135R tankers. Helicopters include 14 Boeing CH-47D/SD Chinook, of which several are based in the USA, 34 AS332M/UL Super Puma for transport and SAR, 27 Bell UH-1B/H Iroquois, eight Bell 205, and 29 AS550C-2/2A Fennec for scout and liaison duties, as well as eight AH-64D Longbow Apache helicopters equipped with Hellfire missiles and Hydra unguided rockets. Trainers include 27 S211s based in Australia. There is a requirement for up to 12 additional Apaches, between nine and 12 naval helicopters for new corvettes, and a new multi-role fighter in the F-16/Typhoon/Rafaele category, as well as the future F-35 JSF. Missiles include AAM AIM-7P Sparrow and AIM-9N/P Sidewinder, with AGM-45 Shrike, AGM-65B/G Maverick and AGM-84 Harpoon in the ASM role.

Currently, the RSiAF is responsible for all service aviation, providing aircraft for naval and army use as required.

Slovakia

POPULATION: 5.4 million

LAND AREA: 18,940 sq m, 49,435 sq km

GDP: $19.6bn (£13.7bn), per capita $8,184 (£5,723)

DEFENCE EXPENDITURE: $347m (£242.7m)

SERVICE PERSONNEL: 33,000 (45% conscript) active, plus 20,000 reserves.

ABOVE: *Slovak L-29 Delfin. (Jeremy Flack/Aviation Photographs International)*

REPUBLIC OF SLOVAKIA AIR FORCE

Formed: 1993

Slovakia became an autonomous state again in 1993 when the former Czechoslovakia split in two, with the former air force as a whole divided 2:1 in favour of the Czechs, the larger and more populous, as well as more prosperous, portion of the country. The 2:1 split applied to the overall strength, but varied slightly on particular items: The MiG-29 force was divided equally while the Czechs received all MiG-23s. Since then, the LPOASR added eight extra MiG-29s in 1995-96 and has retired older aircraft. Although not a member of NATO, the MiG-29s and Mi-24s are being upgraded with Western equipment, including IFF. There are plans to replace the Su-25s, Su-22s, MiG-21s, L-39s and L-29s with a single armed-trainer type to cut costs, with a requirement for up to 50 aircraft, once funding permits.

The LPOASR has 10,200 personnel. Flying hours are low, at an average of 45 per annum. The fighter units operate 24 MiG-29A/UB and 30 MiG-21, while attack units have 12 Su-25K/UBK and eight Su-22M-4/U. Eight MiG-21RF equip a reconnaissance squadron. There are 19 Mi-24 attack helicopters, as well as 20 Mi-8 and Mi-17 transport helicopters and six Mi-2 for liaison duties. Transport is provided by six L-410M and two each of An-24 and An-26. Training is provided on 11 Aero L-29 Delfin and 18 L-39C Albatros.

Slovenia

POPULATION: 2 million

LAND AREA: 7,796 square miles, 16,229 sq km

GDP: $18.6bn (£13bn), per capita $12,518 (£8,753)

DEFENCE EXPENDITURE: $227m (£158.7m)

SERVICE PERSONNEL: 7,600 (60% conscript) active, plus 61,000 reserves.

ABOVE: *The break-up of Yugoslavia has spawned many new air arms, including that of Slovenia. Here is a heavily-armed Pilatus PC-9M, known as the Swift in Slovenian Service. (Marko Malec)*

SLOVENIAN MILITARY AVIATION

Formed: 1991

Formerly part of Yugoslavia, Slovenia has escaped the upheaval and fighting that affected much of the rest of that state. The armed forces are centred on the army, which maintains maritime and air elements, the latter accounting for around 120 of the army's 7,600 personnel. The initial role of the air element has been communications and transport support for ground forces, using aircraft based in Slovenia at the time of independence being declared in 1991. These were augmented by three ex-US Army Pilatus PC-9 armed-trainers, and a further nine PC-9 Mk2 have since been purchased, and all 12 aircraft fitted with a weapons and mission system. Fighters may be acquired in the near future, possibly either ex-Israeli Kfirs or ex-Dutch F-16s, although defence expenditure was cut by almost a third between 1999 and 2000, placing expensive procurement programmes in doubt. Five Bell 412EP helicopters are used for frontier patrols and SAR, with two 412HP and a 412SP for transport, on which a Let L-410V-E Turbolet and two Pilatus PC-6 Turbo Porters are also used. A UTVA 75 is used for communications duties. Eight Zlin 242L and two 143L are used for training, as well as three Bell 206B-3 JetRanger helicopters.

ABOVE: *A number of Slovenian unarmed training types also carry civilian registrations, although the 15th Air Brigade insignia is still carried. (Marko Malec)*

Somali Republic

POPULATION: 10.3 million

LAND AREA: 246,000 square miles, 637,658 sq km

GDP: $900m (£629m), per capita $1,100 (£769)

SERVICE PERSONNEL: Nil.

SOMALIAN AERONAUTICAL CORPS

Formed: 1960

The former protectorate of British Somaliland gained independence in 1960 at the same time as the Italian Trust Territory of Somalia, the two amalgamating to form the present state. The Italian-sponsored air corps provided the basis for an air arm, which developed with Soviet aid. The Somalian Aeronautical Corps reached 2,000 personnel in the mid-1970s, by which time it was operating six MiG-17 and 12 MiG-15 fighter-bombers, six Beech C-45s, a Douglas C-47 and an Antonov An-24 in the transport role, with 20 Yakovlev Yak-11, ten Piaggio P148 and six MiG-15UTI trainers. Many of these aircraft soon became unserviceable, but were replaced by at least eight MiG-21s and, during a brief flirtation with China, more than 20 Shenyang F-6 (MiG-19). During the 1980s, the country was courted by the West, and received military aid in return for making airfields and ports available to US forces. Aircraft were not provided by the USA, but the SAC was able to obtain at least nine Hawker Hunters, mainly FGA76 fighter-bombers, and some Britten-Norman BN-2 Islander communications aircraft from Abu Dhabi.

In 1991, a revolution split the country into fiefdoms dominated by rival clans. From late 1992 until mid-1994, the United Nations attempted to restore the peace and provide aid to ward off the affects of a severe famine, largely using forces supplied by the USA and Italy. These forces had to be withdrawn as the position within the country made a military presence virtually impossible. The country is now in a state of anarchy with upwards of ten different groups fighting. The SAC's aircraft are reported to lie abandoned at Mogadishu Airport.

South Africa

POPULATION: 40.8 million

LAND AREA: 471,445 square miles, 1,224,254 sq km

GDP: $122bn (£85.3bn), per capita $6,281 (£4,392)

DEFENCE EXPENDITURE: $1.9bn (£1.3bn)

SERVICE PERSONNEL: 61,500 active, plus 89,189 reserves.

ABOVE: *With the Suez Canal closed to shipping on a number of occasions, the Cape route around South Africa was important and vulnerable, it was patrolled by SAAF aircraft such as Avro Shackleton MR MkIII -a SAAF restored aircraft and the only MkIII flying today! (Jeremy Flack/ Aviation Photographs International)*

SOUTH AFRICAN AIR FORCE

Formed: 1920

The country with the longest history of military aviation in Africa, South African Army officers received flying training as early as 1913, and in 1915, the South African Aviation Corps was formed. In World War I, the Royal Flying Corps had a number of South Africans amongst its personnel. The SAAC's initial equipment consisted of a number of BE2As and also some ex-RNAS Henri Farmans. At first, the new force was expected to operate against German South-West Africa, but in the event there was little aerial opportunity on that front, but the SAAC did fight with some distinction against German forces in East Africa. It was disbanded in 1918.

The South African Air Force was formed in 1920 as a separate service using the 'Imperial Gift' of war surplus aircraft. In the SAAF's case this amounted to a hundred aircraft, including SE5A fighters, 48 DH4, DH9 and DH9A bombers, and a number of Avro 504K trainers. During the early years, the SAAF was involved in a number of police actions. From 1925, an air mail service was operated between Durban and Cape Town. Other duties included surveying and the training of Citizen Defence Force pilots. In 1929, 20 Avro Avian trainers were purchased to replace the Avro 504Ks, followed by 31 Westland Wapiti general-purpose biplanes, while another 27 Wapitis were assembled locally. During the early 1930s, 60 Avro Tutors were built under licence in South Africa. The SAAF differed from other air forces in the British Empire by not having a 'Royal' prefix, reflecting the different constitutional arrangements and colonial history of the country. Air force ranks were based on those of the army rather than the Royal Air Force system.

In 1936, a major expansion started. Licence-production began of 65 Hawker Hartebeest bombers, as the Hawker Hind was known in South Africa. Aircraft supplied direct from the UK included Hawker Fury fighters, the last biplane fighter, the Gloster Gladiator, and then Hawker Hurricane fighters, Fairey Battle and Bristol Blenheim bombers. When war broke out in Europe in September, 1939, the SAAF had a hundred modern combat aircraft, and was able to put the Junkers Ju86 airliners of South African Airways into service on MR patrols.

Although not formally a member of the Empire Flying Training Scheme, SAAF training facilities expanded considerably and placed at the disposal of the RAF. SAAF bases were also made available for the Royal Navy's Fleet Air Arm, using them for carrier air wings in transit to Egypt and the Mediterranean after the direct route across the Mediterranean became impassable. They then proved invaluable as rear bases for the Fleet Air Arm once the Royal Navy returned in force to the Indian Ocean and then to the Pacific. Many South African personnel also served with the RAF. The SAAF itself operated MR over the South Atlantic and the Indian Ocean, while its fighter and bomber units operated in North Africa against Italian and German forces. SAAF units were stationed in East Africa protecting Kenya and the Sudan from Italian forces, and helped liberate Ethiopia. After taking part in the Allied invasion of Sicily, the SAAF saw further service in Europe. Inevitably, the rapidly expanded SAAF operated a wide variety of aircraft during the war years. Fighters included Curtiss Kittyhawk, Mohawk and Tomahawk, Hawker Hurricane, Supermarine Spitfire and the North American F-51 Mustang. Bombers included Martin Maryland, Marauder and Baltimore, Vickers Wellington, Fairey Battle, Bristol Beaufort and Blenheim, Douglas Boston and the Consolidated Liberator. Vickers Warwick, Lockheed Ventura, Lodestar and Harpoon MR aircraft operated alongside Short Sunderland flying boats and Consolidated PBY-5A Catalina amphibians. Transport aircraft included the Ju86s, as well as the more usual Douglas C-47 and some ex-civil Junkers Ju52/3Ms, the C-47's angular rival! Training used de Havilland Tiger Moth, Airspeed Oxford, Avro Anson, North American T-6 Harvard, Northrop Nomad, Miles Master and Hawker Audax aircraft.

South Africa

Post-war, the SAAF was reorganised as a small force capable of rapid expansion from SAAF Reserve and Active Citizen Air Force personnel. Many wartime aircraft remained in service, working out their useful lives. The SAAF took part in the Berlin Airlift, enabling the city to survive the Soviet siege. In 1950, the SAAF sent a small force of F-51D Mustang fighters to Korea to join the UN forces there, facing MiG-15 jet fighters, and it was during the Korean War, in 1952, that the SAAF received its first jets, North American F-86 Sabres loaned by the USAF. Meanwhile, back in South Africa, de Havilland Vampire FB5 jet fighter-bombers, with Vampire T55 trainers, started to replace the Spitfires. Canadair CL-13 Sabre 6 fighters, de Havilland Dove and Heron transports, and Sikorsky S55 helicopters were delivered in 1956. In 1957, the Short Sunderlands were replaced by Avro Shackleton MR3s for defence of the Cape shipping route, whose importance had been reinforced by the closure of the Suez Canal during the Suez Crisis of the previous year. The bomber force was modernised with English Electric Canberra B12 jet bombers. The South African Navy obtained eight Westland Wasp helicopters to operate from its frigates. US aircraft introduced during this period included seven Lockheed C-130B Hercules transports, and Cessna 185 liaison aircraft for a small Army Air Corps.

A limited expansion programme started during the early 1960s to maintain the balance of power in Africa, where newly independent states hostile towards South Africa were backed by the USSR. The newer members of the British Commonwealth were also hostile, and the country withdrew in 1961. A United Nations resolution against the sale of arms to South Africa specifically excluded armaments for defence against external threats, but by the mid-1960s, a new British government extended the embargo to cover all arms. Before this took effect, under the terms of the Anglo-South African Defence treaty, the Simonstown Agreement, the SAAF was able to obtain 16 Hawker Siddeley Buccaneer low-level strike aircraft. The Buccaneers differed from the standard aircraft in having rocket-boosters to improve take-off performance from 'hot and high' airfields. The ban enabled France to become the main supplier of arms to South Africa, although more than 200 Italian Aermacchi MB326K armed-trainers were built under licence as the Impala in South Africa during the late 1960s. The Impala were used both for training and for Active Citizen Force units. French equipment introduced to SAAF service at this time included 16 Dassault Mirage IIICZ interceptors, 20 Mirage IIIEZ fighter-bombers, four Mirage IIIRZ reconnaissance aircraft, nine C-160 Transall transports, six Alouette II and 50 Alouette III helicopters, as well as 20 SA330 Puma and 16 Super Felon helicopters. During the early 1970s, licence-production of Mirage III and F1 aircraft started in South Africa.

Hopes during the early 1970s that the

ABOVE: *Arms embargoes led to home grown solutions, including the Cheetah C & D - single and twin-seat modified Mirage III airframes for the SAAF. (Jeremy Flack/Aviation Picture Library)*

BELOW: *Aircraft manufacturers struggled to produce an up-dated replacement for the Douglas C-47, but the answer was an up-dated C-47! The South African Air Force was first to use the turboprop C47TP, a concept now found in a number of air forces. (T J Walker/Aviation Picture Library)*

South Africa

ABOVE: *Before arms sanctions began to bite, the South African Navy was able to obtain Westland Wasp helicopters for its frigates. (GKN Westland)*

strict British interpretation of the arms embargo would be eased proved groundless, and other nations also applied stricter controls. South African support for the regime in neighbouring Rhodesia, now Zimbabwe, aggravated these tensions. South Africa started to develop its own armaments industry. In 1986, the Cheetah development of the Mirage III first appeared, and a substantial number of the SAAF's Mirage IIIs were upgraded to this standard. The C-47s, of which there were 40, were re-engined with turboprop engines. Many older types were retired, first the Sabres replaced by Mirage F1s, and then Canberras and Buccaneers, which finally left the SAAF in 1991.

Substantial political changes were taking place inside South Africa during the early 1990s, culminating in majority rule in 1994. This resulted in the end of UN sanctions, but the SAAF was not able to modernise quickly due to financial constraints, while the USA did not relax its arms embargo until 1998. Plans to replace the Cheetahs and older Impalas with new aircraft were cut back, but the SAAF is currently introducing Saab Gripens and BAe Hawks, while Agusta A109s are replacing the Alouette IIIs. The C-130B fleet has been renovated and upgraded, while the first of an initial 12 Denel CSH-1 Roivalk attack helicopters - itself a demonstration of the capability of the South African industry - entered service in 1999.

Southern Africa remains strategically important, but suffers from political instability, while there is also a frequent need for humanitarian aid for the areas to the north. The SAAF remains the most capable air force on the African continent, although the period of sanctions and the financial constraints of the post-sanctions era have meant that certain significant capabilities have been lost, notably MR. Strike capability is now confined to the tactical level. Although the current strength of 9,250 personnel has risen from around 8,000 over the past 30 years, it seems likely that this will reduce given constraints on the procurement programmes.

Today, the SAAF has introduced 19 Saab Gripen single-seat fighters and nine two-seat aircraft, replacing half the Cheetahs, as well as 12 Hawks for lead-in fighter training, with another 12 on option. Up to 34 Impala II (MB326K) attack aircraft and 35 Impala I (MB326M) training aircraft remain in service. The Alouette IIIs have been replaced by 30 Agusta A109 helicopters, while options exist for another ten. Plans exist to buy up to four Westland Lynx for the shipborne ASW and ASuW role, but these are on hold. Transport is provided by five Boeing 707-320, some of which can operate in the tanker role and some in the ELINT role, as well as nine upgraded Lockheed C-130B and three C-130E Hercules, and 12 Douglas C-47 and C-47TP Dakota, which also undertake maritime-reconnaissance. VIP and communications duties fall to a BBJ (Boeing 737), two Dassault Falcon 50 and three Falcon 900, as well as three Beech Super King Air 200 and a 300, and two Cessna Citation II and a PC-12. Utility aircraft include 11 Cessna 208 Caravan I and 13 185A/D/E, and a CASA CN235. There are 12 CSH-1 Roivalk attack helicopters and 51 transport Oryx (licence-built Super Puma), with nine BK117A-1/-3. There are 59 Pilatus PC-7 and PC-7 Mk II trainers, known as the Astra in SAAF service, with the remaining Alouette III helicopters. Missiles in the inventory include the Raptor ASM and V-3C AAM.

BELOW: *More advanced still was the Denel Rooivalk, designed for the SAAF, but a serious contender for many attack helicopter competitions by armies elsewhere in the world. (Jane's))*

Spain

POPULATION: 39.7 million

LAND AREA: 194,945 square miles, 504,747 sq km

GDP: $568bn (£397bn), per capita $18,703 (£13,079)

DEFENCE EXPENDITURE: $7.2bn (£5bn)

SERVICE PERSONNEL: 143,450 active, plus 328,500 reserves.

SPANISH AIR FORCE/EJERCITO DEL AIRE

ABOVE: *The post-war Spanish HA-200 Super Saeta. (Jeremy Flack/Aviation Photographs International)*

Formed: 1939

Spain's long history of military aviation started with a balloon company within the Spanish Army in 1896. In 1910, the balloon company saw action during the Riff uprisings in Morocco, and the following year aircraft were introduced and it became the Aeronautica Militar Espanola, with two Henri Farman, two Maurice Farman, two Bristol and six Nieuport aircraft. Additional Maurice Farman and Nieuport aircraft were acquired, with Morane-Saulnier MS14 and Lohner biplanes, before the outbreak of World War I. A neutral country, Spain had difficulty in obtaining aircraft for the AME and the newly-formed Aeronautica Navale during the war years, resolving this through licence-production of DH4 bombers and Morane-Saulnier Parasols. A Spanish-design, the Flecha, also entered production. The Aeronautica Navale managed to acquire some Curtiss F flying boats.

The return of peace meant that the two Spanish air arms were able to benefit from the release of large numbers of war surplus aircraft by the belligerent powers, acquiring Ansaldo, Bristol F2B, Martinsyde F4A and Spad S13C1 fighters; Farman F-50, Salmson SAL2-A2, Breguet Br14A-2 and additional DH4 bombers; Macchi M9 and Savoia S16 flying-boats, and a number of Caudron GIII trainers. Many of these aircraft were deployed in Morocco where there was renewed fighting in the Riff that continued until 1926. Additional aircraft entered service at this time, with Dornier Wal and Macchi M18 flying-boats, Blackburn Velos seaplanes and Supermarine Scarab amphibians for the Aeronautica Navale; and licence-built Breguet Br19A-2 and DH9 bombers, 20 Fokker CIV AOP and CIII training aircraft, and Avro 504K trainers, for the AME. A number of Wals were also built in Spain under licence, as were Nieuport 52C1 fighters. Spain's own Loring R-1 reconnaissance aircraft entered production. Nevertheless, by the late 1920s, aircraft were entering service in smaller numbers, although some Vickers Vildebeest torpedo-bombers were obtained in 1931, and the combined aircraft strength of the two services fell sharply from 700 to 300.

In 1931, the Spanish monarchy was replaced by a republic, followed by growing civil unrest within Spain itself. In 1936, the Spanish Civil War began between the Nationalists, supported by Germany and Italy, and the Republicans, supported by the USSR. The Republicans managed to obtain most of the AME's aircraft, and at the outset had some 200 aircraft to the 60 or so of the Nationalists. Soviet support for the Republicans encouraged Germany and Italy to provide overt support for the Nationalists. France sold aircraft to the Republicans, including 100 Dewoitine D373, D500 and D510, Liore-Nieuport LN46 and Spad 510C fighters, and Potez 56 and Bloch MB200 bombers. Czechoslovakia supplied Letov S231 fighters and Aero 100 general-purpose aircraft. Some of the supporting nations were able to test tactics and aircraft in this theatre of war, while their 'volunteer' pilots also developed their skills. Bombing was used extensively, while fighter combat moved on from the elementary operations of 1914-18. Of significance was the use of Ju52/3M transports to move the Moroccan troops of the Spanish Foreign Legion from North Africa to Spain, giving the Nationalists an important edge. The war ended in 1939 with a Nationalist victory. The new government immediately reorganised Spanish military aviation, merging the two air arms into the Spanish Air Force, or Ejercito del Aire, a separate service. Despite heavy losses during the war, the EdA on its creation had 1,000 aircraft available.

Spain remained neutral throughout World War II, although aligned politically with the Axis Powers. Once again, the country was effectively cut-off from outside aircraft supplies, leaving the two Spanish manufacturers, CASA and Hispano-Suiza, to produce aircraft under licence. Although a few Messerschmitt Bf109F fighters, Junkers Ju88A bombers, Heinkel He114 seaplanes and Dornier Do24 flying-boats were supplied from Germany, with some Fieseler Fi156 Storch general-purpose aircraft, for the most part Spain's needs were met by licence-production of Germany and Italian designs. These included additional Bf109Fs and He114s, 250 He He111H bombers, 100 Junkers Ju52/3M transports, and Fiat CR32

Spain

ABOVE: *Post-war Spanish naval aviation started with the Bell 47G Sioux helicopter for liaison duties. (AAAE)*

ABOVE: *McDonnell Douglas/ BAe AV-8A Harrier, known in Spain as the Matador. (AAAE)*

fighters. Spanish pilots flew with the Luftwaffe on the Russian front. Post-war, the EdA was to become notable for continuing the operation of many wartime aircraft, some of which remained in production until 1953, with the fitting of Rolls-Royce Merlin engines to the Bf109s, the same engines as their wartime rivals, the Hurricane and Spitfire!

Spain was late in joining NATO, but agreed a separate defence treaty with the United States in 1953, receiving US military aid in return for the use of Spanish air and naval bases. This resulted in the EdA receiving 200 North American F-86D/F Sabre fighters, 15 Douglas C-47 transports, a number of Grumman HU-16 Albatross amphibians, Sikorsky H-19 Chickasaw and Bell 47G Sioux helicopters, 30 Lockheed T-33A and 100 North American T-6G Texan trainers. The Spanish aircraft industry produced the CASA 201B, 207 and 352L transports, and the Hispano Aviacion HA-100 and HA-200 trainers. These aircraft were followed during the early 1960s by 20 Lockheed F-104G Starfighters, enough to equip one interceptor squadron, joined later by CASA-built Northrop SF-5A/B fighter-bombers. The EdA also received Dassault Mirage IIIE fighters at this time. When the Spanish-American Defence Agreement was extended in 1970, Spain received McDonnell Douglas F-4 Phantom fighter-bombers and a number of other US aircraft, including Lockheed P-3 Orion MR aircraft and C-130 Hercules transports. Other aircraft included DHC-4 Caribou tactical transports. The Mirage IIIs were later joined by Mirage F1s.

The restoration of the monarchy and of democracy in Spain in 1975 brought a major change in Spain's international position, joining NATO and, later, the European Union, and also joining aircraft programmes with the country's European partners. Nevertheless, US and French designs continued to dominate the EdA, which introduced 72 licence-built EF-18A/B Hornet fighters during the 1980s, replacing the F-4 Phantom in the fighter-bomber role, while still retaining some of these aircraft for reconnaissance. The SF-5s and Mirage IIIs were modernised during the late 1980s and early 1990s. A plan to develop a Spanish attack aircraft, initially designated the AX, was abandoned during the early 1990s. Spain's hopes that participation in the Eurofighter programme with the UK, Germany and Italy would cover its future needs were dashed as the programme ran into repeated delays, necessi-

BELOW: *Advanced jet training for combat pilots with the EdA takes place on the CASA C101B Aviojet.*

ABOVE: *The EdA modernized its Dassault Mirage IIIEs during the late 1980s and early 1990s, but has now retired them.*

tating the acquisition of additional F/A-18 Hornets from the USN between 1995 and 1998. Additional Mirage F1s also had to be obtained from France and Qatar in 1994 and 1997, and all the F1s have been updated to extend operational life to 2015. The first 20 Eurofighter 2000 Typhoons will enter service in 2003.

The EdA has 24,500 personnel, a fall of a quarter over the past thirty years, partly due to the end of conscription. The EdA is organised into four air commands, Central, Eastern, Strait and Canary Islands, plus a Logistic Support Command. A total of 87 Eurofighters is on order, but currently the main frontline aircraft are 81 EF/A-18A Hornets, known locally as the EF15, in six fighter squadrons spread across the four air commands, with a further 25 EF/A-18B in the OCUs. Another two fighter squadrons in the Strait Air Command have Mirage F1CE/BE, known locally as the C14, while Central Air Command has the single RF-4C Phantom reconnaissance squadron. The remaining 24 SF-5A/B (known as AE/AR9) are in two lead-in fighter squadrons in Strait Air Command. Fighters use AIM-7 Sparrow, AIM-9 Sidewinder, AIM-120 AMRAAM and R-530 AAM. The EdA is responsible for most shore-based maritime aviation, and has a single MR squadron with two Lockheed P-3A and five P-3B Orion, while SAR and EEZ patrols are provided by three Fokker F-27-200MPA normally based in Canary Islands Air Command. A substantial force of around 80 CASA C212-100/200 handles a variety of tasks, including transport, patrol, SAR, ECM and EW, as well as training. Tanker-transports include five Lockheed KC-130H Hercules (known as TK10) and three Boeing 707-320s, with another 707 for ELINT, while there are also seven C-130H/H-30 Hercules (T10/T10L) transports. There are nine C295 transports, and 21 licence-built CASA 127A (Do27) in the communications role. VIP aircraft include a Dassault Falcon 20 (plus four for ECM), a Falcon 50 and two Falcon 900, as well as three Super Puma (HT21). Extensive SAR coverage around the long coastline of Spain and her islands is provided by 19 SA330H/J Puma (HD19), 12 Super Puma (HD21), three AS332BM1 Super Puma (HT21A) and eight Sikorsky S-76A Spirit, some of which are also used for training. There are 15 Canadair CL215T (UD13) amphibians for fire fighting and SAR. Two Cessna V Citations (TR20) are used for photographic surveys. Training aircraft include five Beech B55 Baron and 23 F33A Bonanza, 76 C101B Aviojet and 37 Enaer T-35C Pillan, with Hughes 300s replaced by 15 Eurocopter EC120 Colibri (HE25). ASM include AGM-65G Maverick, AGM-84D Harpoon and AGM-88A HARM, while there are Mistral and Skyguard/Aspide SAM batteries.

SPANISH NAVAL AIR ARM/AARMA AEREA DE LA ARMADA ESPANOLA

Although the history of Spanish naval aviation dates from the early years of the last century, all service aviation was merged into the new EdA at the end of the Spanish Civil War. Spanish naval aviation reappeared with the loan by the USN of an Independence-class light aircraft carrier, the USS *Cabot*, in 1967, although the EdA retained shore-based MR. Renamed *Daedalo*, the carrier was sold to Spain in 1973, and at first mainly operated helicopters as an ASW carrier. The late 1970s saw the Spanish Navy become one of the first to deploy V/STOL aircraft at sea, obtaining five AV-8A Harrier and two TAV-8A trainers through the US Marine Corps. These aircraft provided the *Daedalo* with a strike capability, primarily for action against shore-targets without the fighter/anti-shipping capability of the Sea Harrier. Meanwhile, frigates capable of carrying helicopters also entered service with the Santa Maria-class (US Oliver Hazard Perry) frigates, while many older destroyers and frigates were modified for helicopters.

The *Daedalo* was replaced by a new Spanish-built aircraft carrier designed specifically for the operation of V/STOL aircraft, the *Principe de Asturias*, which usually operates with an air wing of up to eight

Spain

ABOVE: *A Spanish Navy Sikorsky SH-60 Sea Hawk about to land on the US-built frigate Numancia. (AAAE)*

ABOVE: *The Spanish Navy uses its Sikorsky SH-3H Sea King helicopters, known in Spain as the HS9, for SAR (as shown here) and ASW, although there are also three AEW examples. (AAAE)*

EAV-8B/EAV-8B-plus Harrier II/Harrier II-plus aircraft and ten helicopters.

Today, naval aviation accounts for around 700 of the Spanish Navy's 36,950 personnel. There are nine EAV-8 and eight EAV-8-plus Harrier IIs, known in Spain as the VA2. These have been upgraded to use AMRAAM missiles for air defence and Penguin anti-shipping missiles. Helicopters include 13 Sikorsky SH-3H Sea King, known in Spain as the HS9 in the ASW configuration, of which there are ten, and the SH9 in AEW form, with three. There are also 12 Sikorsky SH-60B Seahawk (HS23) for ASW and ASuW operations from frigates, while ten Agusta-Bell AB212ASW (HA18) are used on transport and SAR duties. Three Cessna Citation IIs (U20) are used on communications duties. Ten Hughes 300 (HS13) are used for training.

SPANISH ARMY AVIATION/FUERZA AEROMOVILES DEK EJERCITO DE TIERRA

Spanish Army Aviation developed during the late 1950s and early 1960s, initially using Cessna O-1 Bird Dogs for liaison and AOP duties. These were joined later by six Bell 47G Sioux helicopters, and then by 12 Agusta-Bell UH-1D Iroquois. Over the past thirty years this force has grown substantially, initially as a transport and liaison force but moving into the attack role with 70 BO105ATH/CB, known in the FAMET as HA/HE15, with 27 equipped with HOT anti-tank missiles and the remainder armed with 20mm cannon. The transport capability of the force was also enhanced with CH-47C Chinook, or HT17 in Spain, heavy lift helicopters. The next move is to acquire up to two squadrons of attack helicopters, which should be fully operational by 2010.

Today, the FAMET has 70 BO105ATH/CB helicopters in the combat role, with another nine on training duties, as well as 17 Boeing CH-47D Chinook, recently upgraded from CH-47C standard, in the transport role, with 48 Agusta-Bell UH-1H Iroquois (HU10) and six AB212 (HU18), and 31 AS532UC/UL Cougar (HT21). The AB212 also have a SAR role. Training also uses 11 OH-58B Kiowa (HE15).

BELOW: *An EAV-8B Harrier II takes off from the ski-jump on the aircraft carrier* Asturias. *(AAAE)*

Sri Lanka

POPULATION: 19 million

LAND AREA: 24,959 square miles, 65,610 sq km

GDP: $16.7bn (£11.7bn), per capita $4,600 (£3,217)

DEFENCE EXPENDITURE: $880m (£615m)

SERVICE PERSONNEL: c.120,000 active, plus 4,200 reserves.

ABOVE: *BAC Jet Provost T.51 of the Royal Ceylon Air Force (BAC via Aviation Photographs International)*

SRI LANKA AIR FORCE

Formed: 1950

Although Sri Lanka, or Ceylon, was used as a major British base during wartime, its own history of military aviation dates from 1950, two years after independence from the UK, when the then Royal Ceylon Air Force was formed with assistance from the RAF. After some delay, in 1953, 12 DHC-1 Chipmunk basic trainers and nine Boulton-Paul Balliol trainers arrived, followed by a couple of Airspeed Oxford training and communications aircraft. In 1955, a de Havilland Dove light transport arrived, with Scottish Aviation Pioneers and Westland Dragonfly helicopters. Additional Doves and Pioneers were delivered in 1958, as well as eight Hunting Jet Provost armed-trainers. The RCAF operated against illegal immigrants and smugglers, and on internal security duties. The latter role soon assumed greater importance, and in 1971, Soviet personnel were seconded to the RCAF and six MiG-17 fighter-bombers were supplied to help suppress an armed rebellion.

The country became a republic in 1972, although remaining within the British Commonwealth, adopting the name of Sri Lanka, while the air force also dropped the 'Royal' prefix.

The association with the USSR was short-lived, and by the mid-1970s, the Sri Lanka Air Force had largely returned to transport and communications duties, almost allowing the combat capability to lapse. Within a few years, internal security duties returned, as the Tamil population in the north of the island pressed for autonomy, with a terrorist campaign by the Liberation Tigers of Tamil Eelam, LTTE. From 1983, the SLAF rebuilt its combat capability, initially using armed trainers. In 1991, two Shenyang FT-5 (MiG-17U) trainers were obtained from China with four F-7M Airguard (MiG-21) fighter-bombers - the small numbers reflecting both financial constraints and the availability of suitable personnel. Steady growth in the combat units continued throughout the 1990s, with Mi-17 transport and Mi-24 attack helicopters, while the LTTE, although lacking air support of their own, introduced surface-to-air missiles in 1995. Transport also grew in importance, with nine Harbin Y-12 transports joining the earlier small force of just three HS748s. A civilian aircraft was shot down in 1998, forcing the SLAF to provide internal passenger flights, and aircraft were fitted with infrared warning systems and chaff dispensers in response to the SAM threat. A significant increase in offensive capability occurred in 1996, with ex-Israeli Air Force Kfir C2/CT2s with additional Kfirs in 2000, as well as MiG-27s acquired from the Ukraine. A number of Pucara COIN aircraft were acquired from the Argentine, but are now believed to be unserviceable, as are the F-7s, reportedly retired with wing cracks. The intensity of the conflict is such that in 2000 the SLAF lost four Mi-24s and Mi-35s, having received two Mi-24s in 1996, two Mi-35s in 1999 and three in 2000. Some training has taken place overseas, partly as a result of the internal security situation, but new aircraft have brought this activity back to Sri Lanka.

Today, the SLAF has 10,000 personnel, including a number of reservists called-up for full-time service. Average flying hours are high, at 420 hours. The main offensive capability lies in 12 Kfir C2 fighter-bombers, with a single TC2 trainer. There are four MiG-27s, believed to be flown by Ukrainians, and four Chengdu F-7BS, as well as a single FT-7. There are 16 SF260TP/W armed-trainers that can be used on COIN operations, as well as seven Mi-24/35 attack helicopters. Eleven Bell 212 helicopters can be used for transport and COIN duties. Other transport aircraft include three Lockheed C-130B Hercules, a Shaanxi Y-8 (An-12), an An-32B and five Harbin Y-12, as well as a surviving HS748, ten Mi-17 helicopters as well as six Bell 412 and six 206. The Bell 412s also have a VIP role, as has a single Beech King Air 200, with another four on MR. A Cessna 421 is used for survey work, with five 150s on training, with two Shenyang FT-5. Eight K-8 and six CJ-6 (Yak-18) trainers were introduced recently.

A Chetak helicopter is used on naval liaison duties, but a plan to create a naval air wing within the Sri Lanka Navy in response to LTTE attacks on shipping appears to have been abandoned.

Sudan

POPULATION: 29.6 million

LAND AREA: 967,500 square miles, 2,530,430 sq km

GDP: $9.5bn (£6.6bn), per capita $1,709 (£1,195)

DEFENCE EXPENDITURE: $580m (£406m)

SERVICE PERSONNEL: 117,000 (17% conscript) active.

SUDANESE AIR FORCE

Formed: 1959

Sudan became an independent republic in 1958 after being administered jointly by the UK and Egypt, with the Sudanese Air Force formed the following year. Initial equipment comprised four Hunting Provost T51 armed-trainers, soon joined by Hunting Pembroke and Douglas C-47 transports. During the 1960s, the SAF's role grew, obtaining 16 MiG-21 fighter-bombers and eight BAC Jet Provost T52 armed-trainers and five T45 trainers. Three Fokker F-27M Troopships and five Antonov An-24 transports were also introduced, followed by Alouette III, Mi-4 and Mi-8 helicopters. The SAF has drawn its equipment from a wide variety of sources, with twelve Northrop Grumman F-5E/Fs delivered during 1982-84. Generally, equipment supplies from Western sources have been hampered by concern over the internal political and military situation with concerns over treatment of the Christian minority in the south by the Muslim majority in the north. Donations of equipment by other Muslim states have not helped standardisation, with a Fokker F-27-100 Friendship donated by South Yemen and a DHC-5 Buffalo by Oman in 1986, and Libya donating a squadron of MiG-23s in 1988, although these appear to have suffered heavy attrition. Puma and Agusta-Bell AB-412 helicopters were also placed in service, while combat aircraft in recent years have come from China, with first Shenyang F-5s (MiG-17), and then F-6s (MiG-19) and F-7s (MiG-21), believed to have been paid for by Iran.

Today, the SAF has 3,000 personnel. Many aircraft are believed to be unserviceable, although reports of air attacks suggest that at least some remain operational. Mainstays of the fighter and ground-attack force are up to 20 Chengdu F-7B and two MiG-23s, with small numbers of F-6s and F-5s. There are believed to be three BAC Strikemasters still operational. Attack helicopters include six Mi-24 and five BO105CB. Transport is provided by two Lockheed C-130H Hercules, five Shaanxi Y-8 (An-12), a Fokker F-27-100 and three DHC-5D Buffalo, eight Eurocopter Puma, six Mi-8 and four Agusta-Bell AB-212. A Dassault Falcon 50 and a 20 provide VIP transport. Training uses eight Pilatus PT-6As.

Surinam

POPULATION: 419,000

LAND AREA: 70,087 square miles, 161,875 sq km

GDP: $409m (£286m), per capita $5,200 (£3,636)

DEFENCE EXPENDITURE: $11m (£7.7m)

SERVICE PERSONNEL: 2,040 active.

SURINAM AIR FORCE

Formed: 1982

Formerly Dutch Guiana, Surinam formed a small air force in 1982 within the Army. It provides border and coastal patrols, light transport and MEDEVAC, initially using Britten-Norman BN-2B Defenders and a Cessna TU206G, as well as a Bell 205 and two Alouette III helicopters. Today, it has 200 personnel. Two CASA C212-400s were delivered in 1998 and 1999, with one used for transport and the other, equipped with radar, for offshore patrols. There are four BN-2 Defenders, a Pilatus PC-7 trainer, and a Cessna 310 and a 172.

Swaziland

POPULATION: 600,000

LAND AREA: 6,705 square miles, 17,363 sq km

SWAZILAND DEFENCE FORCE AIR WING

Swaziland is an independent kingdom completely surrounded by South Africa, which donated three Alouette III helicopters in 2000 to join two IAI-201 Arava armed-transports, and a Piper Cherokee training aircraft.

Sweden

POPULATION: 8.9 million

LAND AREA: 173,620 square miles, 449,792 sq km

GDP: $238.6bn (£166.9bn), per capita $24,032 (£16,805)

DEFENCE EXPENDITURE: $5.3bn (£3.7bn)

SERVICE PERSONNEL: 33,900 active (48% conscript), plus 262,000 reserves.

ABOVE: *Saab 18 (© Saab AB)*

ROYAL SWEDISH AIR FORCE/FLYGVAPNET

Formed: 1926

Swedish military aviation started with the Royal Swedish Navy being presented with a Bleriot monoplane by an air-minded citizen in 1911, a year before the Swedish Army was presented with a Nieuport IVG in similar circumstances. Sweden remained neutral during World War I, by which time the RSwN had added two Henri Farmans and a Donnet-Leveque flying boat, while the Army Air Corps had added three aircraft, including a Breguet. The wartime famine in aircraft for non-belligerents led to licence-production of Farman F23, Albatros CIII and Morane-Saulnier Parasol aircraft. When peace returned, the RSwN had 25 aircraft and the AAC around 50. While licence-production continued post-war, including Phoenix 122 fighters and Avro 504K trainers, Swedish designs also appeared, including J23 and J24 fighters, S18, S21 and S25 reconnaissance aircraft, and O1 trainers.

In 1926, the two air arms were amalgamated into a separate service, the Royal Swedish Air Force, Flygvapnet. The RSwAF received some new aircraft, including Nieuport 29C-1 fighters and Fokker CV reconnaissance aircraft, but then entered a period of neglect. It was only when the shortage of aircraft became desperate that new aircraft were obtained. These included 12 Bristol Bulldog fighters, Hawker Hart light bombers and Osprey reconnaissance aircraft, with some of both types built in Sweden, ASJA J6s and ASJA RK26s, and 40 de Havilland Tiger Moth trainers. There was a revival in the RSwAF's fortunes as the political situation in Europe showed the threat to national security. The service was organised into eight wings: F1, F4, F6 and F7 operating bombers; F8 fighters; F2 naval reconnaissance; F3 army reconnaissance; and F5 flying training. Some 60 Gloster Gladiator fighters were built, with 40 Junkers Ju86K and 100 Douglas DB-8A bombers and 40 North American NA-16-4 trainers. These aircraft were joined by 60 Republic EP-1 fighters in 1940. Before the wartime demands on the belligerent nations cut Sweden off from supplies again, 72 Fiat CR42 and CR60, and Reggiane Re2000 fighters were purchased, replacing the obsolescent biplane Gladiators, and 80 Caproni Ca313 bombers. Two new fighter wings, F9 and F10, a new reconnaissance wing, F11, and a new bomber wing, F12, were quickly formed. During late 1939 and early 1940, a small but effective Flygvapnet unit fought alongside Finnish forces during the so-called 'Winter War' that followed an attempted Soviet invasion, but otherwise Sweden remained neutral throughout World War II. Swedish-designed aircraft included the Saab 17 light bomber, Saab-18 bomber and Saab-21 fighter, of which some 300 were delivered.

The final batch of Saab-21 fighters was not delivered until after the war ended, by which time outside supplies were resumed, with an initial 50 North American F-51D Mustang fighters followed by a further 90. The Flygvapnet's first jets, 70 de Havilland Vampire F1 fighters, arrived in 1946. The Saab-21 was a twin-boom design with a single pusher propeller, and in 1949, deliveries started of 60 Saab-21R jet fighters, conversions of the basic piston-engined design. New aircraft entered service in quantity, including 200 Vampire FB50 fighter-bombers and T55 trainers, 60 de Havilland Mosquito NF19 night fighters and 70 Supermarine Spitfire PR19 reconnaissance fighters. Piston-engined aircraft remained an option at this time due to the short range of many of the early jet fighters, but the Mosquito night-fighters were replaced during the early 1950s by 60 de Havilland Venom NF51. A new purpose-designed Swedish fighter, the Saab-29 Tunnan, or 'barrel' (an appropriate name) appeared at this time, along with the first Saab-91 Safir piston-engined trainers.

In 1956, 120 Hawker Hunter F4 jet fighters were introduced, and the Saab-32A Lansen attack aircraft entered service, as did 16 Hunting Pembroke C52 light transports. The first helicopters, Vertol 44s, entered service in 1957. Saab-35 Drakens (Dragon) started to replace the Vampires and Saab-29s in 1959. By this time, the pattern had emerged of Sweden maintaining substantial reserve forces, a position of armed neutrality, with a core of professionals augmented by conscripts under train-

ABOVE: *The Royal Swedish Air Force's first indigenous jet was the Saab-29 Tunnan, which appropriately enough translates as 'barrel'. (© Saab AB/ Ingemar Thurson)*

ABOVE: *The Saab-32 Lansen entered service during the mid-1950s. (© Saab AB/ Ingemar Thurson)*

ABOVE: *At the end of the 1950s, the Saab-35 Draken, or Dragon, entered service, with the Royal Swedish Air Force, and subsequently became the first sophisticated combat aircraft to be exported by Sweden. (© Saab AB/ Hans Olof Arpfor)*

ABOVE: *Although Sweden remained neutral throughout the Cold War, at its peak, the nation's air defences and strike capability rested on the Sabb-37 Viggen. (© Saab AB/ Ingemar Thurson)*

ing. Aircraft numbers were accordingly far higher than the size of the standing air force would indicate. Two other characteristics also helped the personnel/combat aircraft ratio: the lack of a substantial MR force, reflecting Sweden's geographical position, and the small size of the air transport element, reflecting the absence of overseas commitments. Regular replacement of aircraft also became the pattern, with the Saab-37 Viggen (Viking) starting to replace the Saab-32 Lansens during the late 1960s, before replacing the Saab-35 Drakens during the 1970s. Vertol 44s were replaced during this period by Boeing-Vertol 107s, and transport capability increased with the purchase of two Lockheed C-130E Hercules, augmenting a force of seven Douglas C-47s. External purchases of aircraft continued, including 58 Scottish Aviation Bulldog trainers delivered during the early 1970s. An advanced trainer, the Saab 105, also entered service.

In recent years, the Flygvapnet has been through another upgrading exercise, replacing the Saab-37 Viggens with the JAS39 Gripen, which became operational during the late 1990s. As with the Viggen, the Gripen uses a standard airframe as the basis for attack, interceptor and reconnaissance variants, in an attempt to maintain an affordable indigenous aircraft development and production facility, although Sweden is now more active in selling combat aircraft in world markets than in the past. Saab's brief move into the airliner market, with the Saab 340 and 2000, also provided the basis for a low-cost AEW aircraft, the S100 Argus. In 1998, the helicopters of the Flygvapnet and the naval and army air arms were merged into a Helicopter Wing, Helicopterflottij. Despite remaining neutral throughout the Cold War period, Sweden has also reduced the size of its armed forces since the collapse of the Warsaw Pact.

The Flygvapnet has 7,700 personnel, of whom 1,800 at any time are reservists fulfilling their annual training commitment. This is a reduction of more than 50 per cent in personnel strength over the past 30 years, although the true reduction is slightly less due to the transfer of some personnel into the Helicopterflottij. The use of many reserve personnel may account for the relatively low average annual flying hours of between 110 and 140. A total of 204 JAS39 Gripen are entering service with deliveries to be completed by 2006, with many of the

Sweden

ABOVE: *Latest arrival in RSwAF service is the Saab-39 Gripen, on which collaboration has taken place with BAe, largely with an eye on the export market, essential to achieve the volume production to make a modern combat aircraft affordable. (Swedish Defence Images)*

earlier versions being upgraded, and these aircraft undertake interceptor, strike, reconnaissance and combat training, replacing some of the Viggen squadrons. The Viggen force remains substantial, with 65 interceptors recently upgraded to JA37D standard, while there are 48 upgraded AJS37 Viggen in the attack, interceptor and reconnaissance role. The Sk37E variant of the Viggen is used for ECM, with nine aircraft of this type. There are no tanker aircraft, but supporting the interceptor force are six Saab S100B Argus (Saab 340) in the AEW role, with a seventh aircraft for VIP duties, and two ELINT S102B (Gulfstream IV), with a third aircraft as a light transport. The main transport force is now provided by eight Lockheed C-130E/H Hercules, known as the Tp84 in Sweden, some of which are being converted to provide in-flight refuelling, as well as three Beech Super King Airs (or Tp101). A Cessna Citation II is used for communications duties, and two Sabreliners are trials aircraft for new equipment. Training uses 106 Saab 105 (Sk60) jets and 60 Scottish Aviation Bulldogs (Sk61), with both types also used for communications and liaison duties. AAM include AIM-9L Sidewinder and AIM-120 AMRAAM, with Maverick ASM.

HELICOPTER WING/ HELICOPTERFLOTTIJ

Formed: 1998

In 1998, the three Swedish armed forces formed a combined Helicopter Wing with 1,000 personnel on secondment from the other services. Equipment included ex-Flyvapnet Super Puma (HKP10 in Sweden) SAR helicopters, ex-Marinflget, or Naval Aviation, ASW Boeing-Vertol 107 (HKP4) Retrievers and Agusta-Bell 206 (HKP6A/B) JetRangers, and ex-Armeflygkar, or Army Air Corps, Agusta-Bell 204B (HKP3C) and 206A, BO105CB (HKP9) and Hughes 300C (HKP5B). The Royal Swedish Navy had been operating helicopters since 1958, while the Swedish Army had started an aviation element in 1964. The Helicopterflottij initially maintained the 13 squadrons provided by the constituent services, but in 1999 these were reorganised into four peacetime battalions. Standardization has started with the medium-lift helicopter chosen under the Nordic Standard Helicopter Programme, with 18 NH90s (and seven on option) replacing the 107s from 2004, while 20 A109s are entering service, replacing the 204Bs and 206s as well as the 300Cs.

At present, transition is taking place with 20 A109s replacing 15 Hughes 300C and up to 29 206s by 2007 on light transport and training duties, suggesting either further follow-on orders or a reduction in the ultimate size of the Helicopterflottilj. From 2004, 18 NH90s will replace 14 Boeing-Vertol 107 (Hkp4 in Sweden) and eight Agusta-Bell AB204A/B helicopters on ASW, SAR and transport duties. There are also 11 AS332M Super Puma on SAR with five AB412 (Hkp11). Anti-tank helicopters are 20 BO105CB (Hkp9A), but there are plans for up to 24 new attack helicopters. ASW and MR is supported by a CASA C212-200 Aviocar. Basic training uses the Flygvapnet's Bulldogs.

Switzerland

POPULATION: 75 million

LAND AREA: 15,941 square miles, 41,310 sq km

GDP: $245bn (£171.3bn), per capita $30,017 (£20,990)

DEFENCE EXPENDITURE: $3bn (£2.1bn)

SERVICE PERSONNEL: 3,600 active, plus around 23,000 conscripts for 15 weeks, 351,200 reserves.

ABOVE: *The early 1930s saw the Swiss Air Force operating French Dewoitine D26 fighters. (Swiss Air Force)*

SWISS AIR FORCE / SCHWEIZER LUFTWAFFE

Formed: 1939

Swiss military aviation dates from the creation of an Air Troop, Fliegertruppe, in 1914, with eight aircraft of Aviatik, Bleriot, Henri Farman, Morane and Schneider manufacture. Although Swiss pilots flew with the French Aviation Militaire during World War I, Switzerland itself remained neutral. To overcome the shortage of aircraft from the belligerent countries, Swiss designed aircraft were put into production, including the Haefeli DH1, DH2 and DH3 observation aircraft. In 1919 the force was reorganized and renamed as the Militar-Flugwesen, by which time it had a hundred, mainly Swiss, aircraft. Some foreign aircraft were acquired, including war-surplus Fokker DVII fighters, but the 1920s also saw continued production of Swiss aircraft, including Haefeli M7 fighters, and DH5 and M8 bombers. The late 1920s and early 1930s saw foreign designs produced under licence, including Dewoitine D9, D26 and D27 fighters, Potez 25 general-purpose aircraft and Fokker CVE reconnaissance-bombers, while the M-F also obtained Hawker Hind bombers and de Havilland Moth and Tiger Moth trainers.

Immediately before World War II, Potez 63 fighter-bombers were obtained, as well as 90 Messerschmitt Bf109E fighters and 13 Bf108 liaison aircraft. Morane-Saulnier MS406C fighters and Bucker Bu131 Jungmann and Bu133 Jungmeister trainers were built under licence for what had now become the Swiss Air Force, Schweizerische Flugwaffe, although still a corps of the Swiss Army. At the outbreak of World War II in 1939, during which Switzerland again remained neutral, the SF was operating 100 fighters and 100 AOP aircraft. Additional aircraft were obtained from Germany early in the war, including Bf109Es, Fieseler Fi156 Storch AOP aircraft and Bucker Bu181 Bestmann trainers. Despite Switzerland's official neutrality and the purchase of German aircraft, on a number of occasions, there was air-to-air combat with German aircraft, especially during the early years of the war.

The first post-war aircraft were 100 North American F-51D Mustang fighter-bombers and 40 T-6 Harvard trainers. In 1949 and 1950, the first jets were introduced, 75 de Havilland Vampire FB6 fighter-bombers, followed by a further 100 of these aircraft built in Switzerland to replace the Mustangs. The Vampires were followed by 250 licence-built de Havilland Venom FB50 fighter-bombers, and, in 1958, 100 Hawker Hunter F58 fighters entered service. Swiss-designed aircraft continued to enter service, including Pilatus P-2 and P-3 trainers. During the late 1960s, 57 Dassault Mirage IIIS fighters were built under licence for what had now become the Swiss Air Force and Anti-Aircraft Command, Kommando Flieger und Fliegabwehrtuppen. The Mirages were joined by Mirage IIISR recon-

ABOVE: *Morane-Saulnier D380 fighters were licence-built and in service during World War II, although Switzerland remained neutral. (Swiss Air Force)*

naissance aircraft, the KfuF entered the 1970s also operating Hunters and Venoms, as well as 30 Alouette II and 90 Alouette III, and 20 Bell 47G Sioux, helicopters. A few Bucker trainers and three Ju52/3M transports were survivors from the war years.

During the 1980s, the Venoms were retired and replaced by Hunters, upgraded to carry Maverick air-to-surface missiles, while the Hunter was replaced in the line of battle by Northrop F-5E/F Tiger IIs. Vampires were replaced by the BAe Hawk F58 and T68, providing ground attack as well as training capability. Tactical transport was improved with 15 AS332M Super Puma helicopters delivered during the early 1990s. Pilatus PC-7 Turbo Trainers were also introduced, with PC-9s for communications duties. Despite being neutral, the ending of the Cold War saw a reduction in the armed forces starting in 1995, largely by reducing the length of training periods. New equipment continued to enter service, with the F/A-18C/D Hornet assembled in Switzerland. The Hunters were retired in 1995.

The Swiss Air Force has a mobilised strength of 30,600 personnel. Flying hours are between 150-200, but closer to 50 hours for the high proportion of reservists. There are eight fighter squadrons, with two operating 26 F/A-18C and seven F/A-18D Hornets, and six operating 89 F-5E and 12 F-5F Tiger II, some of which may be sold. There are still 16 Mirage IIIRS and four IIIDS in two reconnaissance squadrons, which may be replaced by additional F/A-18Cs, possibly from Kuwait if that country's plan to buy F/A-18Es goes ahead. Transport is provided by a squadron with 18 Pilatus PC-6, two Dornier Do27, a Falcon 50, and a Learjet 35A. Helicopters include three squadrons with 15 AS332M-1 Super Puma and 68 SA316 Alouette III, but 12 Cougar are entering service. Training uses 19 BAe Hawk Mk66 and 38 Pilatus PC-7, with 11 PC-9 target tugs. The force is tactical without tankers or a strategic transport element. Some training takes place in Sweden since supersonic aircraft soon reach this small country's borders! AIM-9 Sidewinder and AIM-120 AMRAAM are used.

ABOVE: *The first of 100 Hawker Hunters entered service in 1958. (Swiss Air Force)*

ABOVE: *The F/A-18C is the latest addition to Switzerland's air defences. (Swiss Air Force)*

ABOVE: *Not surprisingly, this Swiss Super Puma is equipped for landing in the snow. (Swiss Air Force)*

Syria

ABOVE: *MiG-23s equip five Syrian Air Force fighter squadrons. (Jane's)*

POPULATION: 16.5 million

LAND AREA: 71,210 square miles, 184,434 sq km

GDP: $13.8bn (£9.6bn), per capita $7,818 (£5,467)

DEFENCE EXPT: $775m (£542m)

SERVICE PERSONNEL: 321,000 active (65% conscript), plus 354,000 reserves.

SYRIAN AIR FORCE

Formed: 1946

Syria gained independence in 1943 having previously been a French-mandated territory, but foreign forces were not completely withdrawn until 1946. Despite being formed in 1946, the first Syrian Air Force aircraft did not enter service until 1949, including Fiat G46 and G59, and DHC-1 Chipmunk, trainers, with Beech C-45 and Douglas C-47, as well as some French-assembled Ju52/3M, transports. Ex-Armee de l'Air bases were taken over. Despite a British arms embargo between 1951 and 1953, the SAF managed to obtain 30 de Havilland Vampire FB52 jet fighter-bombers via Italy, although these were quickly passed on to Egypt. Syria's first combat aircraft did not enter service until 1953, with 23 Gloster Meteor F8 fighters, NF13 night-fighters and T7 trainers, as well as 40 Supermarine Spitfire 22 fighters.

In 1955, Soviet aid began, promising 25 MiG-15 fighters and personnel to train Syrian air and ground crew. Of the aircraft that did arrive, all were destroyed on the ground in Israeli raids during the 1956 Suez crisis. Syria obtained some MiG-15s and 60 MiG-17 fighters later. An attempt by Egypt and Syria to form a United Arab Republic in 1958 was short-lived as Syria withdrew following a *coup d'etat.* In the 1960s, the SAF received MiG-21 interceptors, Ilyushin Il-14 transports, and Mil Mi-1 and Mi-4 helicopters, as well as Yak-11 and Yak-18 trainers. The June, 1967, Arab-Israeli War saw up to 75 per cent of this equipment destroyed. The USSR made good Syria's losses, then increased the SAF's strength, with 90 MiG-21 interceptors, 80 MiG-15, MiG-17 and Sukhoi Su-7B fighter-bombers, and transports, helicopters and trainers. Syria remained committed to the Soviets and continued to introduce new aircraft throughout the late 1970s and 1980s, including MiG-23, MiG-25 and, in 1989, MiG-29 interceptors, Su-20, Su-22 and Su-24 strike aircraft. A small naval air arm developed, using SAF personnel. While transport aircraft remained an assortment of Soviet types, all in small numbers, a substantial helicopter force was established with more than 100 Mi-8 and Mi-17 transport helicopters, and 36 Mi-24 attack helicopters.

Syria opposed the 1990 Iraqi invasion of Kuwait and committed troops to Saudi Arabia to assist in the liberation of Kuwait, but only helicopters and transport aircraft were deployed, with no Syrian combat missions.

The Syrian Air Force has 40,000 personnel, having quadrupled in size over the past 30 years. A further 60,000 personnel are in a separate Air Defence Command, which is responsible for anti-aircraft defences. Flying hours are very low, with an average of 30 per year, suggesting poor combat readiness. There are 17 fighter squadrons, with eight operating 170 upgraded MiG-21PF/MF/bis, five with 90 MiG-23BN, two with 30 MiG-25 and two with up to 54 MiG-29A. Another ten squadrons operate in the strike role, with five operating 90 Su-22, two with 44 MiG-23BN, two with 20 Su-24 and a recently formed squadron of 14 Su-27s. Reconnaissance units have six MiG-25R and eight MiG-21H/J. There are 36 Mi-24 attack helicopters and 55 SA342L Gazelle anti-tank helicopters, as well as at least 100 Mi-8/Mi-17 transport helicopters, 20 Mi-2 and ten heavy lift Mi-6. Transport aircraft include four Antonov An-26, four Ilyushin Il-76M and two Tupolev Tu-134B, with a VIP flight operating two Yak-40 and two Falcon 20Fs. Training uses 70 L-39A/ZO Albatros, 30 MiG-17F and 15 MiG-15UTI, six MiG-29UB, 40 L-29 Delfin, 48 MBB Flamingo and six Mushshak.

SYRIAN NAVAL AVIATION

Syrian Naval Aviation operates 20 Mil Mi-14 and four Kamov Ka-25 ASW helicopters, all operated from shore bases by Syrian Air Force personnel. The small force of frigates and missile craft does not include ships capable of carrying helicopters.

Taiwan

POPULATION: 22.1 million

LAND AREA: 13,890 square miles, 35,975 sq km

GDP: $314bn (£220bn), per capita $16,800 (£11,748)

DEFENCE EXPENDITURE: $17.6bn (£12.3bn)

SERVICE PERSONNEL: 350,000 (50% conscript) active, plus 1,657,000 reserves.

ABOVE: *Taiwan's Mirage 2000-5s mean that the country is no longer completely dependent on arms supplies from the USA, and additional security comes from its own indigenous Ching-Kuo project. (Jane's)*

REPUBLIC OF CHINA AIR FORCE

Formed: 1949

The late 1940s saw Communist victory in China, with Nationalist forces forced to withdraw to the offshore island of Formosa in 1949. With a number of smaller islands, this formed the basis of the new state of Taiwan, at one time known as 'Nationalist China'. Taiwan is not recognised by the Chinese People's Republic, which still claims sovereignty.

At the end of World War II, the Central Government Air Force had been reorganized and renamed the Chinese Air Force. The surviving wartime aircraft, supplied by the USA, were augmented by ex-USAF aircraft, including additional North American F-51 Mustang and Lockheed P-38 Lightning fighters, and North American B-25 Mitchell bombers. Aircraft types new to the Chinese Air Force included Republic F-47 Thunderbolt fighter-bombers and Consolidated B-24 Liberator bombers, while 250 ex-RCAF de Havilland Mosquito fighter-bombers were purchased. Many of these were lost in the withdrawal to Taiwan, where the new Chinese Nationalist Air Force started with just 160 aircraft. In 1951, US military aid started, providing Republic F-84G Thunderjets, and these were followed in 1954 by North American F-86F Sabres to replace the Thunderbolts, while the remaining Lightnings were replaced by Republic RF-84F Thunderflash reconnaissance-fighters. The next generation of aircraft into service was 50 North American F-100 Super Sabres, soon joined by Lockheed F-104A Starfighters. In 1963, Lockheed F-104G Starfighters replaced the Thunderjets and the earlier F-104As. During the late 1960s, Northrop F-5A/B fighters took over from the F-86 Sabres. A number of transport aircraft were also supplied by the USA, including Fairchild C-119 packets and C-123 Providers, as well as the necessary helicopters and trainers. Grumman S-2 Trackers were acquired for MR, and re-engined with turboprop engines in 1991, before transferring to the Navy in 2000.

Attempts by the United States to seek an accommodation with Communist China have left Taiwan at the mercy of the prevailing diplomatic mood, with arms supplies fluctuating according to the state of relations between the US and China. Overall, the US has always supported Taiwanese autonomy, while the country's successful industrialisation enables it to purchase aircraft, although many countries are wary of upsetting the People's Republic. Northrop F-5E/F Tiger II interceptors were introduced during the 1980s. An Indigenous Defence Fighter (IDF) project during the 1990s led to the A-1 Ching-Kuo interceptor, although orders for this aircraft were cut from the planned 250 to 130 in 1992. IAI Kfir C7 fighters were considered at one stage. In 1991, the People's Republic acquired Su-27 interceptors from Russia, and this prompted the CAF to buy 60 Dassault Mirage 2000-5 attack aircraft and 150 Lockheed Martin F-16A/B interceptors. Older aircraft have been upgraded, and substantial numbers are held in store, for use by reservists or against attrition losses.

The CAF is organized on US lines. An F-16 conversion unit is based in the United States. It currently has 68,000 personnel, its size having remained steady over the past quarter century since here the Cold War continues. There are 90,000 reservists. Flying hours are fairly high, at around 180. There are three fighter squadrons with 47 Mirage 2000-5EI and 11 2000-5DI, and 20 fighter/ground attack squadrons, of which seven have 126 F-16A/B, six have 128 Ching-Kuo, one has 22 AT-3/3B Tzu Chung, and six have 130 F-5E/F Tiger II, with another 70 in store. A number of F-5Es have been converted to RF-5E reconnaissance standard, while one of the F-16 squadrons can undertake reconnaissance missions. Supporting the combat aircraft are six Lockheed E-2C/T Hawkeye AEW aircraft. Transport use 19 Lockheed C-130H Hercules and five Douglas C-47s, with 11 Beech 1900 and three Fokker 50, while two Boeing 727-100 and a 737-800 provide VIP transport. There are 17 Sikorsky S-70C Black Hawk helicopters in the utility role. There are 15 A-CH-1B Chung Sing in the forward air control role. Training uses some 30 AT-3/3B Tzu Chung, 40 Northrop T-38A Talon and 36 Beech T-34C Turbo Mentor. Missiles include AAM AIM-4 Falcon, AIM-9J/P Sidewinder, Shafrir, Skysword I/II,

Taiwan

ABOVE: *Taiwanese Ching Kuo (Mader-Acaes Collection)*

Mica and R-550 Magic II, with AGM-65B/G Maverick ASM.

Plans to procure a tactical transport to replace the C-119 Packets, retired in 1997, could lead to a C-27A Spartan purchase. The threat posed by China's development of the J10 advanced fighter is likely to lead to a new fighter type, such as the Typhoon or the F-22 in due course.

REPUBLIC OF CHINA NAVAL AVIATION

The Republic of China Navy initially used aviation for communications and liaison, but it now operates ASW helicopters from frigates and destroyers, as well as having 32 Grumman S-2T Trackers for ASW and MR. There are 20 Sikorsky S-70 Seahawks and 12 MD500 helicopters. Plans exist for up to 12 SH-2Gs for Knox-class guided missile frigates being leased from the USN. Lockheed P-3 Orions could be acquired, or there could be a navalised version of a tactical transport to be chosen by the Air Force. There are 12 destroyers and 21 frigates capable of operating helicopters.

REPUBLIC OF CHINA ARMY AVIATION

Formed: 1970

The Republic of China Army became involved with aviation in 1970, with the introduction of Bell UH-1H Iroquois helicopters. It has 12 Boeing CH-47SD Chinook heavy lift helicopters, up to 90 UH-1H Iroquois in the utility role, and 62 anti-tank Bell AH-1W Super Cobra helicopters. There are 39 OH-58D Kiowa on observation and liaison duties. Training uses 30 TH-67 and 15 TH-55A helicopters. An Iroquois replacement is under consideration.

Tajikistan

POPULATION: 6.3 million

LAND AREA: 55,240 square miles, 144,263 sq km

GDP: $1.3bn(£909m), per capita $1,000(£699)

DEFENCE EXPENDITURE: $82m (£57m)

SERVICE PERSONNEL: 6,000 active.

TAJIKISTAN AIR FORCE

Tajikstan is a member of the Commonwealth of Independent States, CIS, which has provided troops and aircraft to help the Tajik government in its struggle with Muslim rebels. Forces from Russia, Kazakhstan and Uzbekistan are based in the country. The government is forming an air force, initially with ten Mi-8 and five Mi-24 helicopters, and in the longer term may also acquire Su-25s from Belarus, presumably with contract pilots.

Tanzania

POPULATION: 34.5 million

LAND AREA: 361,800 square miles, 939,706 sq km

GDP: $8bn (£5.6bn), per capita $737 (£515)

DEFENCE EXPENDITURE: $144m (£100m)

SERVICE PERSONNEL: 27,000 (inc conscripts) active, plus 80,000 reserves.

TANZANIAN PEOPLE'S DEFENCE FORCE AIR WING

Formed: 1964

Tanzania was formed on the federation of two former British colonies, Tanganyika and Zanzibar, in 1964. The Tanzanian People's Defence Force Air Wing came into existence with Luftwaffe assistance, including six Nord Noratlas transports, eight Dornier Do28 liaison and communications aircraft, and nine Piaggio P149 trainers. In 1965 the aid ended abruptly before deliveries could be completed after Tanzania recognised the German Democratic Republic and East German aid took over. As a result, the aircraft of the TPDFAW included an Antonov An-2, five DHC-3 Otters and four DHC-4 Caribou transports, and seven Piaggio P149 trainers. The Air Wing bought Chinese combat aircraft supported by Western transports and helicopters, taking first the Shenyang F-4 (MiG-17), then the Shenyang F-6 (MiG-19) and finally the Chengdu F-7 (MiG-21) into service, with a squadron of each successive type, although attrition has reduced the numbers of the older aircraft.

Training of combat pilots takes place in China after initial training in Tanzania. Today, the TPDFAW has 3,000 personnel. It has three combat squadrons operating 11 F-7s, ten F-6s and three F-5s, as well as having two MiG-15UTI. There are four DHC-5D Buffalo transports, three HS748 and a single Harbin Y-7, with two Fokker F28 Fellowships and a BAe125-700B for VIP duties. Four Agusta-Bell 205B, six Bell 206 JetRanger and four ex-South African Alouette III helicopters cover transport and communications duties, with two Cessna 404s, seven 310s and a 206. Five Piper Cherokees provide basic training. There is a substantial SAM force with more than 150 SA-3, SA-6 and SA-7 launchers.

A police air wing has a Cessna 206, two Bell 206L LongRangers and a Bell 47G.

Thailand

POPULATION: 61.6 million

LAND AREA: 198,250 square miles, 519,083 sq km

GDP: $123bn (£86bn), per capita $8,500 (£6,004)

DEFENCE EXPENDITURE: $2.5bn (£1.75bn)

SERVICE PERSONNEL: 306,000 (50% conscript) active, plus 200,000 reserves.

ROYAL THAI AIR FORCE

Formed: 1937

Siamese army officers were sent to France for flying training in 1911. They returned home in 1913 with four Nieuport and four Bleriot aircraft. The then Kingdom of Siam entered World War I on the side of the Allies, sending a contingent of the new Siamese Flying Corps to Europe as part of an expeditionary force. It was not until the end of the war that the first Siamese operational sorties were flown, but more than a hundred Siamese officers and NCOs received flying training in France during the war. The SFC spent a short period as part of the Allied Army of Occupation in the Rhineland before returning home in 1919 with ex-wartime aircraft, including Spad SVII and SXIII and Nieuport-Delage ND29 fighters, Breguet Br14A2 and 14B2 reconnaissance-bombers, and Nieuport trainers. In 1919, the name was changed to the Royal Siamese Aeronautical Service.

In 1920, some of the Br14s were converted to operate a domestic airline, while the remainder operated reconnaissance and liaison duties. Pilots were trained in the USA, UK, France and Italy, as well as in Siam. In 1930, 20 Avro 504 trainers were purchased, and a further 50 produced in Siam under licence. Despite evaluating new aircraft, no further orders were placed until 1934, when 72 Vought V935a Corsair AOP aircraft were built in Siam to replace elderly Br14s. These were followed by 12 Curtiss Hawk II and 12 Hawk III fighters, with another 25 Hawk IIIs assembled in Siam.

Another new name, the Royal Siamese Air Force, was adopted in 1937, changing to the Royal Thai Air Force in 1939 when the national name changed.

Further re-equipment occurred in 1937, with six Martin 139 bombers followed in 1939 by an order for 25 Curtiss Hawk 75N fighters and a number of North American NA-69 bombers and, in 1940, six North American NA-68 fighters. Only the Hawks were delivered, with the later aircraft being taken off their ships at the Philippines and Hawaii respectively to be pressed into US service. Some support for Japan amongst certain sections of the Thai community led to the acquisition of nine Mitsubishi Ki21 bombers and nine Tachikawa Ki55 trainers. A border dispute led to an invasion of French Indo-China by Thailand in January, 1941, with combat between Thai and French aircraft until Japan brokered a truce in May, after which the Vichy French allowed Japan to use bases in French Indo-China. Japan invaded Thailand in December, 1941, and the Thai Government, faced with overwhelming odds, arranged a cease-fire and surrender after a few days.

Thailand was officially an ally of Japan for the remainder of the war, but RTAF per-

Thailand

ABOVE: *Many air forces still use the veteran Cessna T-37 as their intermediate trainer, including the Royal Thai Air Force.*

sonnel were confined to non-combatant roles. A number of RTAF members helped an underground movement, and flew Allied agents into and out of the country. Pro-Allied and pro-Japanese RTAF personnel were segregated. Additional Japanese aircraft were supplied, including a few Nakajima Ki27 and Ki43 fighters, Mitsubishi Ki30 bombers and Mansyu Ki79 trainers. Peace found the RTAF operating abandoned Japanese aircraft and some surviving pre-war aircraft. RAF personnel were seconded to rebuild the RTAF, providing 30 Supermarine Spitfire Mk14s and a number of Fairey Firefly FR13s. These were followed by 20 Miles Magisters, 42 North American T-6G Texans, de Havilland Tiger Moths and DHC-1 Chipmunks for training.

US military aid started in Thailand before the country became a founder member of the South East Asia Treaty Organisation, SEATO, in 1954. Thai personnel were trained in the USA. New aircraft included 50 Grumman F8F-1 and F8F-1B Bearcat fighter-bombers and additional Texan trainers. There were also Stinson L-5 Sentinels, Piper L-18 Super Cubs and Cessna O-1 Bird Dogs for AOP duties; Fairchild 24W, Cessna 170 and Beech C-45 communications aircraft; Westland S51 Dragonfly (S-51), Sikorsky S-55 and Hiller 360 helicopters; and in 1957, another 75 Texans. The first jet aircraft also arrived in 1957, 30 Republic F-84G Thunderjet fighter-bombers and Lockheed T-33A trainers. Thunderjets and Bearcats were replaced in 1962 by North American F-86F Sabres, with additional Sabres in 1966, when the RTAF also received its first Northrop F-5A/B fighters. Transport aircraft were also provided, with Fairchild C-123 Providers and Sikorsky CH-34C helicopters. Cessna C-37B trainers were also supplied.

Thailand's importance grew during the 1960s and early 1970s, a stable country in an unstable region, as the Vietnam War spread into both Laos and Cambodia. COIN became important, with the RTAF becoming one of the few operators of the Fairchild OV-10C Bronco. During the 1980s, Northrop F-5E/F Tiger II and Lockheed Martin F-16A/B interceptors entered service, as well as Lockheed C-130H Hercules transports.

Over the past 30 years, Thailand has placed a high priority on defence. A mixture of surplus aircraft, usually from the US, and new aircraft has produced a modern air force. The Asian economic crisis of the late 1990s seriously affected procurement programmes, including an order for F/A-18C/D Hornets.

Over the past 30 years, the RTAF has almost doubled its personnel from 25,000 to 48,000. Flying hours are relatively limited at an average of 100 despite the high priority accorded defence. Mainstay of the RTAF are four fighter squadrons, three having 43 F-16A and eight F-16Bs, and a fourth with 14 F-5A/B. The F-16s will be upgraded when funding permits. Three fighter/aggressor squadrons operate 33 upgraded F-5Es and five F-5Fs. For COIN duties, 25 Alphajets, upgraded by DASA, have replaced the OV-10C Broncos in one squadron, while of four other squadrons in this role, one has four AC-47 armed-transports, two have 22 Fairchild AU-23 (PC-6B) and one has 19 GAF Nomads. Reconnaissance is provided by three RF-5As, while three IAI-201s are used for ELINT. Transport is provided by three squadrons, one of which has seven Lockheed C-130H and five C-130H-30 Hercules; one with three C-123K and four BAe748, and the third with six Alenia G222. The VIP Royal Flight has an Airbus A310-324, a Boeing 737-400, three Beech King Air 200, two BAe748 and three Fairchild Merlin IV, as well as two Bell 412 and three AS532A2 helicopters. Communications and liaison aircraft include three Commanders, a King Air E90, two Queen Air, three Basler

Thailand

Turbo-67s, and two O-1 Bird Dogs, while survey work occupies two Learjet 35A, three Merlin IVA and three GAF N-22B Nomads. There are two helicopter squadrons, one operating 17 Sikorsky S-58T and the other with 25 Bell UH-1H Iroquois, while there are also five Bell 412 for transport and SAR. Training uses 30 Fantrainer 400/600, ten SF260MT, and six Bell 206, as well as 22 PC-9s, and a number of T-37s and L-39ZA Albatros, and 23 CT-4 Airtrainers.

ROYAL THAI NAVY AIR ARM

Originally a shore-based force with MR Grumman S-2F Trackers and SAR HU-16 Albatross amphibians, it was updated during the 1970s and 1980s with Canadair CL-215 amphibians and Fokker F27s, as well as Dornier Do228s for EEZ and anti-smuggling patrols. Several warships entered service able to carry helicopters for ASW and Bell 212 and Sikorsky S-70 were introduced. Lockheed P-3 Orions entered service during the early 1990s, providing longer-range MR. The most recent advance has been the acquisition of a small aircraft carrier built in Spain, *Chakri Naruebet,* to operate eight AV-8S Matador (Harrier) V/STOL fighters and up to six S-70 helicopters; financial problems have prevented effective operations.

Some 1,700 of the Royal Thai Navy's 68,000 personnel are involved with naval aviation. There are 14 LTV A-7E and four TA-7E Corsair II attack aircraft based ashore, while seven AV-8S and two TAV-8S Matadors can operate from the carrier. There are three Lockheed P-3A/UP-3T Orions for MR, while six Grumman S-2F Trackers remain in service. Additional coastal MR and EEZ patrols are provided by five Fokker F-27-200/400M, six Dornier Do228-212, two Canadair CL-215 amphibians, and five GAF N24 Searchmasters. Helicopters include six Sikorsky S-70B Seahawk and seven Bell 212ASW for ASW, with five Bell 214ST and four UH-1H for SAR and transport, with five Sikorsky S-76N. Four Cessna O-1 Bird Dogs and four U-17Bs support a strong force of marines.

ABOVE: *The introduction of the Sikorsky S-70 Sea Hawk helicopter for shipboard operations has marked a step change in the Royal Thai Navy's capabilities. (Sikorsky)*

ROYAL THAI ARMY AIR DIVISION

Although badly affected by the financial problems affecting Asia in the late 1990s, the Royal Thai Army is developing an air mobility brigade. An attack helicopter squadron was established in 1990, well before the downturn, originally with a nucleus of eight Bell AH-1F Cobra helicopters, with the three surviving machines likely to be joined by a number of ex-US Army aircraft. Two Sikorsky UH-60L Blackhawks were introduced, but the economic crisis has prevented plans to order 36 additional aircraft, and instead 30 additional US-surplus UH-1Hs have been introduced bringing the number of these aircraft to 98, with another 38 Bell 212s, as well as ten 206A JetRangers. There are six Boeing CH-47C/D Chinooks for heavy lift. Fixed-wing aircraft include two Shorts 330UTT, two BAe Jetstream 41, two CASA C212-300, two Beech 1900-C1 and two King Air 200, ten Cessna O-1A Bird Dog and five U-17. Training uses seven T-41D Mescalero (Cessna 172), 18 Maule M7-25 Super Rocket and 40 Schweizer TH-300C helicopters.

The Royal Thai Border Police also operate a number of helicopters, including 27 Bell 205A, 14 Bell 206, three Bell 212 and six Hiller UH-12, as well as a number of utility aircraft.

Togo

POPULATION: 4.7 million

LAND AREA: 22,000 square miles, 54,960 sq km

GDP: $1.6bn (£1.1bn), per capita $1,481 (£1,035)

DEFENCE EXPENDITURE: $31m (£21.8m)

SERVICE PERSONNEL: c.9,000 active (inc selective conscription).

TOGO AIR FORCE/FORCE AERIENNE TOGOLAISE

Formed: 1960

Togo formed a small air force on achieving independence in 1960, and the French Community gained the standard arms 'package' of a Douglas C-47 transport, two Max Holste 1521M Broussard communications aircraft, and an Alouette II helicopter. A 1963 agreement with France on aircrew training survives to the present. A light attack capability was acquired during the 1980s with Aerospatiale TB.30s and Embraer EMB-326 Xavante as well as armed Alphajets. The main roles are transport and liaison, with a small transport and helicopter fleet including two VIP aircraft.

Today, the FAT has 250 personnel. The three remaining TB.30 Epsilon are relegated to training, leaving four Alphajets and four EMB326G Xavante for attack. Just one DHC-5 Buffalo remains in service, with two Beech 58 Baron and two Reims-Cessna F337 on communications duties, with three SA318B Lama II and an Alouette III helicopter. VIP transport is provided by a Boeing 707-320B, a Fokker F28-3000 Fellowship and an AS332 Super Puma helicopter.

Tonga

POPULATION: 90,000

LAND AREA: 270 square miles, 699 sq km

TONGA DEFENCE SERVICES AIR WING

Formed: 1996

The Tonga Defence Services formed an Air Wing in 1996 with a Beech G18S for EEZ patrols and SAR around this South Pacific groups of 150 islands. An American Champion Citabria was acquired for training in 1999.

Trinidad and Tobago

POPULATION: 1.3 million

LAND AREA: 1,980 square miles, 5,128 sq km

GDP: $7.3bn (£5.1bn), per capita $12,800 (£8,951)

DEFENCE EXPENDITURE: $62m (£43.3m)

SERVICE PERSONNEL: 2,700 active.

TRINIDAD AND TOBAGO DEFENCE FORCE AIR WING

Formed: 1977

The Trinidad and Tobago Defence Forces Air Wing is part of the Coast Guard, and has 50 personnel. It operates two Fairchild C-26 (Merlin), a Cessna 310 and a 410, with a 172 for training, and has just recently introduced two Piper Navajo. There are no armed aircraft. A government-owned company, Helicopter Services, operates two Sikorsky S-76s and three BO105CBS, mainly for charter to commercial customers, but the aircraft are also available for SAR when needed.

Tunisia

POPULATION: 9.7 million

LAND AREA: 63,362 square miles, 164,108 sq km

GDP: $21bn (£14.7bn), per capita $7,100 (£4,960)

DEFENCE BUDGET: $356m (£249m)

SERVICE PERSONNEL: 35,000 (75% conscript) active.

REPUBLIC OF TUNISIA AIR FORCE

Formed: 1956

Tunisia became independent in 1956, and shortly afterwards formed the Republic of Tunisia Air Force with Swedish assistance. The first aircraft, 15 Saab-19D Safir trainers, arrived in 1960, and were joined by two Alouette II helicopters in 1962. Eight Aermacchi MB326B jet trainers entered service in 1966, and were followed soon afterwards by the first combat aircraft, 12 North American F-86F Sabre fighters. These were later joined by North American T-6G Texan trainers and Dassault Flamant transports, as

Tunisia

well as additional Alouette II helicopters. Armed versions of the MB326 were acquired later, as well as SF260 armed-trainers, but Tunisia also started to receive US military aid, with Northrop F-5E Tiger IIs for interception and reconnaissance. Agusta-Bell 205 and UH-1H helicopters were also delivered, as well as two Lockheed C-130H Hercules transports. In 1989, additional F-5Es were obtained from surplus USAF stocks. In 1995, further C-130Bs were provided by the USA, as well as 12 UH-1H and three HH-3 helicopters, while the Czech Republic supplied L-59T Albatros armed-trainers.

The main threat to Tunisia lies in action by Algerian rebels, possibly encouraged by Tunisia's ambitious neighbour, Libya. The RoTAF has 3,500 personnel, having grown almost six-fold over the past thirty years. Mainstay of its fighter force are 12 F-5E and three F-5F Tiger II fighter-bombers, augmented by three MB326K and two MB326L, and 12 L-59 Albatros armed-trainers, as well as 12 armed SF260Ws, plus six in the training role. There are also five SA341 Gazelle anti-tank helicopters, while two HH-3E (SH-3) Pelican helicopters are equipped for ASW, with another two on transport duties, where they operate alongside up to 29 Bell and Agusta-Bell AB205A and UH-1H. Transport aircraft include five Lockheed C-130B and two C-130H Hercules, five Alenia G222 provided by Italy in 2000, three Let-410 and a Falcon 20. A helicopter wing operates 15 Agusta-Bell 205, and 17 UH-1H/N Iroquois, six AS350B Ecureuil, an AS365 Dauphin, six SA313 Alouette II and three SA316 Alouette III.

Turkey

POPULATION: 67.7 million

LAND AREA: 301,302 square miles, 780,579 sq km

GDP: $210bn (£146.9bn), per capita $6,101 (£4,266)

DEFENCE EXPENDITURE: $10.8bn (£7.6bn)

SERVICE PERSONNEL: 515,100 (76% conscript) active, plus 378,700 reserves.

RIGHT: *After the end of World War II, the Turkish Air Force introduced a number of de Havilland Mosquitoes - this is a T Mk3. (BAE SYSTEMS)*

TURKISH AIR FORCE

Turkey has the longest history of military aviation in the Middle East, having ordered Bristol, DFW, Deperdussin, Mars, Nieuport and REP aircraft in 1912, to be flown by foreign pilots for the Army. In 1914, the Turkish Flying Corps was formed with German assistance. During World War I, Turkey was aligned with the Central Powers, operating AEG CIV and Albatros reconnaissance aircraft, Halberstadt DII fighters and, for naval cooperation, some Gotha WD13 seaplanes, all flown by German pilots with Turkish observers.

When the war ended, the Treaty of Versailles banned the Central Powers and their allies, operating or building military aircraft. The ban was circumvented by the formation of the Turkish Air League, Turk Hava Kurumu, in 1925, with public subscription buying a Caudron trainer and an Ansaldo A300. The Air League received assistance from France and some pilots were trained in the UK. In late 1926, some Morane-Saulnier MS53 trainers were delivered, by which time the Turkish Air Force, Turk Hava Kuvvetleri, was operating as part of the Turkish Army. In 1928, Rohrbach RoIII flying boats entered service, followed by 18 Curtiss Hawk fighters and some Fledgling trainers. Development of the THK accelerated during the 1930s, with 20 Breguet Br19B2 reconnaissance-bombers and six Supermarine Southampton MR flying boats. As the pool of trained pilots grew, substantial orders were placed for new aircraft in 1937, with orders for 30 each of Heinkel He111D, Bristol Blenheim I and Martin 139 bombers. There were smaller quantities of Supermarine Walrus amphibians; Avro Anson light bombers and Vultee VIIG fighter-bombers; Hanriot 182 and Westland Lysander army cooperation aircraft; Miles Hawk II and Curtiss-Wright CW22 trainers; and Focke-Wulf Fw58 Weihe communications aircraft. The following year, Hawker Hurricanes and additional Blenheims were ordered. These were joined by de Havilland Dragon and Dragon Rapide light transport and navigational training aircraft. Licence-built Gotha trainers entered service, but an order for 40 Gotha G23 fighters was never fulfilled when the aircraft were diverted to Spain.

Turkey remained neutral during World War II, signing a non-aggression pact with Germany and obtaining aircraft from both sides. The new aircraft included further examples of those already in service, as well as Curtiss Tomahawk IIBs, Fairey Battles, Airspeed Oxfords, Morane-Saulnier MS406 and Focke-Wulf Fw190As. The objective was

Turkey

ABOVE: *Turkey remains one of the operators of the McDonnell Douglas F-4E Phantom, although the aircraft have been upgraded to maintain their effectiveness in modern combat conditions.*

to keep Turkey well disposed towards the combatant nations, although had it not been for the failure of Barbarossa, Turkey could have been invaded. The RAF also supplied spares for He111s, salvaged from aircraft shot down over the UK. The Allies also provided Lend-Lease equipment, including additional Hurricanes and Blenheims, Supermarine Spitfire VBs, Bristol Beaufighters and Beauforts, Martin Baltimores and Consolidated Liberators.

Post-war, the THK received large numbers of war-surplus aircraft, including de Havilland Mosquito FB6 fighter-bombers and T3 trainers; Republic F-47D Thunderbolt fighter-bombers; North American T-6 Harvard and Beech T-11B Kansan trainers, Douglas B-26 Invader bombers; Beech C-45 and Douglas C-47 transports. On joining the North Atlantic Treaty Organisation, NATO, in 1952, Turkey became eligible for the US Military Aid Programme, including aircraft and USAF personnel as additional instructors. The THK received its first jet aircraft in 1952, with the first of 300 Republic F-84G Thunderjet fighter-bombers, while 24 Beech T-34 Mentor trainers were assembled in Turkey. In 1953, Canadair F-86E Sabre Mk2 and Mk4 fighters were introduced, at the same time as the first of Turkey's own MKEK Ugar (Lark) basic trainers. Deliveries of US aircraft continued into the 1960s, with North American F-100C Super Sabre fighter-bombers, Convair F-102A Delta Dagger and Lockheed F-104G Starfighter interceptors, RT-33A tactical reconnaissance aircraft and T-33A trainers, and C-130E Hercules transports, as well as Cessna T-37 trainers. Other aircraft have included Dornier Do27 and Do28 communications aircraft, Piper L-18 AOP aircraft, and Northrop F-5A fighter-bombers.

Despite Turkey's important Cold War strategic position, aircraft had to be withheld on occasion to prevent tension between Turkey and neighbouring Greece flaring up into open warfare. Tension centred on territorial rights in the Aegean Sea and over the island of Cyprus, where a substantial Turkish minority objected to Greek Cypriot demands for union with Greece. These tensions erupted in a Turkish invasion of Cyprus in 1973, with the THK operational over the island and also landing troops. Direct aerial combat with Greek forces was avoided as Greek aircraft would have had to overfly Turkey to reach Cyprus.

NATO policy was to maintain a measure of equality in the quality of equipment supplied between Turkey and Greece. This resulted in a mixture of new aircraft and surplus aircraft supplied under MAP to other NATO countries. The original F-5A force was augmented by former RNethAF aircraft, and the THK's F-4 Phantom force was later augmented by ex-USAF aircraft.

In 1975, the THK took charge of the

ABOVE: *US attempts to maintain a balance of power between Greece and Turkey have meant that Turkey has also received F-16s, in this case an F-16D normally used for conversion training.*

Turkish Navy's 33 shore-based Grumman S-2A/E Tracker MR aircraft, but these were retired in 1993.

Following the Iraqi invasion of Kuwait in August, 1990, Turkish bases were used by NATO forces, and this has continued as Operation Northern Watch, enforcing a 'no fly' zone in the north of Iraq to protect the Kurdish community.

The 1990s saw Turkey's most ambitious aircraft programme so far, building 240 Lockheed Martin F-16C/D interceptors in Turkey, but plans to buy 32 additional aircraft have been affected by a serious economic crisis in March, 2001. Meanwhile, 50 CASA CN-235M medium transports were also built under licence. In addition, F-4 Phantoms and F-5A/Bs have been upgraded by IAI, with some aircraft upgraded in Israel and others in Turkey using kits. Current plans centre on six Boeing 737 AEW&C aircraft, with one on option. Turkey is also planning to buy ten Airbus A400M transports and upgrade its C-130Es.

The THK has 60,100 personnel, a slight increase in numbers over the past 30 years. Flying hours are around an average of 180 per year, suggesting a good state of readiness. There are eighteen fighter and ground-attack squadrons: nine, including an OCU with F-16Ds, operate 210 F-16C and 30 F-16Ds; six, including an OCU, use 163 F-4E Phantoms; the remaining three have 110 F-5A and 30 F-5Bs, including an OCU. There are two reconnaissance squadrons with up to 44 RF-4E Phantoms. There are nine upgraded Boeing KC-135R tankers. Transport is provided by five squadrons, one with seven Lockheed C-130B and seven C-130E Hercules, one with 20 C-160T Transalls, two with 48 CN-235s and a VIP unit with two CN-235s, two Cessna Citation IIs and two VIIs, and three Gulfstream IV. Another two CN-235s are used for EW. Transport helicopters include 40 UH-1H Iroquois and six AS532AL Cougar, with another 14 used for CSAR. A Beech King Air 200 is used for communications. Training uses 40 SF260D, 24 Cessna T-37B and 42 T-37C, 69 Northrop T-38 Talon and 29 Cessna T-41D Mescalero (172), with plans to replace the T-37s from 2005. Missiles include AIM-7E Sparrow, AIM-9S Sidewinder and AIM-120 AMRAAM AAM, AGM-65 Maverick, AGM-88 HARM, AGM-142 and Popeye I ASM, as well as four squadrons with Nike Hercules and two with 86 Rapier SAM.

TURKISH NAVAL AVIATION

The Turkish Navy relinquished its 33 Grumman S-2A/E Trackers to the THK in 1975, but retained control over their deployment, to become a purely helicopter force. It has returned to fixed-wing flying with nine CASA CN235MPA Persuader MR aircraft built in Turkey, that have replaced the Trackers. At least 16 frigates can operate ASW helicopters, and the Navy operates 16 Sikorsky S-70B Seahawk ASW helicopters, with plans for an eventual force of around 30, as well as four CH-60S utility helicopters.The S-70s are replacing ten Agusta-Bell AB212ASW and three AB204AS ASW helicopters, as well as three EW AB212s. Training uses eight SOCATA TB20 Trinidads.

TURKISH ARMY AVIATION

Turkish Army Aviation originally developed primarily as a transport and liaison force, but in 1983, it received Bell AH-1 Cobra anti-tank helicopters with TOW missiles. A substantial number of Sikorsky S-70A Black Hawk helicopters brought a significant improvement in the transport capability during the 1990s, but the Army still wants a heavy lift helicopter.

A paramilitary gendarmerie, the Turk Tandarma Teskilati, operates under army control, and operates a number of helicopters, primarily on transport duties.

Today, the Turkish Army has 36 Bell AH-1P/S and nine AH-1W Cobra, as well as the first batch of 50 AH-1Z King Cobra, with the possibility of 145 of these eventually, in the anti-tank and attack roles. An indication of changing times is that this order was won in the face of competition from the Kamov Ka-50. Transport is provided by 95 Sikorsky S-70A Black Hawk, but there could be as many as 200 eventually, when they will doubtless replace many of the 120 Agusta-Bell 205s and 35 UH-1H Iroquois. There are two AB212a and 20 transport AS532UL, as well as another ten on CSAR. AOP duties are handled by 20 OH-58B Kiowa and 30 Cessna U-17A. There are five Beech King Air 200 and four Cessna 421 Golden Eagle communications aircraft. Training uses 24 Cessna T-41D Mescalero (172), five Beech T-42A Conchise (Baron), 30 Bellanca Citabria, and 25 Hughes 300, 12 AB204B and 30 AB206B JetRanger. The TTT has 19 Mi-17 and 14 S-70A, as well as eight Agusta-Bell 204B, six 205A, eight 206A and one 212, in addition to two Dornier Do28D.

Turkmenistan

POPULATION: 4.5 million

LAND AREA: 188,400 square miles, 491,072 sq km

GDP: $4.4bn (£.3.1 bn), per capita $2,600 (£1,818)

DEFENCE EXPENDITURE: $176m (£124m)

SERVICE PERSONNEL: 17,500 (inc.conscripts) active.

TURKMENISTAN AIR FORCE

Formed: 1993

Plans to establish an air force were finalized in 1993; two years after the collapse of the USSR, using equipment left behind by the Russians. Unlike many breakaway and newly independent states, Turkmenistan enjoys Russian support for its military efforts, possibly because it includes the main combat training range of the former Soviet armed forces. The core of the new air force is to be a composite regiment of MiG-29s and Su-17s, while there are plans to raise funds by selling more than 200 advanced combat aircraft, including 80 MiG-23s, 40 MiG-27s and 50 Su-17s. Difficulties are that the aircraft are deteriorating, while the local economy is also contracting.

The Turkmenistan Air Force is believed to have around 3,000 personnel and is organised along Soviet lines as a VVS, or air force, and PVO, or air defence force. Most aircraft are in storage and there are no details about annual flying hours. There are likely to be around 30 MiG-29s and Su-17s in the combat regiment, drawn from a total of 22 MiG-29 and two MiG29U and 65 Su-17M. Other aircraft, believed to be in storage, include 220 MiG-23M and ten MiG-23U and 24 MiG-25s, while the MiG-21s are believed to have been scrapped, along with a few Su-7s. There are also up to 46 Su-25, ten Mi-24 and ten Mi-8, with a handful of Antonov transports and at least two L-39 Albatros trainers. Missiles include around 50 SA-2, SA-3 and SA-5 launchers.

Uganda

POPULATION: 22.3 million

LAND AREA: 93,981 square miles, 235,887 sq km

GDP: $8.5bn (£5.9bn), per capita $2,000 (£1,398)

DEFENCE EXPENDITURE: $251m (£176m)

SERVICE PERSONNEL: 50-60,000 active.

UGANDAN PEOPLE'S DEFENCE FORCE AIR WING

Formed: 1987

Uganda was a British colony and before independence in 1962, created a Police Air Wing, with an Army Air Wing created in 1964. At first, the Army Air Wing shared the two Westland Scout helicopters of the Police Air Wing. Israeli aid at the outset was replaced during the late 1960s by Soviet and Czechoslovak aid, with MiG-15 fighter-bombers joining the earlier 12 Potez Magister armed-trainers. MiG-17s and MiG-21s arrived later, but political unrest, worsened by an Israeli raid on Entebbe in 1976, and then an invasion by Tanzania in 1979, followed by a military coup in 1985, saw all but a few aircraft destroyed. Police and Army aviation merged to create the Ugandan People's Defence Force Air Wing in 1987 with Libyan aid.

Today, the status of the force is uncertain, although it is believed to have received seven Polish MiG-21s, upgraded by IAI. There are up to seven Mi-17 and three Mi-24, but most of these are believed to be grounded, as are three Bell 206 and two 412s. A number of Aero L-39 and SIAI SF260W armed-trainers are thought to be unserviceable. A VIP Gulfstream IV-SP obtained in 2001 is nevertheless serviceable!

Ukraine

POPULATION: 50.4 million

LAND AREA: 231,990 square miles, 582,750 sq km

GDP: $32bn (£22.4bn), per capita $4,762 (£3,330)

DEFENCE EXPENDITURE: $1.1bn (£769m)

SERVICE PERSONNEL: 303,800 (inc.conscripts) active, plus 1,000,000 reserves.

MILITARY AIR FORCE

Formed: 1991

Many of the former USSR's military aircraft were stationed in the Ukraine and were taken over when the USSR broke up. Some have been sold, while others have been transferred back to Russia in order to reduce the Ukraine's debts. Most of the remaining aircraft are either in store, or unserviceable. The Ukraine could have difficulty in sustaining strong armed forces since the country is not heavily industrialized, although it is home to the Antonov concern and 65 An-70 transports are on

Ukraine

ABOVE: *The ubiquitous Mil Mi-8 helicopter remains the mainstay of many helicopter fleets and active in a variety of roles, in this case a special forces training exercise. (NATO)*

order. The Ukraine is restructuring its armed forces in three phases, with the first due to end in 2005 and the third in 2015, which will mean a substantial reduction in both aircraft and personnel numbers. Meanwhile, the Naval Air Arm's combat fixed-wing aircraft have been transferred to the Military Air Forces, VVS, which are organised along Soviet lines.

The VVS has 96,000 personnel and upwards of 900 combat aircraft, although aircraft numbers will be cut by two-thirds under the first phase of restructuring. Two regiments operate around 40 Tupolev Tu-22M bombers. Fighters are operated by eight regiments with 225 MiG-29s, upwards of 50 MiG-23s and around 70 Su-27s. Ground-attack units include, one training, and four operational regiments with 70 or Su-25s, while another training and four operational regiments operate up to 200 Su-24s. Four regiments operate around 60 Su-17s and 30 Su-24s, and possibly some Tu-22s for reconnaissance.

Transport aircraft include up to 78 Ilyushin Il-76s, and 20 Il-78 tanker-transports, more than 50 An-2, 20 An-12, ten An-24, 28 An-26 and 26 An 72, with rationalization likely as the An-70s arrive. Helicopters include more than 100 Mi-8 and Mi-17, as well as 20 Mi-6 and more than 100 Mi-2. There are more than 300 Aero L-39 Albatros trainers, with probably as many again in store, as well as more than 200 Yak-52, and conversion trainers versions of the main combat aircraft. A wide selection of missiles is available, although not necessarily operational. Flying hours are believed to be low.

UKRAINIAN NAVAL AIR ARM

The Ukrainian Navy was formed out of the former Soviet Black Sea fleet, but the major fleet units have been returned to Russia to reduce debts, and to meet treaty obligations to ensure that the Ukraine does not remain a nuclear power. Others have been sold, notably the aircraft carrier *Varyag* to China in 2000. Fixed-wing combat aircraft have been transferred to the VVS. Some 2,500 naval personnel are involved in aviation out of 13,000, but only one frigate is helicopter capable. The Ukrainian Navy has 12 Kamov Ka-27, 18 Ka-25 and five Mil Mi-14PL helicopters in the ASW role, as well as four Ka-29 in the assault role, with eight Mi-8 transport helicopters. There may also be a few fixed-wing transports in service.

UKRAINIAN ARMY AVIATION

The Ukrainian Army has a substantial force of attack and assault helicopters, but most are believed to be either in storage or unserviceable. In theory, it has almost 300 Mil Mi-24 attack helicopters, including some for EW, around 250 Mi-8 and 25 Mi-26 transport helicopters, 40 Mi-6 heavy lift helicopters, as well as 50 Mi-2, although some sources suggest that all of these are non-operational.

United Arab Emirates

POPULATION: 2.6 million

LAND AREA: 32,300 square miles, 82,880 sq km

GDP: $58bn (£40.1bn), per capita $25,600 (£17,902)

DEFENCE EXPENDITURE: $3.4bn (£2.4bn)

SERVICE PERSONNEL: 65,000 (inc.up to 30 per cent expatriates) active.

UNITED ARAB EMIRATES AIR FORCE

The United Arab Emirates Air Force is a composite force funded by the seven emirates, formerly known as the Trucial States: Abu Dhabi, Ajman, Dubai, Fujairah, Ras al-Khaimah, Sharjah and Umm al-Quain. Several maintain separate royal flights, and most aircraft are assigned to one or the other of the two main states, Abu Dhabi and Dubai. Abu Dhabi had previously maintained an Air Wing, whose equipment included ten Hawker Hunter fighter-bombers and four DHC-4 Caribou transports. Pressure to operate jointly came from the UK, anxious to withdraw from the Gulf during the 1970s, but aware of the threats to these territories. Standardization of equipment has been a priority, but Hawk armed-trainers and Puma helicopters, were common to both Abu Dhabi and Dubai.

There are 4,000 personnel. Flying hours are slightly low, at an average of 110. Over the next few years, 80 Lockheed Martin F-16C/D Block 60 multirole fighters will be entering service, complementing the existing force of 30 Dassault Mirage 2000-9 and 30 upgraded 2000-Es, but replacing Mirage VAs, and possibly Aermacchi MB326KD/LD armed-trainers. Two squadrons handle light attack duties, each equipped with 17 BAe Hawks, although one squadron has Mk102s, ex-Abu Dhabi, and the other a mixture of Mk61 and 63, drawn from the two main air forces. There are 30 Boeing AH-64A and 12 SA342L Gazelle anti-tank helicopters, as well as seven AS565SA Panther and two AS532SC Panther ASW helicopters, with another eight Panther in the transport role. There are four MR CASA CN-235MPA, with seven CN-235M transports. Transport aircraft include six Lockheed C-130H/H-30 Hercules, four CN212-200, a Shorts 330OUUT and an SC7 Skyvan, with 18 Eurocopter SA330 Puma, seven BO105CBS, six Agusta-Bell 205A and six 412, as well as two Bell 212, a 222, a 206L LongRanger and eight 206/AB206 JetRanger. Training aircraft include 24 Pilatus PC-7 Turbo Trainers, 12 Grob G115 Acro, 14 AS350B Ecureuil, four MB339A and five SF260, as well as a Bell 407. Royal Flight aircraft include Abu Dhabi's two Airbus A300-620, a Boeing 747SP, a BAe 146-100, two Beech Super King Air 350, three Dassault Falcon 900 and two AS332L Super Puma; Dubai's Boeing 747SP, a Grumman Gulfstream II and a IV, a Sikorsky S-76 and an AS365 Dauphin; Ras al-Khaimah's Cessna Citation I and Sharjah's Boeing 737-200.

BELOW: *United Arab Emirates C-130H Hercules. (Jeremy Flack/Aviation Photographs International)*

United Kingdom

POPULATION: 58.9 million

LAND AREA: 92,000 square miles, 238,278 sq km

GDP: $ 1.4tr (£979bn), per capita $23,422 (£16,379)

DEFENCE EXPENDITURE: $34.6bn (£24.2bn)

SERVICE PERSONNEL: 211,430 active, plus 247,100 reserves.

ABOVE: *The Westland Wapiti was conceived as a replacement for the RAF's DH9A light bomber, but when it entered service with the RAF, and a number of other air forces and air arms, it was as a multi-purpose aircraft. (GKN Westland)*

ROYAL AIR FORCE

Formed: 1918

The world's first autonomous air force when formed on 1 April, 1918, the Royal Air Force was the result of the merger of the Army's Royal Flying Corps and the Royal Naval Air Service. The new service was already in action, in the bomber and fighter campaigns of World War I, which still had more than seven months to run.

The UK has the longest continuous history of military aviation of any nation. The Royal Engineers had started experimenting with balloons in 1878 at Woolwich Arsenal, on the outskirts of London using these in expeditions to Bechuanaland in 1884, and to the Sudan a year later. The official status of this unit was established in 1890, with the formation of a balloon section within the Royal Engineers, and the building of balloon sheds at South Farnborough. It was not until the outbreak of the Boer War in 1899, that the balloons were used for observation and artillery control duties. In 1911, the Balloon Section became the Air Battalion.

Meanwhile, the Royal Navy had started flying in 1909.

The Royal Air Force's direct predecessor, the Royal Flying Corps, was formed in 1912, on the amalgamation of the Air Battalion, Royal Engineers, and the Royal Navy's Air Branch, but it was short-lived, with the Royal Navy withdrawing from the RFC in 1914, and establishing the Royal Naval Air Service. On the outbreak of war in August, 1914, the RNAS had 100 aircraft, plus airships that were to prove useful as convoy escorts. The RFC, on the outbreak of war under the command of Lieutenant-Colonel Hugh Trenchard, had 180 aircraft. The RNAS was volunteered for the task of home air defence by the then First Lord of the Admiralty, Winston Churchill.

Procurement policies of the two services differed. The RFC largely concentrated on the products of its own Royal Aircraft Factory at Farnborough, while the RNAS bought the products from the civilian aircraft industry. Aircraft included British Avro, Bristol, Short and Sopwith designs operating alongside French Bleriot, Deperdussin and Farman products. Aircraft belonging to the RNAS bombed Zeppelin sheds at Hamburg, Cologne and Friedrichshafen, and torpedoed a Turkish warship in the Mediterranean. In addition, reconnaissance missions were flown from warships, including seaplane carriers, on a number of occasions, but most notably at the Battle of Jutland on 31 May, 1916.

Fighter warfare evolved during 1915, and in 1916, German Fokker aircraft, fitted with synchronized machine guns, gained aerial superiority. At first, the answer lay in pusher-propeller Vickers FE2B and Royal Aircraft Factory DH2s, and later the Lewis-gun-fitted French Nieuport biplanes. It was not until the arrival of Bristol Scouts and Sopwith 1 1/2-Strutters with synchronized Vickers machine guns that the balance was restored. The bomber evolved as a distinct aircraft type, with the appearance of the DH9 series, the Handley Page 0/400 and V/1500, and at the end of the war, the Vickers Vimy. These aircraft were involved on raids on German lines of communication, including railway marshalling yards. During the war, bombs were developed from artillery shells fitted with fins to purpose-designed ordnance of as much as 1,650lbs (750 kg), although smaller sizes were more commonly used.

During the closing months of the war, the RAF was in action on fighter, bombing and reconnaissance duties. In France, the service operated as the Independent Air Force. Aircraft from the world's first aircraft carrier, *Furious*, bombed the German airship sheds at Tondern in August, 1918, destroying two airships. In 1919, a Vickers Vimy made the first non-stop transatlantic flight.

The new service was seen as an example to follow by many, not least in the United States Army Air Corps and later in post-war Germany. It was controversial, with a single service responsible for all aspects of military aviation. In the event, a small compromise was agreed, which allowed the fleet spotting aircraft operated by battleships and cruisers to be flown by naval officers, but aircraft

United Kingdom

ABOVE: *A Vought V66E Corsair during demonstrations to the RAF in 1932. (Vought)*

ABOVE: *The famous Spitfire (Jane's)*

aboard the growing fleet of aircraft carriers were flown and maintained by the RAF.

At the end of the war, the RAF had some 360,000 men, 200 squadrons and 23,000 aircraft. Severe budgetary cutbacks, and the desire to suppress the old inter-service RFC/RNAS rivalries, led to a reduction to just twelve squadrons, with one in Germany, two in the United Kingdom, and the remaining nine in the Middle East and India. In 1923, it was decided that fifteen fighter and thirty-seven bomber squadrons should be available for home defence in addition to units overseas, making an overall strength of seventy-four squadrons. The Depression years meant that this strength was not attained until 1936. British military expenditure during this period was also hampered by the 'Ten Year Rule', a belief that there would be ten years in which to prepare for a major conflict!

In 1924, the RAF's carrier squadrons became known as the Fleet Air Arm.

It was not until 1923 that the first post-war design was introduced, the Fairey Fawn. A succession of designs followed, all of which were biplanes and purchased in such small numbers that it was not unknown for a single type to equip just a single squadron. Boulton Paul Sidestrands, Fairey IIIKs, Handley Page Hinaidis and Hyderabads, Hawker Harts and Horsleys, and Blackburn Iris and Supermarine Southampton flying boats followed Gloster Glebes, Armstrong-Whitworth Siskins and Vickers Virginias. Despite the RAF High Speed Flight participation and successes in the Schneider Trophy contests, performance of service aircraft improved but slowly. Nowhere was this more evident than at sea aboard the Royal Navy's fleet of aircraft carriers.

Operations between the wars included a number of police actions, especially in the Middle East, in Afghanistan and on the Northwest frontier of India. 'Air control' was seen as a cost-effective means of maintaining the peace in territories such as Iraq, but aircraft on their own were no substitute for forces on the ground. Colonial and League of Nations policing meant that a network of well-equipped bases was established, complementing the efforts of the new national airline, the rapidly expanding Imperial Airways. The RAF had also operated a number of diplomatically important airmail services in the period before the creation of Imperial Airways in 1924.

In 1936 and 1937, a major reorganization of the RAF took place, with the division of the service into Bomber, Fighter, Coastal, Maintenance and Training Commands in 1936. The following year it was decided to return naval aviation to the Admiralty. It was decided in 1936 to increase the RAF's strength to 134 regular squadrons plus 138 Royal Auxiliary Air Force, or reserve, squadrons, a massive increase over the 13 RAuxAF squadrons of the early 1930s. Development and production of new aircraft started, including the Armstrong-Whitworth Whitley, Fairey Battle, Bristol Blenheim and Vickers Wellington bombers, and the Hawker Hurricane and Supermarine Spitfire fighters. It was not until 1938 that industrial capacity rather than finance became the limiting factor in re-equipping and expanding the RAF, so that by 1939, the service still had only an eighth of the manpower and two-sevenths of the equipment of the Luftwaffe.

On the outbreak of World War II in Europe in September, 1939, the RAF's Bomber Command possessed 55 squadrons. Of these, five had Armstrong-Whitworth IIIs and IVs, six had Handley Page Hampdens, six Bristol Blenheim IVs, another six with Vickers Wellington Is, the most effective of these aircraft, and another ten had the ineffectual Fairey Battle light bomber. The remainder operated outdated aircraft. Coastal Command had ten squadrons of Avro Ansons, one of Lockheed Hudsons, and two of Short Sunderland flying-boats, as well as six squadrons of obsolete Supermarine Stranraer and Saunders-Roe London flying-boats and Vickers Vildebeest torpedo-bombers. Fighter Command had 22 squadrons of Hawker Hurricanes and Supermarine Spitfires, and another 13 of the obsolete Gloster Gladiator biplanes.

Initially, 27 squadrons were deployed to France, including the light bombers of the

Advanced Air Striking Force, AASF, which accompanied the British Expeditionary Force, BEF. Just 600 British and French aircraft faced more than three thousand Luftwaffe aircraft. The RAF could only provide token support for the battle for Norway, partly because of the limited number of suitable airfields, but did manage to send a small number of Hawker Hurricane fighters. After the massive German advances during spring, 1940, the RAF withdrew many of its aircraft from France, ready for Luftwaffe air attacks against the British Isles, the prelude to invasion. Another 27 squadrons were based around the Mediterranean, where, despite operating mainly obsolescent aircraft, they managed to maintain air supremacy and helped defeat the Italian attack on Greece in 1940, although the German attack the following year was successful.

The summer that followed saw the RAF face its most testing period in what became known as the Battle of Britain, as the Luftwaffe concentrated on attacking British airfields. The main phase of the battle was between 11 August and 30 September, 1940. In the heat of the battle, with sometimes more than one pilot claiming an enemy aircraft as a 'kill', the initial claims for both sides were heavily exaggerated. At the time, the RAF claimed 2,698 enemy aircraft destroyed, while the Luftwaffe claimed 3,058, but the true figures are now estimated to be 1,733 Luftwaffe losses against 1,140 of the RAF. The shadow system of aircraft production, which saw, amongst others, much of the motor industry converted to aircraft production, meant that the supply of new fighter aircraft took second place to the shortage of aircrew and especially experienced pilots. Desperate for pilots, the RAF borrowed heavily from the Fleet Air Arm and used French, Polish and other airmen who had fled their countries to continue the fight against Germany, while American volunteers eventually manned three squadrons. There were substantial numbers of personnel from the countries of the then British Empire. The 'Chain Home' network of radar stations was a secret weapon that enabled fighters to be scrambled in time to face German attacks.

ABOVE: *The most effective heavy bomber in the European theatre during World War II was the RAF's Avro Lancaster, modified versions of which could drop a 20,000-lb bomb.*

The Battle of Britain was followed by a sustained heavy bombing campaign against British cities, known as the 'Blitz'. The Luftwaffe believed that they had destroyed the RAF, and that they could hamper its rebuilding by attacking factories and communications - but had the attack on the air fields been maintained, the RAF might eventually have run out of pilots. The German blitzkrieg, so successful when operating in support of ground forces in Poland, Norway, France and the Low Countries was defeated by heavy anti-aircraft defences. Later these defences were augmented by increasingly sophisticated night fighters, initially the Bristol Beaufighter and then the de Havilland Mosquito. The Luftwaffe also suffered from the absence of an effective bomber. Severe damage, with heavy casualties, was inflicted on British cities, but the war effort was maintained. The British had also cracked the German 'Enigma' codes and learnt how to 'bend' the radar beams guiding German bombers.

After the fall of France and Norway, and then the entry of Italy into the war on the side of Germany in 1940, the only means of carrying the war to the enemy lay in a heavy bomber offensive. The RAF's early experiences of offensive raids at the outset of the war had produced unsustainable casualties, due to inadequate aircraft and poor tactics. A generation of genuinely heavy bombers, the Short Stirling, Avro Lancaster and Handley Page Halifax, capable of delivering heavy warloads in excess of 8,000lbs and with a good defensive armament, meant that from 1942 onwards the balance began to tip in the RAF's favour. Navigation and bomb-aiming improved with aids better suited to fast-moving aircraft, and assisted by the creation of crack Pathfinder units to mark targets in advance of the main bomber force. No less important, the forces deployed against each target increased, attaining a critical mass that overwhelmed defending fighters and anti-aircraft defences. This lead to the famous 'Thousand Bomber Raids', with the first on the night of 30-31 May, 1942, against Cologne, with 1,050 aircraft deployed, including Hampdens and Whitleys bought out of their working retirement on training and other duties. Losses on that raid amounted to 40 aircraft, or 3.8 per cent of the aircraft, far below that on many less successful operations in which a third or more of the aircraft deployed had been lost. With the longer ranges of the new aircraft, targets in Italy could also be reached.

One of the most famous bomber operations was against the Moehne and Eder dams in May, 1943, by Lancasters of 617 squadron, using specially designed bouncing bombs. The squadron was later to become a specialist in precision bombing using ever larger weapons, including the 12,000lb (5,400kg) 'Tallboy' and 22,000lb (10,000kg) 'Grand Slam' against a variety of targets, including the battleship *Tirpitz*, the V-weapons sites, and railway junctions.

The entry of the USA saw RAF night raids and USAAF day raids alternate. Many American aircraft entered RAF service, with the Douglas C-47, known to the RAF as the Dakota, compensating for the absence of a good British transport aircraft. Transport Command was formed in March, 1943, providing specialized transport units to support paratroop operations such as that at Arnhem, and to provide assault and supply operations in many theatres of war, including towing Hamilcar and Horsa gliders used in the Normandy invasion and the crossing of the Rhine.

There had been little that the RAF could do in the Far East against the Japanese assaults of late 1941 and early 1942. In the Mediterranean, the RAF was outmatched in the defence of Malta, for example, at one time left to just three Gloster Sea Gladiator biplanes. Nevertheless, as the war progressed, tank-busting Hurricane fighters were put into service in North Africa and then later, after D-Day, the Hurricane's successor in this role, the Hawker Typhoon, undertook this role in France.

The RAF played a major role in the fight against the German U-boat campaign against Allied shipping around the British Isles, in the Atlantic, and on the Russian convoys, Shore-based MR aircraft, notably the Short Sunderland flying boat and the Wellington bomber, maintained patrols. Later, American Consolidated Catalina flying boats, Lockheed Hudson and Consolidated Liberator bombers provided a welcome increase in Coastal Command's capability. A novel RAF unit formed for convoy protection was the Merchant Service Fighter Unit, operating Hurricanes catapulted from merchant vessels to protect them from German aircraft, with the pilot having to ditch his aircraft or bail out afterwards. This unit was made redundant by the advent of the escort carrier.

At the end of the war, the RAF was operating Hawker Typhoons and Tempests, Lockheed P-61 Lightnings, North American P-47 Mustangs and Avro Lincoln heavy bombers in addition to many of the aircraft types introduced earlier. While the Luftwaffe had received Messerschmitt Me262 jet fighters during the war, Hitler's insistence that the aircraft be used as a bomber meant that the RAF was first to operate jet fighters when the Gloster Meteor entered service in 1944.

Peace in Europe found the RAF with 1.1 million personnel, and 487 squadrons, of which about a hundred came from the Dominions and the 'free' air forces, and 9,200 aircraft. Heavy post-war commitments meant that the service could not immediately reduce its numbers to the peacetime target of 300,000.

During the war, the RAF had shown that its strengths lay in the projection of strategic air power, with the capability of providing strong tactical support when called upon to do so. This was in contrast to the Luftwaffe, which had been developed primarily as a tactical force.

The RAF was heavily involved in the Berlin Airlift, which started in June, 1948, when the USSR blockaded West Berlin, cutting access between the city and the British, French and American zones of occupation in what later became West Germany. A massive RAF and USAF transport operation was augmented by chartered civilian aircraft, moving food, clothing and fuel, including coal, to the beleaguered city, Upwards of a hundred RAF transport aircraft were used on this operation.

Despite receiving new aircraft, such as the de Havilland Venom fighter and the world's first operational purpose-built jet bomber, the English Electric Canberra in 1950, the RAF was fully stretched with colonial style police operations. The first of these were in Palestine, under a former League of Nations mandate, and in Malaya, while also confronted with the need to face the growing threat from the USSR. Only limited RAF participation was possible in the Korean War because of other demands, leaving the main British contribution to United Nations air power to the Royal Navy.

The United Kingdom was a founder member of the North Atlantic Treaty Organisation, NATO, with much of the

BELOW: *RAF Hurricane IIb-J of No 261 Squadron. (Jeremy Flack/Aviation Photographs International)*

United Kingdom

RAF's frontline strength in the United Kingdom, West Germany, Gibraltar, Malta and Cyprus assigned to NATO. The UK was also involved in many other regional defensive alliances, including the South East Asia Treaty Organisation and Baghdad Pact.

During the 1950s, the RAF received a considerable quantity of new aircraft of advanced performance for the period. These included the de Havilland Comet C2 jet transport and the Bristol Britannia turboprop transport, before which it gained the unwieldy Blackburn Beverley, a heavy transport with a good short field performance. The Westland Whirlwind helicopter was soon replaced by the Wessex, like its predecessor a licence-built development of the Sikorsky original, while a heavier helicopter for the RAF at this time was the twin-rotor Bristol Belvedere. The RAF's first swept-wing fighter, the Supermarine Swift, entered service in 1954, only to be withdrawn because of technical failings. By contrast the Hawker Hunter which followed was a great success both for the RAF and for its manufacturer, as was the English Electric Canberra light jet bomber. Gloster Javelin delta-wing interceptors were also introduced, while the famous trio of 'V' bombers to carry Britain's nuclear deterrent, the Vickers Valiant, Avro Vulcan and Handley Page Victor, became operational. A major step forward in the RAF's strategic capabilities came with the use of Valiants as a substantial fleet of tanker aircraft for in-flight refuelling.

The most significant action for the RAF during the 1950s was the Anglo-French operation against Egypt after that country nationalized the Suez Canal in 1956. RAF, Fleet Air Arm, Armee de l'Air and Marine Nationale aircraft participated, with the RAF using British bases in Cyprus. Although a tactical success, the operation ended abruptly due to international pressure.

During the early 1960s, the RAF supported ground forces in a successful campaign against Indonesian attempts to undermine the newly created Federation of Malaysia.

The period saw further sharp reductions

ABOVE: *One of the most successful British aircraft was the Hawker Hunter, in this case a ground-attack variant. Other versions included a world speed record setting aircraft (BAe systems)*

in the RAF's strength, from 250,000 during the early 1950s to 110,000 two decades later. These reductions were accompanied by the disappearance of the RAuxAF, officially deemed to be impractical given the growing complexity of modern aircraft, but units of which had been called upon for up to three months at a time to maintain RAF strength. A significant element in cutting RAF numbers was the ending of conscription during the early 1960s, making the United Kingdom the first major power to have all-regular armed forces. Further reductions in both personnel numbers and aircraft were to follow throughout the next three decades, with the exception of a small and short-lived reversal of the process during the early 1980s.

New aircraft continued to enter service, reflecting the rapid developments taking place. The English Electric Lightning interceptor was followed by the McDonnell Douglas F-4K Phantom, itself accompanied by the Short Belfast, Vickers VC10, Lockheed C-130K Hercules and Hawker Siddeley Andover transports. After a number of policy changes and aborted projects, the Hawker Siddeley Buccaneer became the mainstay of the bomber force. Both the Phantom and Buccaneer forces received additional aircraft with the transfer of former Fleet Air Arm aircraft when the Royal Navy retired the last of its conventional aircraft carriers. Two strike aircraft types entered service during the 1970s, the Anglo-French Sepecat Jaguar and the Hawker Siddeley Harrier, a vertical take-off aircraft and the first of its type to enter operational service with any air force. Aerospatiale-Westland Puma helicopters replaced the remaining Whirlwinds while later Westland Sea Kings (S-61) replaced the Wessex in the SAR role. For MR, Hawker Siddeley Nimrods, developed from the Comet airliner, replaced elderly Avro Shackletons. By the 1970s, with a few exceptions, procurement projects were made in collaboration with one or more European partners, or purchased from the United States. A multi role combat aircraft project led to the Anglo-German-Italian Panavia Tornado, available in interceptor and interdictor variants.

The slimming down of the RAF led to the disappearance of its 1936 command structure, which had been augmented by the post-war creation of area commands such as RAF Germany and RAF Middle East. Two new commands were created during the late 1960s, Strike Command and Support Command.

Major operations for the RAF, and for the

United Kingdom

ABOVE: *RAF Hawk trainers were originally painted red and white, but have now been painted black so that they can be more easily seen, as have all British training aircraft. (BAe)*

British armed forces as a whole, continued to arise. RAF units were deployed to Zambia for a short period after Rhodesia's Unilateral Declaration of Independence during the late 1960s, while units were deployed to Belize after a threat to the country's sovereignty. The RAF also became increasingly involved in providing humanitarian aid, most notably with air drops of supplies by Hercules aircraft during the Ethiopian famine of 1985. It has provided tactical transport for the British Army in Northern Ireland since the late 1960s, currently using Puma and Chinook helicopters. Other conflicts involving the RAF included the Falklands campaign in 1982, and a major effort during the Gulf War of 1991, followed in 1999 by operations against Serbian forces and installations in the former Yugoslavia in support of the Kosovo Albanian population.

The Falklands campaign to recover the islands after an Argentine invasion was spearheaded by the Royal Navy's aircraft carriers, but RAF transports dropped mail and supplies to the Task Force. On the outbreak of hostilities, a solitary Vulcan made the first bombing raid, attacking the airfield at Port Stanley, although little damage was caused. RAF helicopters and Harrier GR3 ground-attack aircraft were also deployed to assist in the recovery of the islands, although many of the large twin-rotor Boeing Chinook helicopters were lost when the ship carrying them was lost to enemy action. After the surrender of the Argentine forces, the RAF secured an air supply service linking the United Kingdom and the Falklands. RAF Phantoms were deployed on the islands as a deterrent against any further interference by Argentine forces.

The Gulf War followed the invasion of Kuwait by Iraqi forces in 1990. Under the auspices of the United Nations, a Coalition of countries under American command provided air, ground and naval forces for the liberation of Kuwait. The RAF moved

United Kingdom

ABOVE: *Throughout 2000 and 2001, a question mark hung over the RAF's three Harrier GR7 squadrons, as the need for economy suggested losing an aircraft type, but its future may have been assured by increasing deployments aboard the Royal Navy's aircraft carriers. (BAe)*

British troops to Kuwait during the military build-up to the conflict, although a shortage of heavy lift capability following the sale of the Belfast fleet some years earlier, meant that much heavy equipment had to be moved by sea or in chartered aircraft. The main role played by the RAF's interdictor Tornado squadrons once hostilities started with an intensive air campaign against Iraq on 17 January, 1991, was to attack Iraqi airfields, flying at low level with their runway denial weapons. The RAF's Tornado GR1 aircraft suffered a number of casualties, casting doubt on the effectiveness of the technique in combat conditions.

The air campaign over the former Yugoslavia started on the night of 24/25 March, 1999, and continued until June. For this, the RAF mainly operated its Harrier GR7 force, augmented by a number of Tornados, deployed to the Italian Air Force's Gioia del Colle base in southern Italy.

In 2001, the RAF closed its last base in Germany. The service is in demand for many 'out of area' operations. Squadrons are deployed to Turkey, to monitor a no flying zone in northern Iraq, and to Italy to monitor a peace agreement in Bosnia, part of the former Yugoslavia, as well as operating in the Gulf region to confront continuing difficulties with the Iraqi regime. During 1997-98, RAF Harriers were once again deployed alongside Royal Navy Sea

BELOW: *The Tornado GR4 is the latest upgrade for this mainstay of the RAF's strike force, likely to remain in service until well after 2010. (BAe)*

United Kingdom

ABOVE: *Tornado F3s are the air defence variant, ADV, of the international project, but are likely to be replaced by Eurofighter Typhoons once the Jaguar force has been replaced.*

Harriers aboard the aircraft carrier HMS *Illustrious* in the Gulf. The RAF's tankers provided air-to-air refuelling for the USN and USMC during operations over Afghanistan in the winter of 2001/2 as the USN uses a similar inflight refuelling system to that of the RAF - and different from that of the USAF. During spring, 2002, three RAF Chinooks supported a force of Royal Marines operating in the mountains close to Afghanistan's border with Pakistan.

Additional Chinooks and the new Agusta-Westland Merlin helicopter are being introduced, but replacements are needed for the VC10 and ex-British Airways Lockheed TriStar 500 transport and in-flight refuelling fleets, which may include a variant of the Airbus A310 or Boeing 767.

Today, the RAF has 53, 950 personnel, a force that is spread thinly. Flying hours are being cut by around 20 per cent, and for many types will fall below 200 hours per annum. Interceptor squadrons are being reduced from five to four, as well as a flight based in the Falklands, with fewer than 90 Panavia F3 Tornado operational, while there are five interdictor squadrons and two reconnaissance with 142 upgraded Tornado GR4 strike aircraft. Ground attack is handled by two squadrons with some of the 53 BAe Harrier GR-7, operated jointly with the Fleet Air Arm's Sea Harriers and liable for service at sea under the Joint Force Harrier, JFH, project. The aircraft are

ABOVE: *The RAF's new medium lift helicopter is the Westland-Agusta EH101 Merlin. (GKN Westland)*

being upgraded to GR-9 standard, and the RAF and FAA will standardize on this aircraft. Two squadrons, and a reconnaissance unit, have 43 extensively upgraded SEPECAT Jaguar GR3/3A. Five Canberra PR9 form a fourth reconnaissance squadron, with sophisticated cameras and communications equipment. The Tornado F3 and Jaguars will be progressively replaced from 2003 by the first of up to 232 Eurofighter 2000s, while in the longer term the Harrier GR-9s are due to be replaced by up to 150 Lockheed Martin F-35s, operated jointly with the Royal Navy. Two squadrons operate seven Boeing E-3A Sentry AWACs aircraft. There are three MR squadrons with 23 BAe Nimrod MR1, being progressively re-winged and re-engined to MR-2 standard, by which time numbers will be reduced to 18 aircraft. An ELINT squadron operates the Nimrod R-1, while five Global Express ASTOR aircraft are entering service. Transport and tanker aircraft include a squadron with Lockheed TriStar K-1/KC-2A transport tankers, while there is still a squadron of BAe VC-10C1K/K-3, although these aircraft are due to be retired. Four squadrons operate 26 Lockheed C-130K and 25 C-130J Hercules/Hercules II, with the former likely to be replaced by 25 Airbus A400M by around 2008, while a fifth operates four leased Boeing C-17 Globemaster II. Helicopters include nine squadrons operating 49 Boeing CH-47 Chinook HC2/HC2A/HC3 heavy lift helicopters, 22 Agusta-Westland Merlin HC3, and 41 Puma HC1, as well as two SAR squadrons with 25 Westland Sea King HAR3/3A. There are two BAe 146 CC2 and two Eurocopter Twin Squirrel HCC1 on VIP duties. Training uses a wide range of aircraft, including 99 contractor-owned Grob 115 Tutor trainers and 18 Slingsby T67M-2 Firefly; with further training on 73 Shorts-built Tucano T1 and 97 BAe Hawk T1/T1A, with the latter also equipping the famous 'Red Arrows' aerobatic team. Other training uses ten Dominie (BAe125) and 11 Jetstream T1 (Jetsream 31) navigational trainers, 38 Squirrel HT1/HT2 and nine Griffin HT1 helicopter trainers. Missiles include AIM-9L/M Sidewinder, Sky Flash AMRAAM, ASRAAM AAM, with AGM-65G2 Maverick and AGM-84D Harpoon ASM, and ALARM ARM, with Rapier SAM.

ABOVE: *The Royal Navy's Fleet Air Arm provided most of the British contribution to UN air power during the Korean war. This is the light carrier, HMS* Ocean*, with her Hawker Sea Fury fighters ranged on deck. One of her aircraft shot down a MiG-15 jet fighter. (IWM)*

FLEET AIR ARM

Formed: 1939

Although the Royal Navy's Fleet Air Arm in its present form dates from 1939, as a result of the 1937 decision to return control of naval aviation to the Admiralty, the history of British naval aviation is far longer. In 1909, the sum of £35,000 (US $175,000 at the then rate of exchange) for an airship was included in the Naval Estimates. The pre-World War I Royal Navy made considerable strides, experimenting with take-offs from ships and converting warships as seaplane tenders. In 1912, the air branches of both the British Army and the Royal Navy were merged to form the Royal Flying Corps, but this was reversed with the formation of the Royal Naval Air Service in 1914. During the war, the RNAS was offered by the Admiralty to provide fighter protection for English cities threatened by German Zeppelin and, later, heavy bomber attack. The need to provide protection for convoys led to innovation, with landplane fighters flown off lighters towed at high speed behind destroyers, and seaplanes launched on trolleys from primitive flight decks on seaplane carriers. In 1916, fleet spotting by aircraft preceded the clash of the British and German fleets at the Battle of Jutland. The greatest innovation of all was the series of piecemeal conversions of the battlecruiser, HMS *Furious*, that led to the appearance of the first aircraft carrier. The first air strike from the sea was from HMS *Furious* against the German balloon sheds at Tondern in August, 1918. But by this time the RNAS had lost its independence, merged in April with the RFC to form the Royal Air Force.

In 1924, the Fleet Air Arm first came into

United Kingdom

ABOVE: *The last conventional aircraft carrier to enter service with the Royal Navy was HMS* Hermes, *seen here shortly afterwards during the early 1960s, with a deckload including Sea Vixens and Scimitars. (Harry Wragg)*

existence as the carrier-borne arm of the Royal Air Force, although a number of naval aviators served with the fleet, mainly flying seaplanes and amphibians from battleships and cruisers. By 1924, there were four aircraft carriers, HMS *Furious*, HMS *Argus*, HMS *Eagle* and, the first to be designed as such from the keel upwards, HMS *Hermes*. These were later joined by HMS *Courageous* and HMS *Glorious*, converted battlecruisers, that had been sisters of HMS *Furious*. A seaplane carrier, HMS *Ark Royal*, was later renamed HMS *Pegasus* to release her name for a new aircraft carrier. The Fleet Air Arm between the wars suffered from neglect of its aircraft needs as the RAF struggled within a restricted defence budget to maintain its strategic forces. The late 1930s saw a plan to build a new carrier force, largely to replace the earlier ships. The first of these, HMS *Ark Royal*, joined the fleet before the outbreak of war. The next four were fast, armoured carriers, with two more ordered as the threat of war loomed, and with improvements included in the later ships. The new ships were HMS *Illustrious*, HMS *Victorious*, HMS *Formidable*, HMS *Indomitable*, HMS *Implacable* and HMS *Indefatigable*. Instead of replacing the earlier vessels they became additions to the fleet. These were to prove to be the most successful carriers during World War II.

While the Royal Navy found itself with new aircraft carriers, its aircraft on the outbreak of war were obsolescent, at best. Mainstay of the strike squadrons was the Fairey Swordfish, a biplane with open cockpits for its three-man crew and barely able to maintain 100 knots! There were no high performance fighters, with Gloster Sea Gladiator biplanes and the monoplane Fairey Fulmar in service. Poor tactics saw the loss of HMS *Courageous* during the first month of war. The Fleet Air Arm proved its worth during the Norwegian campaign of spring, 1940. It made the first sinking of an operational warship, when shore-based aircraft attacked and sunk the German cruiser *Koenigsberg*. During the withdrawal from Norway, HMS *Glorious* was lost to shellfire from two German battlecruisers.

One of the Fleet Air Arm's greatest achievements was the crippling of the Italian Fleet at Taranto in November, 1940, using just 21 Fairey Swordfish flown from HMS *Illustrious*, with just two aircraft lost in the night attack. Carrier-borne aircraft were also used against the Vichy French fleet in North and West Africa, at the Battle of Matapan and in the pursuit and sinking of the battleship *Bismarck*, as well as in many attacks against the *Tirpitz*, in a Norwegian fjord. The Fleet Air Arm operated aircraft from MAC-ships, merchant aircraft carriers, converted tankers and grain ships, before receiving large numbers of escort carriers, mainly from the United States, for protection of convoys in the Arctic, North Atlantic and across the Bay of Biscay. Malta convoys were covered by large attack carriers, which also flew off aircraft to the beleaguered island. After early losses in the Indian Ocean, including HMS *Hermes*, the Royal Navy returned to the East as the war progressed, operating against targets in the Dutch East Indies, and then working with the USN in the Pacific as the war edged towards Japan. New aircraft provided improved performance, including the Hawker Sea Hurricane and Supermarine Seafire, although the mainstay of the FAA was to prove to be the American Vought Corsair, Grumman Wildcat (known to the RN as the Martlet) and Hellcat fighters, and Avenger bombers.

As the war ended, the Royal Navy had 52 aircraft carriers of all kinds. The US-owned escort carriers were returned. Older ships were scrapped or went into reserve. New light carriers of the Colossus-class and their derivatives were to prove both effective, and also attractive to other navies, including the Dutch, French, Brazilian, Argentinian, Indian, Canadian and Australian. Successful deck-landing trials were carried out with the de Havilland Sea Hornet, a fast long-range twin-engined fighter, and with the de Havilland Sea Vampire jet fighter, the first jet to land on a ship, HMS *Ocean*, and with a Sikorsky R-4 onto the battleship, HMS *Vanguard*. The Royal Navy was to the forefront with the invention, but not always the introduction, of aids to modern naval aviation such as the steam catapult, mirror landing aids and the angled flight deck. The latter allowing take-offs and landings at the same time, and enabling landing aircraft to fly round safely if a first attempt did not succeed. Equipment also improved, with Douglas AD-4W Skyraider anti-submarine and, later, airborne-early warning aircraft; Westland Dragonfly (S-51) helicopters for communications and rescue, and de Havilland Sea Venom, Supermarine Attacker and Armstrong-Whitworth Sea Hawk fighters.

Heavy commitments prevented the RAF from participating fully in the Korean War, leaving British air operations to the RN. The FAA supported ground forces using the carriers HMS *Triumph*, HMS *Theseus*, HMS *Glory* and HMS *Ocean*, with one of the latter's piston-engined Hawker Sea Furies shooting down a jet MiG-15.

In 1956, the first rotary-wing assault on an enemy coast came with Royal Marines being landed during the Suez campaign from HMS *Theseus* and HMS *Ocean* using a combination of Westland Whirlwind (S-55) and Bristol Sycamore helicopters. Throughout the post-war period, the carrier force also supported ground forces in the so-called 'bush fire' wars. This included bandit activity in Malaya during the early 1950s, a mutiny by troops in East Africa in 1961 and the confrontation between Indonesia and the newly independent state of Malaysia during the early 1960s. Later, the carriers were to be involved enforcing sanctions against Rhodesia after that country's unilateral declaration of independence.

ABOVE: *The Royal Navy's frigates and destroyers each carry a Westland Super Lynx - this is a Mk8 - and a number of the vessels can carry two, for ASW and ASuV duties. (GKN Westland)*

New ships entered service, including Britain's two largest carriers, HMS *Ark Royal* and HMS *Eagle*, and the ships of the Hermes-class, HMS *Albion*, HMS *Bulwark*, HMS *Centaur* and, last in service, HMS *Hermes*. Aircraft for these ships included Fairey Gannet turboprop anti-submarine, AEW and COD aircraft, Westland Wessex (S-58) and Sea King (S-61) helicopters, Supermarine Scimitar and de Havilland Sea Vixen fighters, Blackburn Buccaneer bombers, and McDonnell Douglas F-4K Phantom fighters. Amongst the first to send helicopters to sea aboard destroyers and frigates, the Fleet Air Arm operated first Westland Wasp and then Westland Lynx helicopters, with all destroyers and frigates built after 1960 either converted to carry helicopters or, in the majority of cases, designed for helicopters. The fleet supply train, the Royal Fleet Auxiliary, RFA, also introduced ships capable of operating helicopters.

The middle and late 1960s saw a steady reduction in the strength of the Royal Navy, and especially of the Fleet Air Arm, with plans to end fixed-wing naval aviation completely. The creation of the 'Harrier Carrier' concept saved the Fleet Air Arm, although to overcome political opposition at first these had to be described as 'through deck cruisers'. The first, HMS *Invincible*, entered service in 1980, although a defence review shortly afterwards proposed selling the ship. The Royal Navy selected the BAe Sea Harrier for its new carriers, a true fighter with secondary air-to-surface attack capability rather than the Harrier or AV-8 series favoured by the USMC.

British fixed-wing naval aviation, and the three carrier force, was finally saved when the Argentine invaded the Falkland Islands, 1,000 miles off the coast of Argentina and 8,000 miles from the UK, in spring, 1982. HMS *Invincible* and HMS *Hermes* were despatched with a task force for the recov-

United Kingdom

ABOVE: *The Royal Navy's new ASW helicopter is the Westland Agusta Merlin.*

ery of the islands, despite being heavily outnumbered by Argentine air force and naval aircraft, and just 20 Sea Harriers. The successful outcome of this campaign, despite losing two destroyers and two frigates to Argentina air attack, would not have been possible without carrier-borne aircraft, with the islands beyond the realistic operating range of shore-based aircraft, which required up to eleven tanker aircraft for each sortie! The loss of so many ships highlighted the lack of carrier-borne AEW, rectified afterwards with the conversion of a number of Sea King helicopters.

Post-Falklands, the Fleet Air Arm has suffered from further cuts, with one of its three carriers in maintained reserve or refit, although a helicopter carrier, HMS *Ocean*, has been introduced that can also act as an aircraft transport. Both HMS *Ocean* and HMS *Illustrious* operated in the Arabian Sea as part of the Coalition fleet in Operation Enduring Freedom during winter 2001/2. Two assault ships, each capable of operating helicopters, HMS *Fearless* and HMS *Intrepid*, are being replaced by HMS *Albion* and HMS *Bulwark*, again both are helicopter-capable. Operational Sea Harrier squadrons have been reduced to two, and are now operated jointly with the RAF's Harrier II squadrons, also being reduced to just two in number, under the Joint Force Harrier, JFH, project. They will lose their Sea Harriers when these are replaced by Harrier GR-9s after 2004, also marking the end of fighter cover for the fleet. The frigate and destroyer force has been reduced to just 30 ships, against an official figure of 32, half the number in the 1960s. The Government has announced that two larger carriers, of up to 50,000 tonnes each, will replace the present three, probably using the V/STOL variant of the F-35, in which the UK is a full partner, with Harrier GR-9s due to be replaced by up to 150 Lockheed Martin F-35s, operated jointly with the Royal Air Force. An order for conventional carrier variants remains a possibility as it is still to be decided whether the new ships will have catapults or 'ski-jumps'.

Today, the Fleet Air Arm has 6,740 of the Royal Navy's 43,770 personnel. It has 29 BAe Sea Harrier FA2 aircraft in two squadrons and a training squadron. Frontline RN squadrons are numbered in the 8XX series, and support squadrons in the 7XX series to distinguish them from RAF squadrons. There are seven Harrier T7s for conversion training. A typical carrier air wing is a squadron of eight Sea Harriers, and 12 Westland Sea King helicopters, with nine for ASW and three for AEW. There are 60 Sea King HAS5/6 for AEW and SAR, and these are been gradually replaced by an initial 44 EHI Merlin HAS1/2. There are nine Sea King AEW2A, and there are three squadrons with 33 Sea King HC4 transport and assault helicopters. Up to 59 Westland Lynx HAS3/HMA8 operate mainly from frigates and destroyers, with another six AH7 and nine Gazelle AH1 operating in a Royal Marine helicopter squadron, with RM aircrew. Training uses nine BAe Jetstream T2, with three T3 used on communications duties, as well as 12 BAe Hawks and five Grob 115 Heron.

ARMY AIR CORPS

Formed: 1957

Although the British Army's experience of military aeronautics dates from 1878, this led to the creation of the Royal Air Force. Present day Army aviation dates from the formation of the Glider Pilot Regiment during the height of World War II, with its members taking part in a number of major operations, including the invasions of Sicily, Normandy and the Rhine Crossing, using troop-carrying Horsa and Hamilcar gliders. The Glider Pilot Regiment disbanded - the shortest-lived British Army corps - towards the end of the war. During the war, most flying in support of ground forces was conducted by the RAF, but 12 AOP squadrons were operated with mixed Army and RAF personnel. Some of these units survived the war years using Auster AOP. The eventual formation of the Army Air Corps in 1957 saw the disappearance of RAF personnel.

During the 1960s, the emphasis moved from light aircraft to helicopters, including the Saro Skeeter and licence-built Westland 47G Sioux helicopters, with the Westland Scout introducing anti-tank operations with SS11 wire-controlled missiles. Units were established in the UK and to support the British Army of the Rhine, BAOR, in Germany, as well as in Hong Kong, the Persian Gulf and in Malaysia; the last two being run down first as British forces were gradually reduced and then withdrawn from East of Suez. During the early 1970s, Westland Lynxes replaced the Scout helicopters, while Westland-Sud SA340 Gazelles replaced the Sioux. At one time, DHC-2 Beaver utility transports were operated.

In 2001, the Army Air Corps started to introduce Westland-built versions of the Boeing WAH-64 Longbow Apache, with the Lynx moving into the combat scout and light transport roles. They may be upgraded with mast-mounted sights and up-rated engines. The sole fixed-wing aircraft are now five Britten-Norman BN2 Islander AL1s on communications duties. Heavier battlefield transport lies with the RAF's Pumas, being replaced by Merlins in the near future, and heavy-lift Boeing CH-47D Chinooks. A recent reorganisation has retained individual Army, RAF and Fleet Air Arm helicopter squadrons, but placed these in a Joint Helicopter Command, in which the AAC is 16 Air Assault Brigade.

ABOVE: *Although not the first helicopter for the British Army's Air Corps, the Westland Scout was the first to enter service in substantial numbers for anti-tank and other duties. (GKN Westland)*

Today, the AAC has 67 Westland WAH-64 Longbow Apache attack helicopters; 130 Westland AH7/AH9 Lynx helicopters, capable of anti-tank operations, but used mainly in combat scout and utility roles; 119 Gazelle AH1 on reconnaissance duties, four Agusta A109A on special forces duties, and three Bell 212 utility helicopters, and the five Islanders.

ABOVE: *Westland Lynx Mk7 anti-tank helicopters of the Army Air Corps. (GKN Westland)*

ABOVE: *Boeing's Apache helicopter has been built under licence for the British Army by Westland as the WAH-64 Longbow Apache, marking a significant advance in the Army's attack capability. (GKN Westland)*

United States of America

POPULATION: 281.4 million

LAND AREA: 3,775,602 square miles, 9,363,169 sq km

GDP: $9.9tr (£6.9tr), per capita $34,300 (£23,986)

DEFENCE EXPENDITURE: $310.5bn (£217.1bn)

SERVICE PERSONNEL: 1,367,700 active, exc. USCG, plus 1,200,600 reserves.

UNITED STATES AIR FORCE

Formed: 1947

As early as the 1860s, balloons were used for observation duties during the American Civil War, but the history of American military aviation really dates from 1892, when a balloon section was formed within the United States Army during the Spanish-American War, and saw service in Cuba during 1898. The US Army supported several unsuccessful attempts at building heavier-than-air aircraft after the turn of the century, notably the so-called 'Aerodrome A' of Professor Samuel Langley, but rejected offers of the Wright brothers' aircraft. In 1907, the Aeronautical Division of the Signals Corps was formed to take responsibility for military aviation, and in 1908, this finally purchased a Wright aircraft and a dirigible. By 1911, four Wright aircraft and two Curtiss aircraft were being operated on training, reconnaissance and experimental duties. Ten Curtiss aircraft flown during the war with Mexico in 1913 failed to impress, and just five aircraft were operational in 1914 when World War I broke out in Europe. In 1916, the Aeronautical Division became a separate corps, the US Army Aviation Section. The USA entered the war in 1917 against the Central Powers with 200 aircraft, mainly of Curtiss, Martin, Standard and Wright manufacture.

Ambitious wartime expansion schemes were initially heavily dependent on European designs, many built under licence, including Bristol F2B 'Brisfit' and Spad fighters, as well as a number of DH4 bombers. European aircraft included almost 5,000 French-built aircraft, mainly of Breguet, Caudron, Farman, Morane-Saulnier, Nieuport, Spad and Voisin manufacture, with 300 British-built Avro, Airco, Bristol and Sopwith. Rapid expansion was helped by the return of those Americans who had volunteered to fly with other Allied air arms before US entry into the war, especially in the French *Escadrille Lafayette.* Massive expansion plans meant that orders for 61,000 aircraft had to be cancelled within days of the Armistice in 1918.

Post-war, the strength of the US Army Aviation Section was established at 2,500 aircraft in 1920, but few new aircraft were forthcoming. Attempts to form an autonomous air service on the lines of the RAF failed, with the resulting courts martial and demotion of Brigadier-General William

BELOW: *A USAAC Vought 02U-3 Corsair scout and observation aircraft in 1929. (Vought)*

ABOVE: *The USAAC's aircraft included the Beech AT-10, a twin-seat, twin-engined, transition trainer for the aspiring bomber or transport pilot. (Raytheon)*

Mitchell as a result of intense inter-service rivalry with the USN, and the unauthorised bombing of the surrendered German battleship *Ostfriesland*. There was a further change of name to the US Army Air Service in 1920, with an initial strength of 27 combat squadrons each of 18 aircraft. The USAAS did grow, reaching 48 squadrons by 1923, although based on the obsolescent DH4B bomber. To maintain morale and gain publicity, long-distance flights were made. In 1926, the name changed again, to the US Army Air Corps. New aircraft started to enter service during the late 1920s, including Curtiss PW-8, P-1 and P-2 fighters and O-1 AOP aircraft; Martin MB-2 bombers; Douglas O-2 AOP aircraft, Consolidated PT-1 trainers and Douglas C-1 transports. The USAAC was hit by the Depression, curtailing a planned programme of expansion originally set to begin in 1926. Even so, a number of new aircraft continued to arrive, including 150 Curtiss A-3 Falcon ground-attack aircraft; 12 Curtiss B-2, and numbers of Keystone LB-5, LB-6, LB-7, B-3, B-4 and B-6 bombers; 12 Fokker FVIII and 13 Ford C-3 and C-4 tri-motor transports. Berliner-Joyce P-16, Boeing P-12B and Curtiss P-16 fighters, followed these. Four balloon squadrons were retained for AOP duties.

The USAAC steadily achieved the size and equipment quality of the air arm of a major power. As the economic situation began to improve, the USAAC grew more rapidly. New aircraft included the Boeing P-26A fighter of 1933, and the advanced Martin B-10 bomber of 1934. The USAAC also undertook internal air mail services at this time, transferred from the commercial airlines, but a series of accidents in 1935 led to an investigation, which gave the USAAC a greater degree of autonomy, although still remaining part of the US Army. The year also saw the first of 50 Consolidated P-30A and 76 Seversky P-35 fighters, 32 Martin B-12 and 250 Northrop A-17 bombers and attack aircraft respectively, and 90 Douglas O-46A AOP aircraft. In 1936, the Douglas B-18, the bomber version of the C-47 transport, entered service. Meanwhile, inter-service rivalry had flared up again, over responsibility for maritime-reconnaissance.

Following the Munich crisis of 1938, expansion speeded up considerably. Very large orders were placed, especially after war finally broke out in Europe in September, 1939. Aircraft ordered included additional Boeing P-26, Seversky P-35 and Curtiss P-36 fighters; Northrop A-17 and Curtiss A-12 and A-18 ground-attack aircraft; Martin B-10 and B-12, Boeing B-17 Fortress and additional Douglas B-18 bombers; Douglas OA-3 and OA-4, Sikorsky OA-8 and Grumman OA-9 amphibians; Bellanca C-27, Douglas C-33, C-39 and C-47, Lockheed C-36 and C-40 transports; North American O-47 and Douglas O-38, O-43 and O-46 AOP aircraft; and North American AT-6 and BT-9, Consolidated PT-11, Stearman PT-13 and Seversky BT-8 trainers. Experimental aircraft included nine Kellett G-1 autogiros.

Many of the aircraft that followed benefited considerably from experience gained with service with the British and French armed forces before full-scale entry into US service. These included Lockheed P-38 Lightning fighters, and Consolidated B-24 Liberator, Martin B-26 Marauder and North American B-25 Mitchell bombers. These showed that in wartime the USAAC would use air power strategically, in contrast to the Germans and Russians, who used air power

United States of America

ABOVE: *The World War II Consolidated Liberator was operated both by the USAAF and the USN, and by the RAF. The aircraft was notable for its exceptional range. (RAFM)*

tactically in close support of ground forces; a doctrine that worked well when supported by fast-moving tank armies, but otherwise suffered from serious limitations. In 1941, the USAAC became the United States Army Air Force, with four main constituent air forces, the 1st and 2nd in the north, and the 3rd and 4th in the south, deployed primarily for the defence of the United States. When the Japanese attacked Pearl Harbor on 7 December, 1941, bringing the USA into World War II, the USAAF had some 2,500 aircraft in service. The latest were the Bell P-39, Curtiss P-40, Republic P-43 and P-47 (later changed to F-47 when classifications were changed from P-pursuit to F-fighter) Thunderbolt, and North American F-51 Mustang fighters; Douglas A-20 Boston and A-24, and Curtiss A-25 attack aircraft; Beech C-45, Curtiss C-46 Commando and Douglas C-47, C-53, C-54 and Lockheed C-59 and C-60 transports. Most aircraft were still based in the continental United States, while many in Hawaii and the Philippines were destroyed on the ground by air attack.

US entry into the war was provoked by the Japanese attack, although the USA had given strong support to the UK in the war with Germany and Italy, doing almost everything short of an open declaration of war. Debate raged over whether the war against Japan should be given priority, defeating that country first before settling the war in Europe. In the end, the USA decided to wage war on two fronts. This decision helped to shorten the war. The USAAF would have had great difficulty in taking the offensive against Japan until bases had been secured within range of the Japanese islands. In a major propaganda coup that influenced future Japanese strategy, the USAAF mounted Operation Shangri-La, a raid on Japanese cities, including Tokyo. This used 16 North American B-25 Mitchell bombers flown off the USS *Hornet* on 18 April, 1942, with the aircraft flying on to bases in China after the raids, proving that Japan was not immune from air attack.

From 1942 onwards, the USAAF operated in increasing strength in North Africa, and from the British Isles over Europe, from Hawaii and Midway over the Pacific, and from India over China and Burma. Few combat aircraft remained in the United States. Many new concepts were developed, including 'shuttle bombing' in which bombers engaged targets deep inside enemy-occupied Europe and then flew on to bases in North Africa. The USAAF co-operated with the RAF, notably in the 'Point Blank' initiative, with the RAF attacking targets by night with a follow-up USAAF raid by day. Day bombing demanded heavier defensive armament on bombers, and the use of long-range escort fighters, such as the Lockheed P-38 Lightning and the North American F-51 Mustang. As experience grew, bomber wing commanders devised elaborate formations so that aircraft could provide covering fire for each other. In addition to combat, transport grew in importance, with the USAAF flying supplies 'over the hump' to China. Individual aircraft types were deployed to maximise their strengths, with the long range of the Liberator proving useful for raids on Burma from India, and on raids on the Romanian oil refineries from bases in England and North Africa.

The USAAF took part in the invasion of Sicily in 1943, and the following year in the Normandy landings, as well as in landings in Italy and the South of France. Transport continued to increase in importance, with troop-carrying aircraft and gliders with troops and light tanks or artillery. For the most part, aircraft in service towards the end of the war were developments of those available at the start of US involvement, but there were significant new arrivals, including the Northrop P-61 Black Widow night-fighter and the Boeing B-29 Superfortress bomber. B-29s destroyed Japan's major cities in incendiary raids, while Japanese fighters, incapable of matching the aircraft's speed and altitude, were reduced to *Kamikaze* suicide attacks using fighters stripped of weapons to ram USAAF aircraft. The B-29 effectively ended the war in the Pacific by dropping the first atomic bombs on Hiroshima and Nagasaki on 6 and 9 August, 1945.

The USAAF ended the war as the world's strongest air force, with some 60,000 aircraft and more than 2,250,000 personnel. In 1946, it was reorganised into Air Defence, Tactical and Strategic Air Commands, with support commands including Air Material, Air Proving, Air Training and Air Transport. The long cherished ambition to become a separate service was realised in 1947 when the USAAF became the United States Air Force. A hangover from the period of Army control remained with the main reserve units organized on a state-by-state basis as the Air National Guard. That same year saw the USAF receive its first jet fighter, the Lockheed F-80 Shooting Star, and the switch from 'P', 'Pursuit', to 'F', 'Fighter' designations. When the US services eventually standardised aircraft designations, it was the straightforward USAF system that was adopted. The Republic F-84 Thunderjet joined the F-80s in 1948. New aircraft entering service at this time also included the Boeing B-50, a higher-powered version of the Superfortress, and the six-engined Convair B-36, designed to bomb Germany from bases in the United States had the UK fallen; both entered service in 1947. Many B-29s were converted to the tanker role for in-flight refuelling.

ABOVE: *The heavy defensive armament of the Boeing B-17 Fortress was vital for survival during daylight raids over Europe. (IWM)*

Structural changes also occurred. In 1948, Air Transport Command absorbed the Naval Air Transport Service to create the Military Air Transport Service, MATS, which assumed responsibility for all US strategic air transport requirements. The other services were left with a small number of tactical transport aircraft, just as Tactical Air Command also operated some theatre transport and communications aircraft. This removed duplication of effort between the armed services.

The return of peace was relatively short-lived. An indication of what was to come had already occurred in 1944, when three USAAF B-29s had forced-landed on 'friendly' Soviet territory, but had been confiscated and copied, producing the Tupolev Tu-4. Germany was already being divided into zones, with the American, British and French zones eventually to become West Germany, but leaving the capital, Berlin, also divided into zones, well inside what was to become East Germany. In 1948, anxious to take power over the entire city, the USSR blockaded the land routes into Berlin from the West, forcing the Allied Powers to mount an airlift to keep the West Berliners supplied. Everything, including coal, had to be moved by air. USAF Fairchild C-82, Douglas C-47 and C-54 transports played their part in moving supplies, but the demand was such that military transports had to be augmented by chartered civilian aircraft. New aircraft continued to arrive, including the North American F-86 Sabre and Lockheed F-94 Starfire fighters, and North American B-45 Tornado light jet bombers. Progressive modernization proved its worth when war broke out in Korea in 1950 after Communist North Korea mounted a surprise invasion of South Korea. USAF forces in Okinawa and Japan were well placed to counter the invading forces. Their location was an advantage as bases in South Korea were overrun. Air National Guard units were mobilized during the Korean War, doubling the USAF's strength at a time of international tension. During the conflict, USAF forces fought under the auspices of the United Nations. The Korean War lasted until 1953, and saw some of the first jet fighter battles, with Lockheed F-80C Shooting Stars matched against MiG-15s.

Meanwhile, the first heavy jet bomber, the Boeing B-47 Stratojet, had entered service with Strategic Air Command in 1951. Heavier still was the eight-engined Boeing B-52 Stratofortress that replaced the Convair B-36s in 1955, starting a long period of service that lasted into the new century. Other aircraft of this period included the Martin B-57, a licence-built English Electric Canberra light jet bomber. The number of new aircraft types entering service was increased, reflecting the rapid change from piston to jet technology for combat aircraft and the growing importance of airborne radar. New aircraft included the North American F-100 Super Sabre, McDonnell F-101 Voodoo and Northrop F-89 Scorpion fighters and interceptors; Douglas B-66 Destroyer bombers; Lockheed RC-121 Constellation AEW aircraft; Douglas C-124 Globemaster and C-133 Liftmaster, and Fairchild C-119 Packet and C-123 Provider transports. These were followed during the late 1950s by Convair F-102 Delta Dagger and F-106 Delta Dart, and Lockheed F-104 Starfighter interceptors; the Convair B-58 Hustler supersonic jet bomber and the Lockheed C-130 Hercules transport; still in production today in uprated form. Vertol and Sikorsky helicopters were also introduced during the decade. The need for global reach was reflected not only in increasingly heavy transport aircraft, but in a large fleet of several hundred Boeing KC-135 tankers.

The formation of the North Atlantic Treaty

Organization, NATO, in 1952, and the Organization of American States, also led to the USAF assisting many allied air forces, while retired aircraft were welcomed by the poorer member states. The United States was the major partner in NATO, and its regional counterparts, the South East Asia Treaty Organization, SEATO, and the Baghdad Pact, later the Central Treaty Organization, or CENTO, in the Middle East.

Most of the late 1950s influx of aircraft continued in production into the 1960s, often suitably upgraded. The 1960s saw the Northrop F-5 tactical fighter-bomber, mainly exported under the US Military Aid Programme, MAP; the McDonnell F-4 Phantom II fighter-bomber; the LTV A-7 attack aircraft; the General Dynamics F-111 variable-geometry strike aircraft; and the Lockheed C-141 Starlifter and C-5 Galaxy transports. The F-4 and A-7 were naval aircraft with a performance that earned them a place in the USAF inventory. The decade was marked by a growing US involvement in Vietnam, where the Communist North was intent on occupying South Vietnam. The USAF was involved in countering Communist infiltration, bombing build-ups of Communist Viet Cong guerrillas in the South and also attacking targets in the North. The B-52 heavy bombers were deployed, although subsequently proving vulnerable to heavy SAM fire, with the North Vietnamese firing missiles in salvoes to ensure hitting a bomber. More successful were aircraft such as the F-4 when operating with 'smart' bombs, able to make a precision strike against a target while enabling the aircraft to remain clear of the worst AA and SAM fire. Political uncertainties in the US meant that the intensity of the campaign varied, and eventually South Vietnam was overrun.

ABOVE: *The A-7 Corsair was that rarity, a naval carrier-borne aircraft with sufficient performance to interest air forces - here is an A-7D dropping munitions at Davis Monthan AFB. (Vought)*

Changing technology has been constantly adopted by the USAF, often adapting existing aircraft. The B-52 bomber fleet was upgraded and refurbished, gaining a new lease of life as platforms for the launch of cruise missiles. The ability to deploy cruise missiles proved an effective counter to a new generation of Soviet intermediate-range missiles deployed in Europe. Aircraft such as the F-111 could exploit the gap below the scope of ground-based radar. To counter such techniques, AEW aircraft moved forward with the introduction of Airborne Early Warning and Control, AEW&C, aircraft, the Boeing E-3 Sentry, based on the Boeing 707 airliner. In some cases, older techniques have survived, with reconnaissance aircraft retaining a role even in the days of satellite surveillance. New aircraft reflected these changes, with the supersonic North American Rockwell B-1 bomber, and then the 'stealthy' Northrop Grumman B-2, with a much reduced radar signature. Unlike its pre-war counterpart, the post-WWII USAF deployed worldwide, and especially in the UK, West Germany, Spain and Turkey, and Japan. The Military Air Transport Service became Military Airlift Command, MAC, and by the late 1980s this had acquired a daily airlift capability of 66 million ton/miles per day. One of MAC's duties became SAR, and also Combat Search and Rescue, CSAR, using Sikorsky S-65 helicopters, and on many occasions during the Vietnam War downed US pilots were rescued from behind enemy lines.

BELOW: *Although designed to deliver nuclear weapons, the Boeing B-52G Stratofortress saw extensive action in Vietnam, and then enjoyed a new lease of life as a cruise missile carrier during the Gulf War. (RAFM)*

New aircraft entering service during the final years of the Cold War included the McDonnell Douglas F-15 Eagle air superiority fighter and the Lockheed Martin F-16

ABOVE: *Top left: A flight of three F-15 Eagle interceptors, an aircraft type that has seen increased use following the terrorist attacks of September 11, 2001. (USAF)*

fighter-bomber. While the F-15 was intended as an F-4 replacement, the bulk of the replacement programme fell to the F-16 as the F-15 proved too costly.

The collapse of the Soviet Union and of the Warsaw Pact meant a reassessment of priorities. This was further emphasised in August, 1990, when Iraq invaded Kuwait. The immediate priority was to protect Saudi Arabia and that country's vital oilfields, Operation Desert Shield, followed later by Operation Desert Storm, the liberation of Kuwait. The forward basing of so many units in Europe meant that the USAF was able, within two months, to base the equivalent of five tactical fighter wings in the Middle East, drawing aircraft from across the USA and Europe. Other aircraft operated from bases in Turkey and in the Indian Ocean. MAC was the main force in moving not only the heavy equipment and many of the personnel of the US armed forces, but those of its allies, known as the Coalition, as well. The bulk of the Coalition air offensive was provided by the USAF, mainly using F-111s and cruise-missile carrying B-52Gs, while the new Lockheed F-117 Nighthawk stealth fighter was also deployed for the first time. The F-117 units, deployed against important targets and using laser weapons, achieved success rates as high as 80-85 per cent, compared to the 30-35 per cent managed by advanced weaponry in Vietnam.

The success of the Coalition Forces was limited by the decision not to occupy Iraq, which would have needed a fresh mandate from the United Nations, which had sanctioned the liberation of Kuwait in the first UN operation since the Korean War. Nevertheless, the USAF remained in the region and in Turkey, enforcing two 'no fly' zones, one in the south of Iraq and the other in the north.

The collapse of the Warsaw Pact and the experience of the Gulf War led to a reassessment of the USAF's structure. Strategic Air Command was disbanded and its aircraft, personnel and bases reallocated, mainly to Tactical Air Command, which in turn became Air Combat Command, ACC, from January, 1992. On the same date, the SAC tanker fleet passed to a new Air Mobility Command, AMC, which absorbed MAC. The new structures recognized the reduced threat from Russia, although this remains a factor in US defence planning, the reduced likelihood of a massive nuclear response,

ABOVE: *The last of the cold war warriors, the F-111 was the world's first operational variable-geometry, or 'swing-wing' aircraft. Despite the fighter designation, it was intended as a bomber. (USAF)*

ABOVE: *The high cost of the F-15 meant that it had to be complemented with large numbers of affordable F-16 fighters. (USAF)*

BELOW: *The first of a new generation of stealth aircraft, the Lockheed Martin F-117A Nighthawk, designed to be almost invisible to radar, actually operates in the strike role. (USAF)*

United States of America

ABOVE: *The Northrop Grumman OA-10 series has an unconventional configuration but is optimized for intensive anti-tank operations, flying at low altitudes. (USAF)*

and the growing demand for tactical missions. The USAF changed to become a rapid response service, rather than one primarily concerned with massive retaliation. Air mobility was aided by the new Boeing C-17 Globemaster II heavy lifter.

Post-Cold War reductions in the strength of the USAF were used to enhance the equipment of the USAF Reserve and the Air National Guard (ANG) units, with the latter now manning the majority of the USAF's air defence squadrons. During operations over Kosovo in 1999, Operation Allied Force, the USAF performed a far larger number of the missions than the entire effort of the other allies. Aircraft deployed during these operations included the Lockheed F-117A, Lockheed Martin F-16 Falcon and the Fairchild OA-10 Thunderbolt or Warthog. The B-52 force was deployed again during 2001/2 in operations against the Northern Alliance and Al Q'aeda network, mainly operating from Gan in the Indian Ocean. The use of US-based B-2s to attack targets in Serbia demonstrated the ability to attack over extremely long distances. In 1999, further changes were made, introducing the Aerospace Expeditionary Force, AEF, concept, with almost the entire USAF, Reserves and ANG, organised into ten AEFs, each with between 10,000 and 15,000 personnel, and up to 200 aircraft, including air superiority, air-to-ground precision strike and air mobility capability. Each AEF will be on call for 90 days every 15 months, with at least two AEFs on call at any one time, so that the USAF should be able to handle two medium intensity conflicts at any one time.

The USAF has 353,600 personnel, a reduction of more than 50 per cent over the past 30 years. There are also 73,700 personnel in the Air Force Reserve, and a further 106,600 in the Air National Guard units. Flying hours are an average of 212 annually. In the tactical role, there are 52 fighter squadrons, each with between 12 and 24 aircraft. The fighter squadrons include 14 with F-15 and six with F-15E, with a total of 615 aircraft, two with 52 F-117, 23 with 782 F-16C/D, and seven with around 63 A-10 and 45 OA-10. In due course, 339 F-22 Raptors will replace some of the aircraft in this force, followed later by up to 1,783 USAF versions of the Joint Strike Fighter, JSF, although there has been speculation that the F-35 might be replaced by additional F-22s and updated versions of the F-16. Support aircraft include three reconnaissance squadrons with 31 U-2 and 18 RC-135U/V/W, as well as AEW&C with six squadrons, including one for training, with 32 E-3s, while another two squadrons have 22 EC-130 for AEW and ELINT. Other support aircraft include forward air control with about 63 A-10As and 45 OA-10As in seven squadrons. Bomber aircraft include 73 B1B Lancer and 21 B-2A Spirit, as well as 85 B-52H Superfortress. There are 28 transport squadrons, of which 17 are strategic, with five, including a training squadron, equipped with 80 C-5A/B Galaxy; three with C-17 Globemaster II, still in course of delivery with 120 planned; and these will eventually replace many of the aircraft in the nine squadrons (with two on training duties) of 139 C-141 Starlifters. The other eleven transport squadrons provide tactical airlift, equipped with 191 Lockheed C-130 Hercules, including the new C-130J Hercules II. There are also 23 tanker squadrons, of which 19 operate 255 KC-135, including a training squadron, while the other four have 59 KC-10A Extender (DC-10) tankers. Eight squadrons operate 47 HH-60 helicopters and 24 HC-130N/P Hercules on CSAR. Three squadrons operate 29 MEDEVAC C-9A (DC-9) Nightingales. Weather reconnaissance uses three WC-135s. There are a number of miscellaneous aircraft in the transport role, with small numbers of C-9, C-12, C-20, C-21, C-135 and VC-137s; a number of these operate VIP flights. Training aircraft include a squadron of F-16s in the aggressor role, as well as another 35 squadrons with conversion trainer variants of frontline aircraft. Basic and intermediate training is provided by contractors using Slingsby T-3 Firefly and Boeing T-1 Jayhawk (BAe Hawk) as well as

BELOW: *Pre-production models of the F-22A Raptor, a new generation of stealth fighters, fly in formation. (USAF)*

United States of America

ABOVE: *Two stealth bombers, B-2A Spirits, with one refuelling from a McDonnell Douglas KC-10A Extender tanker, a variant of the DC-10 airliner. (USAF)*

T-37 Tweet prominent amongst the aircraft in use.

There are 35 wings in the Air Force Reserve. These include a bomber squadron, with nine B-52H. Seven fighter squadrons have four squadrons equipped with 71 F-16C/D and three with 74 A/OA-10 Thunderbolt II. There are 19 transport squadrons, with the seven strategic squadrons including two operating 32 C-5A and five with 48 C-141 Starlifter. The eleven tactical transport squadrons operate 130 Lockheed C-130E/H Hercules, while there is a weather-reconnaissance squadron with WC-135E/II. Seven tanker squadrons operate 72 KC-135E/R. SAR squadrons number three, with 23 HH-60 Black Hawk and seven HC-130 Hercules. There are another 26 squadrons without aircraft but with the personnel to augment the operations of active squadrons of transport and training aircraft in an emergency.

The Air National Guard has ten air defence fighter squadrons, with F-15s and F-16s, as well as 41 in the fighter/strike role, of which 32 have F-16s and five F-15A/Bs, and six have A-10/AO-10, with a total of 116 F-15, 600 F-16 and 100 A-10/AO-10 aircraft. There are 27 transport squadrons, of which three are strategic, with one operating 13 C-5A and two 18 C-141. The 24 tactical squadrons, including a training unit, have 225 C-130E/H Hercules. There are 23 tanker squadrons with 224 KC-135E/R. A special operations squadron has eight EC-130E EW Hercules. Three SAR squadrons have 17 HH-60 Black Hawk and 13 CH-130 Hercules. There are seven training squadrons.

Missiles used include the AAM AIM9P/L/M Sidewinder, AIM-7E/F/M Sparrow and AIM-120A/B AMRAAM, and the ASM AGM-65A/B/D/G Maverick, AGM-88A/B Harm, AGM-84B Harpoon and AGM-86B ALCM and AGM-154 JSOW

UNITED STATES NAVY AND MARINE CORPS

Formed: 1911

Although US naval aviation dates officially from 1911, in November, 1910, a young naval officer, Lt Eugene Ely had taken off from a wooden platform built over the forecastle of the light cruiser USS *Birmingham*, while the ship lay at anchor. Ely repeated this exploit the following January, flying from a platform constructed over the stern of the cruiser USS *Pennsylvania*, but on this occasion flew his aircraft onto the ship first! Later that year, the first funds were voted for naval aircraft, a Wright and two Curtiss machines. Within two years, the United States Navy was operating eight aircraft, including some seaplanes operating from warships in a brief war with Mexico. It was to take wartime pressures to achieve rapid expansion. At the start of US involvement in 1917, the USN had just 21 aircraft, but by the end of the war, the USN and USMC had

United States of America

ABOVE: *At sea with the fleet in 1926, Vought O2U-1 Corsair scout and observation floatplanes on their catapults aboard a battleship. (Vought)*

more than 1,000 seaplanes and flying boats as well as 250 landplanes. At the start of the war, many personnel served with other Allied air arms before the USN was ready to mount anti-submarine patrols. In 1918, Curtiss R-6 and HS2-2 seaplanes and H-16 flying boats predominated, but there were substantial numbers of aircraft built by the new Naval Aircraft Factory and foreign imports.

Post-war organization and equipment of the USN and USMC was based on Curtiss JN-4H, Martin MBY and Thomas-Morse M1 aircraft, Curtiss N-9 and R-6, and Boeing CL-4 seaplanes, with Curtiss NC series, licence-built Felixstowe F5 and F5L, Aeromarine 40L and HS2L flying-boats, and Airco DH4B. Sopwith Scouts were flown from platforms constructed over the turrets of battleships. An attempt to fly across the Atlantic with four Curtiss NC flying-boats in 1919 resulted in one of the aircraft, NC-4, reaching Lisbon via the Azores in late May, before flying on to reach England.

In 1921, the USN formed its Bureau of Aeronautics, known as BuAer, to advise the Chief of Naval Operations on all aspects of aeronautical affairs. A collier was converted to become the USN's first aircraft carrier, the USS *Langley*, in 1922, with the first take-off by a Vought VE-7-SF biplane. Some 300 new aircraft entered service at this time, including the Naval Aircraft Factory PT, Davis-Douglas, 30 Martin MO-1 observation floatplanes and 60 Vought OU-1 biplanes, as well as aircraft from many European manufacturers for evaluation and analysis. The USN also experimented with airships. A number of landplanes, including DH4s, Douglas DTs and Vought UO-1s, were fitted with arrester hooks so that they could operate from the *Langley*. USN aviation was organized into Pacific Fleet and Atlantic Fleet air forces, each with six squadrons for bombing, torpedo-bombing, reconnaissance and fighter duties, while the USMC had four air squadrons. Large-scale exercises in 1922 included torpedo attacks against battleship targets.

Squadron designations were specified in 1922 as VF, fighters; VO, observation; VS, scout; VT, torpedo and bombing; ZK, balloons and kites.

In 1922, the Washington Naval Treaty limited the tonnage of all classes of warships in the major navies. The USN had too many battlecruisers, so converted the USS *Saratoga* and *Lexington*, into aircraft carriers. These were the largest warships of their kind when they entered service in the late 1920s. The USN also had 20 battleships each capable of carrying three catapult-launched seaplanes, and ten cruisers, each capable of launching two seaplanes.

Progress continued throughout the 1930s, with a fourth aircraft carrier, USS *Ranger*, joining the fleet in 1934, the first US carrier to be designed as such and intended to be the first of five smaller carriers, but she was too small for successful operation in the Pacific. New aircraft included Boeing F3B-1 and F-4B-2 fighters, Chance-Vought O2U-4 and O3U-2 Corsair and Curtiss O2C-1/2 observation aircraft, and Consolidated P2Y-1 flying boats, followed by Curtiss Helldiver dive-bombers and Sparrowhawk fighters, some of which operated from airships as an experiment.

BELOW: *The first American carrier was the USS Langley, a converted collier shown here with the original Vought Corsairs in 1926. Her humble origins meant that she could be treated as an experimental ship and was not covered by the Washington Treaty limitations. (Vought)*

United States of America

ABOVE: *The Douglas A-4 Skyraider entered service too late for World War II, but saw extensive service right through to the Vietnam War, with a number of variants in between, including this AEW AD-4W. (USNI)*

The USMC also started carrier operations at this time. Two new aircraft carriers, USS *Yorktown* and USS *Enterprise*, entered service during the late 1930s, and were followed by a smaller ship, USS *Wasp*. As the end of the decade approached, the USN and USMC were building up towards a total of 3,000 aircraft, with many of the new arrivals in the early 1940s destined to achieve fame during World War II. New aircraft entering service included Brewster F2A, Vought F4U Corsair II, and Grumman F4F Wildcat and F6F Hellcat fighters; Douglas SBD Dauntless and Curtiss-Wright SB2C Helldiver dive-bombers; Douglas TBD Devastator and, later, Grumman TBF Avenger torpedo-bombers; Vought Sikorsky OS2U Kingfisher observation aircraft; Consolidated PBY-1 Catalina and Martin PBM Mariner flying-boats. The Catalina was also available as an amphibian. In 1941, a new carrier, USS *Hornet*, entered service. Recognizing the prospect of a major war, eleven aircraft carriers were already under construction at the time of the Japanese attack on Pearl Harbor in 1941. Plans had also been made for the conversion of merchant vessels to provide low cost escort carriers, known to the USN as CVEs, with more than a hundred built for both the USN and for loan to the Royal Navy. Between the escort carriers and the large, fast attack carriers, lay the Independence-class light carriers, or CVLs, converted from cruisers.

The Japanese attack on the US Pacific Fleet base at Pearl Harbor on 7 December, 1941, was designed to cripple the US fleet, destroy the base and give the Imperial Japanese Navy twelve months as undisputed masters of the Pacific. It failed. The base was never rendered unusable, and the carriers were at sea at the time. The USN then started the task of taking the war across the Pacific and back to Japan, which had quickly established a vast Asian empire in the weeks following Pearl Harbor. In April, 1942, USS *Hornet* flew off 16 North American B-25 Mitchell medium bombers of the USAF to attack Japanese cities, including Tokyo, and while the raid achieved little in tactical terms, it changed Japanese strategy. The following month, the two navies engaged in the first carrier-to-carrier battle, the Battle of the Coral Sea, in which the USN lost USS *Lexington* and saw USS *Yorktown* badly damaged, but sunk the Japanese *Shoho* and severely damaged *Shokaku*. In June, just six months after Pearl Harbor, the US Pacific Fleet sank four Japanese carriers in one day at the Battle of Midway, putting paid to Japanese ambitions. The vast distances of the Pacific meant that, at first, only carrier-borne aircraft could take the lead in the war. A pattern emerged of the USN supporting invasions, with the USMC and US Army taking islands, and then the USAF would move in and use the bases once the islands were secured. At the same time, the USN had to help protect shipping in the Atlantic and also assisted at the invasion of North Africa and in ferrying aircraft to Malta using the USS *Wasp*. Escort carriers and longer-range MR aircraft, notably the Consolidated Liberator, helped to secure the North Atlantic for merchant shipping, while the United States Coastguard and its aircraft

United States of America

ABOVE: *Grumman F9F-5 Panther jet fighters entered service with the USN in time for the Korean War. (USNI)*

ABOVE: *The LTV F-8A supersonic jet fighter used an unusual variable-incidence wing to reduce the impact of the 'nose-up' position of an aircraft approaching a carrier flight deck. These were amongst the first examples in 1955. (Vought)*

came under USN control for the duration of the war. As the Battle of the Atlantic against German submarines was gradually won, escort carriers started to find new roles, often carrying aircraft for ground-attack duties and covering invasion forces, and as aircraft transports. Japanese resistance was strong, and included the use of Kamikaze suicide aircraft against warships, including aircraft carriers, as well as bombing and torpedo-bombing attack. Significant milestones in the Pacific included the campaign known to the USN as the 'Great Marianas Turkey Shoot', in the Battle of the Philippine Sea, in June, 1944, with 300 out of 365 Japanese aircraft based on Truk, an island in the Marianas, shot down. This was followed in October by the Battle of Leyte Gulf, the biggest naval battle in history. Japan lost three battleships, four aircraft carriers, six heavy cruisers, four light cruisers and many smaller vessels; many of them to aerial attack, and the USN lost just one light carrier, USS *Princeton*.

Post-war, the USN was the largest navy, with 41,000 aircraft, 20 large aircraft carriers and five under construction, eight light carriers and 69 escort carriers. Most of the large carriers were new ships of the Essex-class, although USS *Enterprise* and USS *Ranger* survived the war. In addition to the Liberators, land-based aircraft in USN service included the Boeing B-17 Fortress and B-29 Superfortress, North American B-25 Mitchell and Lockheed PV-1 and PV-2 Harpoon bombers, as well as many transport aircraft, including Curtiss C-46 and Douglas C-47, and Lockheed Lodestars. The war had seen rivalry with the USAAF over allocations of long-range aircraft.

The USN started to reduce to 13 large aircraft carriers. Aircraft numbers were rapidly reduced to around 10,000, but many of these were to be new aircraft, of types developed too late to play a part in the conflict. These included Grumman F7F Tigercat, F8F Bearcat and Ryan FR-1 Fireball fighters, the latter being a compound aircraft, with a main piston engine and an auxiliary jet engine to boost performance. These were accompanied by Douglas A-1 Skyraider and Martin AM-1 Marauder strike aircraft, while Lockheed P2V Neptunes gradually took over MR from Harpoons, Fortresses, Superfortresses and Liberators. The USN introduced its first jets, the McDonnell FH-1 Phantom. New transport aircraft included Fairchild C-119 Packets, Douglas C-54 Skymasters and transport versions of the Grumman VF-1 Albatross amphibian and Martin JRM Mars flying-boat, both of which also operated SAR. While awaiting sufficient carrier-capable jets, land-based Lockheed F-80 Shooting Stars were introduced to increase the pool of experienced jet pilots. Starting in 1948, all but the smaller aircraft in the Naval Air Transport Service were transferred to the USAF's Military Air Transport Service.

Carrier-capable jets eventually started to arrive in large numbers to augment and then replace the Phantoms. The new arrivals were fighters, including the

ABOVE: *Tankers for in-flight refuelling are not always large and based on transport aircraft - here a US Grumman A-6 provides fuel for an LTV A-7E Corsair in 1975, using the system known to the RN and USN as 'buddy-buddy'. (Vought)*

Grumman F9F Panther, McDonnell F2H Banshee, and Chance-Vought F6U Pirate, and the naval version of the Sabre, the North American FJ-1 Fury. Sikorsky S-51 helicopters were introduced for plane-guard and communications duties, relieving destroyers of this role. The plane-guard helicopter was more efficient than a destroyer, and in combat it reduced the risk of a destroyer being sunk while stopping to pick up survivors.

The Korean War meant the recall of many reservists to increase USN and USMC manpower. The two air arms provided support for ground forces. On 20 September, 1951, 12 USN Sikorsky S-55 helicopters lifted a company of 228 fully equipped US Marines to the top of a strategically important 3,000-ft high hilltop in central Korea. The helicopters then carried nine tons of food to the Marines, before completing the operation by laying a field telephone system back headquarters. The entire exercise took four hours, instead of the two days it would have taken on foot. Within a month, 1,000 combat-equipped Marines were moved in full view of enemy forces, taking 6 hours 15 minutes, which was 25 minutes less than the Marines had planned. USMC helicopters joined those of the US Army in such roles as CASEVAC, and survival rates of seriously wounded soldiers were vastly improved because of the helicopter's speed and through being spared a bumpy ambulance journey over rough terrain.

While the war was in progress, the North American AJ-1 Savage attack aircraft started to enter USN service, a piston-engined design, but the first USN aircraft designed to deliver an atomic bomb. Variants of existing aircraft included airborne-early-warning versions of the Skyraider. Other new aircraft included the Douglas F4D-1 Skyray, Grumman F9F-6 Cougar and McDonnell F3H Demon fighters, as well as land-based Lockheed R70s, MR versions of the Super Constellation, and Fairchild R49. North American T-28B Trojan and Lockheed TV-2 jet trainers were also introduced.

New aircraft carriers also appeared, with the USS *Forrestal* and a new USS *Saratoga*. American carriers were the first to incorporate the many advances pioneered by the Royal Navy. Aircraft of the late 1950s included Chance-Vought F8U-1 Crusader and F7U-3M Cutlass fighters and the Douglas A3D-2 Skywarrior carrier-borne bomber. Grumman produced the S2F-1 Tracker anti-submarine aircraft, which was joined aboard the carriers by its derivatives, the WF-2 Tracer AEW aircraft and the Trader carrier onboard-delivery, COD, aircraft. Traditionally, carrier-borne aircraft have had an inferior performance to their land-based counterparts, usually due to the extra weight of folding wings, arrester hooks and strengthened undercarriages. During the 1960s, two new arrivals, the McDonnell F-4 Phantom II and the LTV A-7 Corsair II, were a match for their shore-based counterparts, and both entered service with the USAF and many other air forces. Other aircraft at this time included the Douglas A-4 Skyhawk attack aircraft, the LTV F-8 Crusader, in effect a fighter development of the A-7 and eclipsed by the F-4, and larger and more capable AEW and COD aircraft in the Grumman E-2A Hawkeye and C-2A Greyhound. Many Crusaders found a role as reconnaissance aircraft, as did North American RA-5C Vigilante reconnaissance versions of the USN A-5 bomber. In the late 1960s, Grumman A-6 Intruders started to replace many of the Skyhawks, while the Lockheed S-3 Viking replaced the Trackers. Shore-based MR was not neglected, with the Lockheed P-3 Orion developed from the Electra airliner. Throughout this period, new and more capable helicopters entered service, including the Sikorsky S-

United States of America

ABOVE: *More than 500 F/A-18E/F Hornets are due to enter service aboard the USN's carrier fleet over the next few years, with conversion training given on the F/A-18F as shown here. (Northrop Grumman)*

58, followed by the SH-3A Sea King and, for small ships, the Kaman HH-43 Husky and the UH-2 Seasprite, also often based aboard carriers for plane-guard duties. The SH-3A was primarily an anti-submarine helicopter, but a number were converted for mine-sweeping duties until it was felt that a larger helicopter would be more suitable, resulting in a variant of the Sikorsky CH-53 Sea Stallion. Sea Stallions and Boeing Vertol CH-46 Sea Knights were also used by the USMC.

The 1960s were marked by the Vietnam War, and the USN normally kept at least one aircraft carrier off the coast of Vietnam, joining the USAF in attacks on key targets and also dropping mines in the main North Vietnamese port of Haiphong. US withdrawal from the war, leaving South Vietnamese forces to face the Viet Cong, was not matched by a reciprocal fall in support by the Soviet Union, which continued to ensure ample supplies for the North Vietnamese. In the closing stages of the war, in April, 1975, nine aircraft carriers from the US Seventh Fleet converged on South Vietnam to evacuate US nationals and prominent South Vietnamese political and military personnel, landing 7,000 US Marines. During this time, the fleet had risen to fifteen large attack carriers, CVA, as well as seven older carriers of the Essex-class re-designated anti-submarine carriers, CVS.

The advent of assault ships of the Iwo Jima-class and Tarawa-class, capable of carrying large numbers of aircraft was helped by the introduction of the vertical take-off strike aircraft, with the USMC introducing the McDonnell Douglas AV-8A Harrier, a licence-built Hawker Siddeley Harrier, into service. Nuclear-powered aircraft carriers also started to enter service, increasing the time spent on patrol and also allowing larger and more, capable aircraft to be operated, such as the Grumman F-14A Tomcat, another innovation with its variable-geometry wings. These were augmented by the McDonnell Douglas (later Boeing) F-18, later F/A-18 Hornet, a highly capable fighter-bomber which also proved much less expensive than the Tomcat, and eventually replaced the A-6 Intruder in the strike role as well as taking over from the F-4s. The advent of the nuclear age came in 1961 with the USS *Enterprise*, with a range of some ten years between refuelling, and capable of carrying up to 84 large aircraft. This also led to reclassifications, and with the gradual retirement of the anti-submarine carriers, the USN began to classify all carriers as CV or CVN in the case of nuclear-powered ships. The USN has now standardized on nuclear-powered carriers, and further standardization followed the introduction of the Nimitz-class in 1975.

The USN and USMC have been active in the UN operations in Somalia and in the Gulf War, with the first carrier reaching the area within days of the Iraqi invasion of Kuwait. The USN eventually had six aircraft carriers involved in the Gulf War, about half the number in active service at the time. Oldest of these ships was USS *Midway*, and the newest, USS *Theodore Roosevelt*, one of two Nimitz-class ships present. The operation marked the final combat sorties for the A-7 and A-6 strike aircraft, and the first for the F-14 Tomcat and the F/A-18 Hornet, which undertook bombing and fighter sorties. On one occasion, an F/A-18 carrying four 2,000-lb bombs shot down an Iraqi aircraft en route to the target, without shedding its warload. Operations over Afghanistan during the winter of 2001/2002 saw aircraft operating from carriers, including the USS *John C Stennis* and the USS *Theodore Roosevelt*, often receiving inflight refuelling from RAF tankers in the target zone since the RAF refuelling system is compatible with that of the USN.

Today, the USN has 370,700 personnel, of whom 63,200 are involved with aviation. There are 169,800 personnel in the USMC, with 36,400 involved with aviation. The USN is officially historically divided into just two fleets, the Atlantic and the Pacific, but for practical purposes, there are five fleets. These are: Second Fleet, Atlantic; Third Fleet, Pacific; Fifth Fleet, Indian Ocean; Sixth Fleet, Mediterranean; Seventh Fleet, West Pacific. There are 12 aircraft carriers, of which nine are nuclear-powered vessels. One of the three conventional vessels is in maintained reserve, while of the remaining 11 carriers, one is in refit at any one time. Eleven naval air wings exist, of which one is in reserve, while the average air wing includes nine squadrons, usually one with 14 F-14s, three with 12 F/A-18s each, one with eight S-3B and two ES-3 Vikings, 1 with six SH-60 helicopters, one with four EA-6B, one with four E-2C, and a support squadron with C-2 Greyhounds. Five of the USN's 27 cruisers can accommodate two helicopters, as can 28 Arleigh Burke-class destroyers, with the 24 Spruance-class destroyers able to take one helicopter each, while the 35 (of which ten are in maintained reserve) Oliver Hazard Perry-class frigates can take one or two helicopters each. The USMC AV-8B Harrier II operate from seven Wasp-class LHD, each of which can take five as well as up to 48 helicopters; five Tarawa-class LHA, each of which can take six, as well as up to 21 helicopters; while six Austin-class LPD can take six helicopters each. The USMC uses F/A-18A/B/C/D Hornets, with the -18D for forward air control and reconnaissance as well as training, as well as AV-8B Harriers. There are also Lockheed KC-130/130J Hercules tankers, with the C-130J force likely to reach 50. The CH-46E Sea Knight and CH-53D helicopters will eventually be replaced by 425 Bell/Boeing MV22B Osprey tilt-rotors, while the USN will receive 48 HV-22Bs, providing problems experienced with this aircraft can be overcome. Bell has been upgrading 100 UH-1N Iroquois and 180 AH-1W Super Cobra helicopters, making these the UH-1Y and AH-1Z respectively.

USMC aircraft are included in the USN figures, which include 381 F-14A/B/D Tomcats, 545 F/A-18E/F Hornets, which are replacing the 320 F/A-18A/B and some of the 483 F/A-18C/D Hornets, and likely to be joined in due course by up 480 carrier versions of the F-35. There are 200 USMC AV-8B+ Harrier IIs in the attack role, and 19 TAV-8B conversion trainers, likely to be replaced in due course by up to 609 STOVL variants of the F-35, although the

ABOVE: *Electronic-counter-measures is a role played by the remaining Grumman E-6s, converted to EA-6B Prowler standard. (Northrop Grumman)*

BELOW: *The eyes and ears of the USN are the Grumman E-2C Hawkeye operated from the carriers. France is the only other country to have these aircraft at sea, but a number of countries operate these aircraft from shore bases. (Northrop Grumman)*

ABOVE: *The USN and USMC operates the Sikorsky S-65 series in a number of forms, this is one of the USMC's CH-53E transports. (Sikorsky)*

USMC order may be cut back due to funding problems, despite heavy increases in the US defence budget. A substantial number of A-6 derivatives survive, including 16 EA-6A Mercury long-endurance communications relay aircraft, now being upgraded, and 128 ECM EA-6B Prowler. Carrier-borne aircraft also include 94 AEW E-2C Hawkeye, 137 S-3A/B Viking ASW aircraft, and 38 COD C-2A Greyhound. There are 293 Lockheed P-3A/B/C Orion for MR, although only the P-3C is in regular service with 12 squadrons. Another 10 EP-3E/J Orion operate on ELINT, with 12 RP-3A/TP-3A on survey and training duties, while there are 14 UP-3A/B/VP-3A Orion on VIP duties. MCM helicopters include 19 Sikorsky RH-53D Sea Dragon and 25 Kaman SH-2F/G Seasprite. ASW helicopters include 236 Sikorsky SH-60B Seahawk, with up to 243 MH-60R multi-mission helicopters on order to replace older Seahawks starting in 2004. There are 52 Sikorsky SH-3D/G/H Sea King, which join the Seahawks on SAR. CSAR is provided by 24 HH-60H Black Hawk, while the 150 UH/HH-1N Iroquois can also be used for SAR. Other utility helicopters include 58 Boeing UH-46D/HH-46D Sea Knight, 47 Sikorsky UH-3A/H Sea King, and 45 MH-53E Sea Stallion. Apart from the Super Cobra helicopters and Osprey tilt-rotors already mentioned, transport helicopters include 269 CH-46D/E Sea Knight and 226 CH-53D/E Sea Stallion. Fixed-wing transport aircraft include six Boeing C-40A Clipper (737-700), 17 C-130F/T and seven C-130F/R Hercules, 79 KC-130F/R/T tanker versions, many of which are being replaced by KC-130J, three DC-130A Hercules on drone control duties, and 29 C-9B Nightingale MEDEVAC aircraft. Apart from the special Orions, VIP aircraft include eight VH-60N and 15 VH-3A/D Sea King helicopters. Communications aircraft are fixed-wing, with ten CT-39E/G Sabreliners, two Fairchild C-26A Metro III and seven Grumman C-20D/G (Gulfstream). Training uses 38 Northrop F-5E/F Tiger II in an aggressor squadron. There are 319 Beech T-34C Turbo Mentor, 339 T-6A Texan II, 170 McDonnell Douglas T-45A Goshawk (BAe Hawk), 57 Beech T-44A Pegasus, and a small number of Sabreliner and TC-18F trainers, as well as 94 Bell TH-57B/C Sea Ranger (206) helicopters.

The USN is planning for a new MR aircraft to replace the P-3C Orion after 2010.

UNITED STATES COAST GUARD AVIATION

While many air arms had earlier experience of balloons, the United States Coast Guard can lay claim to the longest connection with aviation, as Surfman Daniels took a photograph of the first flight by the Wright brothers in December, 1903. It was not until 1915 that two young officers experimented with a Curtiss F flying boat for USCG operations. They were sent to train with the USN at Pensacola in 1916. Later that year, Congress authorised the USCG to establish ten air stations, but did not provide funding. To gain experience, a number of USCG personnel flew with the USN during World War I. The USCG is counted as part of the US armed forces, but is under USN control only in wartime, otherwise being administered by the Department of Transportation. Post-war, Read's Curtiss NC-4 was piloted on its historic transatlantic flight by a member of the USCG.

An attempt to establish USCG aviation came in 1920, taking over an abandoned USN air station and borrowing USN Curtiss H2-2L flying boats. Despite proving that aircraft were useful in search and rescue, the continued lack of funding led to the station closing in 1921. Prohibition ensured that Congress funded USCG aviation, after the service proved that it could prevent the smuggling of whisky. Congress allowed $152,000 (£38,000 at the then rate of exchange) for three Loening OL-5 amphibians and two Chance Vought UO4s, flown from two bases in New England. SAR was not overlooked, and the USCG set about buying 'flying lifeboats', aircraft that could land in the open sea to rescue survivors. The first of these were two Douglas RD-2 Dolphins and the specially designed General Aviation Flying Life Boat PJ-15, which operated successfully throughout the early 1930s. The early 1930s also saw Grumman JF-2 amphibians operated from the larger cutters, against opium smugglers off the West Coast and on fisheries protection off Alaska, as well as later on 'plane guard' duties for experiments in transatlantic air services. By 1936, the USCG had six air stations and 42 aircraft. Before US entry into World War II, the USCG surveyed the coast of Greenland, looking for sites for airfields, using a cutter carrying a Curtiss SOC-4.

US entry into World War II saw the USCG tracking down and capturing German weather stations in northern Greenland, which were providing forecasts for U-boats. Rescue missions were flown over the sea and the ice cap to aid downed Allied aviators, and survivors from ships sunk by U-boats. During the war, USCG aircraft located a thousand survivors and directed rescuers to the scene, and rescued a hundred survivors by landing aircraft in the open sea, on occasion having to taxi ashore as the weight to those rescued meant that the aircraft could not take-off. A dedicated Air Sea Squadron was formed in 1943 in California, initially using Consolidated PBY-5A Catalina amphibians. By 1945, SAR accounted for 165 aircraft and nine air stations, and the USCG responded to 686 plane crashes. Offensive operations were also mounted, with the USCG's first U-boat sunk in the Gulf of Mexico in 1942.

ABOVE: *Amongst the early aircraft operated by the USCG on anti-smuggling patrols were Vought UO-4 observation floatplanes. (Vought)*

In 1943, the USCG was given responsibility for trials with the helicopter to assess its role for ASW. It used a merchant vessel, SS *Daghestan*, and a joint trails unit was set up with the Royal Navy after the USCG had trained British pilots, using Sikorsky HNS-1 and HOS-1 helicopters. Evaluation also took place aboard the USCGV *Cobb*, a converted passenger steamer. The first rescue by a helicopter came in 1945, after two ski-equipped aircraft ran into difficulties attempting to rescue the crew of an RCAF aircraft crashed in Labrador. A USCG HNS-1 was dismantled and loaded into a Douglas C-54 transport, flown to Goose Bay, and reassembled, and flown 185 miles to the crash site via a refuelling stop.

Post-war, the USCG returned to civil control. It replaced the PBY-5As with Martin PBM-5Gs. A new role was the International Ice Patrol, a task that started in 1946 and continues, using Lockheed HC-130 Hercules with side-looking airborne radar. Civil disasters saw more than 300 people being rescued by helicopter during floods in 1955. Some 7,000 USCG personnel served in Vietnam, including aviators flying with USAF SAR units. In 1980, the USCG rescued many refugees fleeing from Cuba. Successive generations of helicopters displaced the flying boat and amphibian, using Sikorsky HH-52 (S-63) helicopters, and later replacing these with Eurocopter

ABOVE: *A USCG Sikorsky HH-60J approaches a Coast Guard cutter. (Sikorsky)*

ABOVE: *A US Army UH-60L lifts an artillery piece into position. (Sikorsky)*

HH-65 Dolphins.

The USCG has 36,230 uniformed personnel, of which 1,050 are directly involved in aviation, although this low figure probably only includes actual aircrew. It operates 23 HU-25C Guardian (Falcon 20), its first jet, with another 21 on support duties or in storage, as a rapid response multi-mission aircraft. There are also 30 HC-130H Hercules, but the main strength lies in its helicopter force, with 35 Sikorsky HH-60J Jayhawk, with another seven on support roles, and 80 HH-65A, plus 13 Dolphin (Dauphin) in storage or support. There are 12 cutters of the Hamilton-class able to operate HH-60 or HH-65 helicopters, while the latter can also operate from the 13 Bear-class cutters. The 16 Reliance-class also have helicopter decks.

UNITED STATES ARMY

Formed: 1942

Despite the USAAF being under overall US Army command until it became the autonomous USAF in 1947, from 1938 onwards its aircraft had not been under the control of local Army commanders, leaving the service free to pursue its strategic goals. This situation meant that Army aviation had to be re-invented, to provide the vital AOP and liaison duties. In 1942, shortly after the United States entered World War II, the US Army purchased some 3,000 light aircraft, including Piper L-4 Cub, Stinson L-5 Sentinel, Vultee-Stinson L-1, Taylorcraft L-2 and Aeronca L-3 AOP aircraft. Additional deliveries of these aircraft and subsequent models meant that US Army aviation reached 4,500 aircraft by the end of the war.

The war was notable for the appearance of the first helicopters, with the Sikorsky R-4, R-5 and R-6, entering service. Post-war, Sikorsky H-5, H-19 Chickasaw and Bell H-13 Sioux helicopters were used extensively by the US Army. New fixed-wing aircraft continued to enter service, with the most numerous being the Cessna L-19 Bird Dog, which first appeared in 1950. By the mid-1950s, the US Army had about 1,800 fixed-wing aircraft, and about the same number of helicopters, operating AOP, liaison, communications, light transport and reconnaissance duties. The end of the decade saw Sikorsky H-34A Choctaw and H-37A, Vertol H-21C and H-25C, Hiller H-23C and still more Bell H-13 enter service. Other new aircraft included 370 DHC-2 Beaver, known to the US Army as the L-21, U-1A Otter (DHC-3) light transports, and Beech L-23 (Twin Bonanza), Cessna L-27A (310) and

Aero L-26 Commanders for communications duties. The 1960s saw Bell UH-1A Iroquois and Boeing Vertol CH-47 Chinook helicopters enter service. The early UH-1s had been designated HU (helicopter utility), giving rise to the nickname 'Huey'.

The Vietnam War saw the development of armed helicopters, both to suppress enemy fire and to act as gunship escorts for troop-carrying helicopters. US Army personnel in Vietnam were initially as 'advisors' but eventually took over the full duties of soldiers. In 1970, these actions spread into Cambodia. The growing importance of the helicopter, its improved performance and the demands of the conflict in Vietnam meant that by 1970, the US Army was the world's largest helicopter operator, with 10,000 machines. It also had a number of fixed-wing attack aircraft, including the North American Rockwell OV-1D and RV-1D Mohawk COIN aircraft. After Vietnam, the main theatre for helicopter deployment was in Europe, where the US Army faced possible conflict with Warsaw Pact forces in Germany. New helicopters continued to enter service, with new and more powerful versions of the Chinook joined, at the lower end of the scale, by the Bell OH-58 Kiowa (206), and attack helicopters such as the Bell AH-1 Cobra. The next major US Army involvement using helicopters was the Gulf War, with several hundred machines deployed in Operation Desert Storm, liberating Kuwait. This was the debut for the McDonnell Douglas (now Boeing) AH-64Apache, with the air war started by the US Army whose Apache helicopters destroyed a key Iraqi air defence radar installation, easing the way for the massive air strikes by the air forces and the USN and USMC.

During the late 1980s and 1990s, almost all new aircraft were helicopters, but a few fixed-wing types continued to enter service, notably the Short C-23 Sherpa (330) light transport. By the mid-1990s, more than a third of the helicopter force, by this time down to around 8,000 aircraft, consisted of armed combat helicopters. The most significant new arrival during this period was the Sikorsky UH-60 Black Hawk utility helicopter. A squadron of AH-64 Apache helicopters was deployed during operations in Kosovo, but not used.

ABOVE: *A possible future attack helicopter for the United States Army is the Sikorsky AH-66 Comanche - this is an RAH-66 prototype. (Sikorsky)*

The US Army is in the midst of a programme of change, having retired its AH-1F Cobras in 2001 and retiring the remaining OH-58 Kiowas and UH-1 Iroquois by 2004. In their place, more than 1,200 Boeing-Sikorsky RAH-66 Comanches will enter service, while the UH-60s are upgraded, and the AH-64 fleet will be upgraded and many fitted with the Longbow radar system. The active units of the US Army are backed-up by the state-orientated National Guard and by the US Army Reserve, with strong aviation elements in both.

The United States Army has some 5,000 helicopters, of which 1,500 are armed, a 50 per cut in numbers over the past 30 years, but there have been significant improvements in aircraft capability. Mainstay of the present force are 550 Boeing AH-64A/D Apache anti-tank and attack helicopters, being joined by the Boeing-Sikorsky RAH-66 Comanches. Almost 600 Bell OH-58A/C/D Kiowa are in process of being retired from service. The main utility helicopter force comprises more than 900 Sikorsky UH-60A/L Black Hawks, with another 60 on tactical ECM duties and 60 on special operations. Heavy lift is provided by 418 Boeing CH-47D Chinook helicopters, with another 36 on special operations. SIGINT is gathered by fixed-wing aircraft, 47 Beech RC-12 Guardrail (King Air). Communications and light transport fixed-wing aircraft include 47 Beech C-12 Huron (King Air) and seven Shorts C-23 Sherpa (330), as well as 13 DHC-7 RC-7 and DHC-6 UV-18A Twin Otter, two Fokker C-31 (F-27), three Cessna UC-35 (Citation V), and 11 Fairchild C-26 Metro. Training uses 137 Bell TH-67 (206).

The National Guard and US Army Reserve still has 291 Bell AH-1E/F/G/P/S Cobra attack helicopters, as well as 184 Boeing AH-64Apache in the same role. Heavy lift is provided by 183 Boeing CH-47 Chinook. There are more than 800 Bell UH-1H/V Iroquois utility helicopters with more than 440 Sikorsky UH-60A/L Black Hawk, as well as a handful of these on ECM duties. Scout helicopters include 300 Bell OH-58A/C Kiowa and OH-58D Kiowa Warrior. There are 67 C-12 Huron and 44 C-23B Sherpa fixed-wing aircraft, as well as a handful of C-20 Gulfstream and C-21 Learjet communications aircraft.

Uruguay

URUGUAY

POPULATION: 3.4 million

LAND AREA: 72,172 square miles, 186,925sq km

GDP: $14.1bn(£9.9bn), per capita $9,000 (£6,294)

DEFENCE: $364m (£255m)

SERVICE PERSONNEL: 23,900 active.

ABOVE: *Grumman S-2G Tracker (Jane's)*

URUGUAYAN AIR FORCE/FUERZA AEREA URUGUAYA

Formed: 1952

Uruguay established a Department of Military Aviation in 1916, and despite being neutral during World War I, managed to obtain three Morane-Saulnier aircraft. In 1919, four Avro 504Ks were obtained, followed in 1921 by six Nieuport trainers. The new air arm became the Aeronautica Militar. During the 1920s and 1930s, it operated Ansaldo A300 fighters, Potez XXVA and Romeo Ro37 reconnaissance-bombers, a Farman F190 and Stinson AOP aircraft, a Breguet Br19 transport, Waco D-7 general-purpose aircraft, and de Havilland Moth trainers.

The country remained neutral during World War II, but in return for putting bases at the disposal of US forces, military aid was provided. This included Grumman F6F-5 Hellcat fighter-bombers, TBM-1C Avenger torpedo-bombers and J4F-1 Gosling amphibians; North American B-25J Mitchell bombers, AT-6 and SNJ-4 trainers; Chance-Vought OSU-3 Kingfisher AOP aircraft; Curtiss C-46 Commando and Douglas C-47 transports; Beech T-11B, Fairchild PT-23A and PT-26 trainers. In 1948, Uruguay became a member of the Organisation of American States. North American F-51D Mustang fighter-bombers entered service during the late 1940s, and were not replaced until the early 1960s, when Lockheed F-80C Shooting Star jet fighter-bombers and T-33A jet trainers arrived. The FAU title was adopted in 1952.

During the 1980s, the FAU gave up fast jet fighters and concentrated on aircraft suitable for COIN operations, transport and communications. The Shooting Stars were not replaced with fighter-bombers, but with additional AT-33A armed-trainers, and then these were joined by ex-USAF Cessna A-37Bs in 1989. It also became one of the few export customers for the Argentinian FAMA IA-58 Pucara light attack aircraft. Now, fighter-bombers are being sought, possibly up to eight A-4 Skyhawks or Northrop Grumman F-5E Tiger IIs.

Today, the FAU has 3,000 personnel. There are two combat squadrons, with one operating ten Cessna A-37B Dragonflys and the other five FAMA IA-58 Pucara on COIN duties. The second squadron may be strengthened in an emergency by six armed Pilatus PC-7U Turbo Trainers. A helicopter squadron operates two Bell 212 and six UH-1H, as well as six ex-RAF Westland Wessex (S-58). Three transport squadrons operate three CASA C212 Aviocar on transport and SAR, three Embraer EMB-111C Bandeirante, a Fokker F-27 Friendship, three Lockheed C-130B, a Cessna 206 and a VIP Cessna 310. The FAU also operates an airline, a common practice in Latin America, Transporte Aero Militar Uruguayo, with C212s. A number of Cessna and Piper aircraft are employed on liaison duties. Apart from the PC-7Us, trainers include 12 Beech T-34A/B Mentor and five Cessna T-41D Mescalero (172).

URUGUAYAN NAVAL AVIATION/ COMANDO DE AVIACION NAVAL/ARMADA DE URUGUAY

Formed: 1920

Originally founded in 1920, the Aviacion Naval remains a shore-based force. At one time Grumman F6F-5 Hellcat fighters were operated, as well as Martin PBM-5 Mariner flying boats. Today, 310 of the Navy's 5,500 personnel are involved with aviation, but there are no ships capable of carrying helicopters. There are four Grumman S-2A/G Tracker ASW aircraft, with EEZ patrols maintained by a Beech Super King Air 200 and two BAe Jetstream T2. Six ex-RAF Westland Wessex helicopters provide transport and SAR. Communications duties are carried out by two Piper Seneca II and two Beech T-34B Mentor and two Turbo Mentor, which also have a training role. Stored Lockheed P-3Bs are being considered as Tracker replacements, although such a leap in capability seems remote.

Uzbekistan

POPULATION: 24.6 million

LAND AREA: 173,546 square miles, 412,250 sq km

GDP: $18.9bn (£13.2bn), per capita $3,000 (£2,098)

DEFENCE EXPT: $1.5bn (£1.05bn)

SERVICE PERSONNEL: 53,000 (inc.conscripts) active.

UZBEK AIR FORCE

Formed: 1991

On the break-up of the Soviet Union, the Uzbek Air Force was formed, declaring that it owned 300 former Soviet aircraft on its territoryt, although of these just 200 were believed to be operational. The situation was complicated by the country becoming a member of the CIS, and with CIS, mainly Russian, units based in the country. Reports indicate that the UAF may have provided forces to fight anti-government forces in Tajikistan, including combat helicopters.

Today, the UAF has around 9,000 personnel. It is still organised on Soviet lines, with air regiments. A fighter regiment has 30 MiG-29/29UB, and another has 25 Su-27/27UB. Two fighter-bomber regiments have one with 20 Su-25/25BM and 26 Su-17MZ/UMZ, with the other having 23 Su-24 and 11 Su-24MP. There is a helicopter regiment with 42 Mi-24 attack helicopters, some of which are being upgraded, and 29 Mi-8 transports, while a second regiment operates 26 Mi-6 and 29 Mi-8, as well as two Mi-6AY command posts. A transport regiment has 26 An-12 and An-12PP ELINT, and 13 An-26 and An-26RKR ELINT, while there is also a Tu-134 and an An-24. Up to ten Il-76s are in course of delivery, with CFM56 engines to improve reliability. Training uses 14 L-39 Albatros. A relatively wide selection of missiles is in use, including AA-8, AA-10 and AA-11 in the AAM role, with AS-7, AS-9, AS-11 and AS-12 ASM and around 45 SA-2, SA-3 and SA-5 SAM launchers.

Venezuela

POPULATION: 24.6 million

LAND AREA: 352,143 square miles, 912,050 sq km

GDP: $91bn (£63.6bn), per capita $8,300 (£5,804)

DEFENCE EXPENDITURE: $1,405m (£982m)

SERVICE PERSONNEL: 72,300 (37.7% conscript) active, plus 8,000 reserves.

ABOVE: *Venezuelan Canberra B.82. (Jeremy Flack/Aviation Photogrpahs International)*

VENEZUELAN AIR FORCE/FUERZAS AEREAS VENEZULANAS

Venezuela formed a Military Aviation Service in 1920, with a total of 16 Caudron GIII and GIV and Farman F40 delivered the following year. Shortly afterwards, the Venezuelan Navy also formed an air arm. Little further progress was made until 1936, when the Military Air Service became the Military Aviation Regiment and a number of Consolidated aircraft were bought. This was the start of a modest expansion programme, with the MAR starting a domestic airline, and following a visit by an Italian aviation mission, introducing Fiat CR32 fighters and BR20 bombers.

During World War II, Venezuela allowed the US to use bases. In return, military aid was provided, including Republic F-47D Thunderbolt fighters, North American B-25J Mitchell bombers and NA-16-3 and T-6, and Beech T-7 and T-11B trainers. Post-war, Venezuela became a member of the Organization of American States, and the two air arms merged to the Fuerzas Aereas Venezolanus. Modernization occurred during the early 1950s with de Havilland Vampire FB5 jet fighter-bombers and English Electric B2 jet bombers, followed in 1955 by de Havilland Venom FB4 fighter-bombers and Vampire T55 trainers, North American F-86F Sabre fighter-bombers, additional Canberras and Fairchild C-123B Provider transports. In 1967, ex-Luftwaffe F-86K Sabres were obtained. The 1970s saw Canadair-built CF-5A/D fighter-bombers and Dassault Mirage IIIs and Vs introduced. In addition, Fairchild OV-10E Bronco COIN aircraft, with a second batch later purchased from the USAF as Canberra replacements. The FAV also introduced the Lockheed Martin F-16A/B interceptor during the 1980s. The Mirage IIIs and Vs were

Venezuela

ABOVE: *Venezuelan Army's M28 Skytruck (Jane's)*

upgraded to Mirage 50 standard in 1988, with additional Mirage 50s obtained from Dassault. Ex-Netherlands NF-5As boosted the CF-5A/D force in 1992.

Today, the FAV has 7,000 personnel, a reduction of almost a quarter over the past 30 years. Average annual flying hours are around 155. There is little standardization, but second hand aircraft have maintained the strength of individual types. There are six squadrons; two operating a total of 23 F-16A/B; two operating 20 Embraer EMB-312 armed-trainers, with another ten on training duties; one operating 16 Mirage 50EV/DV; and one operating 16 CF-5A/B, upgraded in Singapore to VF-5 standard, and seven NF-5A/B. A reconnaissance squadron operates 24 OV-10E Bronco. ECM is provided by three Dassault Falcon 20DC. There is an armed helicopter air group with ten SA-316 Alouette III, 12 Bell UH-1D and five UH-1H Iroquois, and six AS532 Cougar. Transport uses six Lockheed C-130H Hercules, seven C-123 Packet and eight Alenia G222, as well as two BAe748 and two Boeing 707 tankers. Transport helicopters include 18 Mil Mi-8/Mi-17, eight AS332B Super Puma, two Bell UH-1N, two 214 and four 412. VIP aircraft in the Presidential Flight include a Boeing 737-200, a Gulfstream II and III, a Learjet 24D and a Bell 412. Cessna and Beech light aircraft are used for liaison and communications, including nine Cessna 182, a Citation I and a II, seven Beech Queen Airs and five Super King Air 200, as well as nine SA316B Alouette III helicopters. Training uses 10 EMB-312, 20 Beech T-34, 17 North American T-2D and 12 SF260E.

Uncertainty surrounds an order for AMX-T light strike aircraft and Aermacchi MB339FD armed-trainers, originally intended to be for eight of each with possibly another 24 MB339FD. Growing links with China suggest that Chinese aircraft might be obtained instead. Missiles include R-530 Magic and AIM-9L/P Sidewinder AAM; Exocet ASM and a small number of Roland and RBS-70 SAM launchers.

VENEZUELAN NAVAL AVIATION/AVIACION DE LA MARINA VENEZOLANA

Naval aviation accounts for 1,000 of the Venezuelan Navy's 15,000 personnel, with six Mariscal Sucre-class (Lupo) frigates capable of operating a Bell 212 ASW helicopter. Twelve Bell 212ASWs are operated, with four HeliDyne modified Bell 412s for SAR. There are nine CASA C212-200AS/400 Aviocar transports, with some having radar for SAR. A DHC-7 Dash 7, two Cessna 310R and a 402C and a Commander 695 provide transport and communications, with a VIP Beech King Air E90 and Super King Air 200. Eight additional ASW helicopters are being sought.

VENEZUELAN ARMY AND NATIONAL GUARD/EJERCITO VENEZOLANA E GUARDIA NACIONAL

The Venezuelan Army and National Guard use aviation on transport, communications and liaison with a wide variety of aircraft. The Army's transport aircraft include four IAI-201 Arava, as well as four Agusta AS-61A (S-61), four Bell UH-1H Iroquois and three Bell 205A helicopters. Communications are covered by a Cessna 172 and two U206G and a T207, a Bell 206 JetRanger and 206L LongRanger, and six Agusta A109A helicopters. Two Cessna 182s are used for training. National Guard aircraft include 18 PZL M-28 Skytrucks delivered during the late 1990s, three Arava IAI-201, a Bell 214ST and ten AS365F Ecureuil helicopters. Communications aircraft include a Beech Queen Air, a King Air and a Super King Air, a BN-2A Islander, a Cessna 206 and five Bell JetRangers with one 206L LongRangers, and eight Agusta A109A helicopters.

Vietnam

POPULATION: 81 million

LAND AREA: 129,607 square miles, 335,724 sq km

GDP: $31bn (£21.7bn), per capita $1,300 (£909)

DEFENCE EXPENDITURE: $950m (£664m)

SERVICE PERSONNEL: 484,000 (mainly conscript) active, plus reserves of upwards of 3 million.

RIGHT: *MiG-15 in North Vietnam Air Force's markings, in the Pima Museum. (Jeremy Flack/ Aviation Photographs International)*

VIETNAM PEOPLE'S ARMY AIR FORCE

Vietnam was reunited in 1976, after being divided following the French withdrawal from Indo-China in 1954. For most of this period, the Communist North backed by the Soviet Union was infiltrating forces into the pro-Western South, backed by the United States. Both countries maintained air forces, although the main thrust of their defences lay in their ground forces. During the Vietnam War, the Vietnamese Air Force operated US-supplied Northrop F-5A/B, Cessna A-37B armed-trainers and at one time a substantial number of Douglas A-1 Skyraiders, all on tactical fighter-bomber duties, as well as many helicopters and C-47 transports. The Vietnamese People's Air Force operated Soviet-supplied MiG-15, MiG-17 and MiG-21s, and a few Ilyushin Il-28 bombers. The VAF provided COIN operations and tactical support for ground forces, while the VPAF concentrated on air defence.

After reunification, Vietnam used its armed forces to influence events in Cambodia, and then to back Communist guerrillas in Thailand. Abandoned US-built aircraft were used until spares became a problem. Supplies of Soviet equipment continued at first, including 40 MiG-23ML interceptors and 30 Mi-24 attack helicopters, with Soviet and North Korean advisors. In 1994 and 1997, Su-27 interceptors entered service, replacing the older MiGs, but new aircraft have been much reduced since the collapse of the USSR. Vietnam is still committed to Russian aircraft, and has obtained spares for its MiG-21s, which were upgraded by MAPO/Sokol in 2000-2002. The VPAAF is likely to be cut since Vietnam now has to pay for its equipment, hence the small numbers of new aircraft delivered recently.

Personnel number 30,000. There are five Su-27Ps and six Su-27UBs. The 35 Su-22 attack aircraft are being upgraded. The 120 MiG-21bis are supported by 25 MiG-21UM trainers. In addition to 30 Mi-24 attack helicopters there are five Ka-25s, seven Ka-28s and six Ka-32 for ASW. Transport is provided by 30 Antonov An-26 and 60 Mi-8/Mi-17 helicopters, with ten Mi-6 in the heavy lift role. Five Yak-40 provide VIP transport. Training uses 20 Yak-18 as well as the Nanching BT-6 variant of this aircraft, and 25 L-39 Albatros.

Yemen

POPULATION: 18.9 million

LAND AREA: 136,890 square miles, 486, 524 sq km

GDP: $6.4bn (£4.5bn), per capita $1,500 (£1,048)

DEFENCE EXPENDITURE: $498m (£348m)

SERVICE PERSONNEL: 54,000 (mainly conscript) active, plus 40,000 reserve.

UNIFIED YEMEN AIR FORCE

Formed: 1990

Until 1990, Yemen was divided into North and South, although unification had been agreed in 1981. South Yemen had gained independence in 1967, having previously been a British protectorate. North Yemen, normally referred to as Yemen, had been independent, ruled by an Iman, and had formed a small Yemen Air Force during the mid-1950s, initially as a transport force until Ilyushin Il-10 ground-attack aircraft were supplied by the USSR, with helicopters and trainers, and Soviet and Egyptian pilots. The country was invaded by Egyptian troops during the mid-1960s, and the Iman overthrown.

The Unified Yemen Air Force mainly uses Russia equipment. Yemen's failure to support action against Iraq following the invasion of Kuwait led to Saudi Arabia cutting off aid, while the collapse of the Soviet Union ended military aid. A civil war in 1994 also undermined the economy and the efficiency of the armed forces. Probably only 50 per cent of its aircraft are operational, with many destroyed in the civil war or grounded by a shortage of spares.

The UYAF has 3,500 personnel. It operates ten MiG-29A/U, now being augmented by a further 24 aircraft ordered in 2001 and up to 70 MiG-21 fighters, some of which are being upgraded, 13 Northrop Grumman F-5E and 17 Su-20/Su-22 in the fighter-bomber role. Helicopters include 15 Mi-24 attack helicopters, around 45 Mi-8. Transport aircraft include an Il-76s, four Il-14s, two C-130Hs, three Antonov An-12 and six An-26s. Used Su-27s may be purchased in the near future.

Yugoslavia (Serbia and Montenegro)

POPULATION: 10.6 million

LAND AREA: 39,450 square miles, 101,691 sq km

GDP: $18.3bn (£12.8bn), per capita $4,931 (£3,448)

DEFENCE EXPENDITURE: $1.8bn (£1.26bn)

SERVICE PERSONNEL: 105,500 (40% conscript) active, plus 400,000 reserves.

ABOVE: *The armed jet trainer, G-2 Galeb (Jane's)*

FEDERAL REPUBLIC OF YUGOSLAVIA AIR FORCE

Formed: 1991

The Federal Republic of Yugoslavia Air Force is the rump of the former Yugoslav Air Force, and is part of the armed forces of the two states, all that is left of the former Yugoslav Federation. Since civil war started in 1991, the other states have declared independence, or attempted to do in the case of Kosovo. In 2002, it lost its autonomy and became one of the nine corps within the army.

Yugoslavia was part of the Austro-Hungarian Empire, but independence came with the end of World War I. When the Yugoslav Army Aviation Department was formed in 1923, it was able to use Serb, Croat and Slovene officers who had flown with the Serbian Military Air Service, formed in 1912. Early equipment included Spad S-7C1 and Dewoitine D-1C fighters, Breguet Br19A/B2 reconnaissance-bombers, and a number of Brandenburg trainers built under licence. These aircraft were followed by Avia BH33 fighters, Potez XXV reconnaissance aircraft and Hanriot A32 trainers, as well as a number of H41 seaplanes for a naval co-operation flight. In 1930, semi-autonomy was attained as Yugoslav Army Air Corps.

The YAAC continued to grow, and by 1935 had 44 squadrons and 440 aircraft. Modernization started with orders for Dewoitine D500 and Hawker Fury fighters, Bristol Blenheim I, Dornier Do17K and Savoia-Marchetti SM79 bombers, with some Dornier Do22 general-purpose aircraft. By September, 1939, it was introducing Hawker Hurricane, Curtiss P-40B Tomahawk and Messerschmitt Bf109E fighters, and Me108 and Westland Lysander army co-operation aircraft. Yugoslavia was occupied by German forces in 1941, following a revolution that had overthrown the monarchy and a pro-Axis government. A number of personnel managed to escape, reaching the UK where Yugoslav squadrons flew Supermarine Spitfire fighters in the RAF. The Axis Powers partitioned Yugoslavia, with Italy governing the Croatian areas, and forming a Croatian Air Force with Ćaproni Ca310, Fiat G50 and Messerschmitt Bf109G fighters, Morane-Saulnier MS40 and Dornier Do17 bombers, and some AOP aircraft.

After the war ended in 1945, Yugoslavia became a republic, initially within the Soviet sphere of influence, and the Yugoslav Air Force, Jugoslovensko Ratno Vazduhoplostvo, was formed. Initially, its former RAF equipment was used, but Soviet aircraft, including Yakovlev Yak-3 and Yak-9 fighters, Ilyushin Il-2 and Il-10 ground-attack aircraft, Petlyakov Pe-2 bombers, Lisunov Li-2 (C-47) transports, and UT-2 and Po-2 trainers soon arrived. There were some 400 aircraft by 1950. The country was the first to break away from Soviet influence, remaining outside the Warsaw Pact although still a Communist state, and this encouraged the US to provide aid. During the 1950s, the JRV equipped with 150 Republic F-47D Thunderbolt and 140 de Havilland Mosquito FB6 fighter-bombers, plus a few Mosquito night-fighters. A nationally designed fighter, the S-49A, entered service, but was soon withdrawn. The first jets, Lockheed T-33A trainers, were introduced in 1953, and were soon joined by 200 Republic F-84G Thunderjet fighter-bombers, Westland Dragonfly (S-51) helicopters, and both Douglas C-47 and Ilyushin Il-14 transports, as well as Aero 3 trainers. Canadair Sabre Mk2 and North American F-86D Sabres were obtained during the late 1950s, and in the 1960s, these were followed by MiG-21F interceptors, reflecting a policy of buying aircraft from both sides of the Iron Curtain. Soko Jastreb armed-trainers and Galeb jet trainers entered service, with Yugoslav manufacturers creating a niche in the trainer and armed-trainer market.

The 1970s and 1980s saw the JRV come to operate Soviet combat aircraft and transports, with Western helicopters. This situation reflected Western concerns about selling the most advanced combat aircraft to a Communist country, and the higher cost of Western aircraft. Additional MiG-21s replaced Western aircraft from the 1950s

Yugoslavia (Serbia and Montenegro)

and early 1960s, but were not joined by anything more sophisticated until 1988, when the first MiG-29A/Bs were introduced.

The collapse of the Yugoslav Federation in 1991 and 1992 saw Serbia retain a firm grip on the assets of the former JRV, with little equipment passing to the breakaway states. The worsening economic situation, an arms embargo and, finally, in 1999, air strikes by the Western Powers, all contributed to a weakening of the new FRYAF. The MiG-29 force of 16 aircraft was destroyed, mainly on the ground but six aircraft were shot down. It was believed that 150 other combat aircraft were destroyed, but this is now believed to have been an exaggeration, with part of the air force operational.

ABOVE: *The Avione IAR-93/SOKO J-22 Orao used for ground attack and reconnaisance (Jane's)*

On paper, personnel strength has remained steady over the past 30 years having fallen very slightly from around 20,000 to just 19,500, but given that this is the 'rump' of the former Yugoslav federation, in reality a far larger proportion of the country's manpower is involved. Flying hours are likely to be low, due to shortages of fuel and spares. At least 18 Super Galeb, 12 Gazelle and three UTVA-75s are known to be operational. Other surviving aircraft could include half the 60 MiG-21bis, some ground-attack IJ22/J22/NJ22 Orao as well as the Super Galebs. There may be as many as 30 Mi-8/Mi-17 transport helicopters. Transport aircraft include a number of Antonov An-26, Yak-40 and VIP Dassault Falcon 50. There may be some UTVA-66 for liaison and the UTVA-75s trainers.

Zambia

POPULATION: 9.2 million

LAND AREA: 288,130 square miles, 752,618 sq km

GDP: \$3.7bn (£2.6bn), per capita \$1,000 (£699)

DEFENCE EXPENDITURE: \$66m (£46m)

SERVICE PERSONNEL: 21,600.

ZAMBIA AIR FORCE AND AIR DEFENCE COMMAND

Formed: 1964

Formerly, the British colony of Northern Rhodesia, Zambia became independent in 1964 and immediately established the Zambia Air Force. The first aircraft were four Douglas C-47 and two Hunting Pembroke C1 transports. Initially, the RAF provided assistance, but an Italian company later helped. The Pembrokes were replaced by six DHC-2 Beavers, while two C-47s were replaced by four DHC-5 Caribou. Agusta-Bell 205 helicopters were introduced, with a mixture of DHC-1 Chipmunk and eight Scottish Aviation Bulldog trainers. In 1971, the first jets arrived, two Soko Galeb and four Soko Jastreb armed-trainers, while eight Savoia-Marchetti SF260 trainers also entered service.

In 1980, the USSR supplied 16 MiG-21 interceptors, while 12 Shenyang F-6s (MiG-19) fighter-bombers were brought from China. Antonov and Yakovlev transports were added, with Mi-8 helicopters, but the ZAF continued to buy Western equipment, Aermacchi MB326 and Saab MFI-17 armed-trainers, and a substantial number of AB205 and 212 helicopters. Poor serviceability soon reduced the operational numbers, with the MiG-21 force also depleted by serviceability from 16 to seven by 1996, while the MB326s fell from 18 to six between 1990 and 1996. Many of the MiG-21s were upgraded in Israel during 1999 and 2000.

Today, the ZAF has 1,600 personnel, a fourfold increase over the past 30 years. There are eight Hongdu K-8 armed-trainers delivered from China in 1999 in one fighter-bomber squadron, while another has eight MiG-21MF and another eight Shenyang F-6, although the number of aircraft fully operational may be lower. A single helicopter squadron operates ten Agusta-Bell 205A, five 212 and seven Mi-8. Transport aircraft include four An-26, four C-47, 4 DHC-5D Buffalo and four HAMC Y-12 (II). Seven Do28s are used on liaison with 12 AB-47G helicopters. Training uses a few surviving MB326s, up to ten Nanchang BT6, two MiG-21U, and a handful of surviving Galebs and Jastrebs. There are also some SF260 as well as the K-8s.

Zimbabwe

POPULATION: 11.8 million

LAND AREA: 150,333 square miles, 389,329 sq km

GDP: $6.6bn (£4.6bn), per capita $2,300 (£1,608)

DEFENCE EXPENDITURE: $401m (£280m)

SERVICE PERSONNEL: 39,000.

AIR FORCE OF ZIMBABWE

Formed: 1980

Originally the British colony of Southern Rhodesia, the country's history of military aviation dates from the late 1930s, when the colony's government offered three squadrons for the Royal Air Force. This was in addition to playing a major role in the planned Empire Air Training Scheme, training pilots for the British and other air forces. The RAF's three Rhodesian squadrons operated Supermarine Spitfire and Hawker Typhoon fighters, as well as Avro Lancaster bombers. Flying training took most of the available personnel, with the country offering excellent flying conditions so that courses took less time, in skies free from the threat of enemy aircraft. Post-war, the Southern Rhodesian Air Force was formed, with a single squadron initially for transport and communications, with Douglas C-47s, an Avro Anson and a de Havilland Rapide, as well as de Havilland Leopard Moths. The SRAF acquired combat aircraft in 1951, with 11 Supermarine Spitfire 22 fighters, replacing these in 1953 with de Havilland Vampire FB9 fighter-bombers, with Vampire T11 trainers following in 1955. In 1958, a bomber squadron was formed with English Electric Canberra jets, and Hawker Hunters joined the Vampires shortly afterwards.

Southern Rhodesia was part of the Federation of Rhodesia and Nyasaland (now Malawi) from 1953 until 1963. During this time the SRAF became the Royal Rhodesian Air Force, retaining this title for the air force of Southern Rhodesia when the federation was dissolved, with Northern Rhodesia (Zambia) and Nyasaland becoming independent. In 1965, a Unilateral Declaration of Independence was declared to avoid any grant of independence on terms unacceptable to the country's government. This led to United Nations sanctions, and in 1969, a republic was declared and the 'Royal' prefix dropped from the air force's title. By this time, the RhAF was operating a squadron of 12 Hawker Hunter FGA9 fighter-bombers, one of 12 de Havilland Vampire FB9 fighter-bombers, and another with of 12 Canberra jet bombers. A squadron of 12 BAC Provost T52 armed-trainers handled COIN duties. Transport used four Douglas C-47s, and eight Sud Alouette III helicopters. Despite sanctions, a Beech Baron was obtained in 1967 for communications work. Guerrilla activity intensified during UDI, and the RhAF was heavily involved, remaining fully operational despite sanctions. Alouette III helicopters frequently ferried troops while overloaded, achieving this through making a 'running' or STOL take-off.

ABOVE: *Hawk Mk.60 (Jeremy Flack/Aviation Photographs. International)*

Rhodesia became independent in 1980, and after a spell as Zimbabwe-Rhodesia became Zimbabwe. The new state changed its allegiances, receiving from China more than 20 X'ian F-7 (MiG-21) fighter-bombers as well as FT-7 (MiG-21U) and FT-5 (MiG-17U) trainers. Air and ground crew for these aircraft were trained in China. BAe Hawk 60 armed-trainers were also introduced, while secondhand Alouette IIIs were obtained, with almost 40 operated at one stage. Zimbabwe became involved in regional conflicts, sending forces to Mozambique to support the government against guerrillas, and later to the Democratic Republic of the Congo, where rebel forces claim to have shot down seven of the F-7s. Rapid expansion to eight squadrons was unsustainable, and in recent years serviceability has been poor and along with attrition has caused operational aircraft numbers to fall sharply, not helped by the country's deteriorating economic situation in recent years.

Today, the ZiAF has 4,000 personnel. Average annual flying hours are around 100. There are just nine F-7s operational in one squadron, while another squadron operates ten Hawker Hunter FGA9s and one T-81, and a third has eight BAe Hawk 60 armed-trainers. A reconnaissance squadron has 15 Cessna 337 Lynx and two O-2s. Training, reconnaissance and liaison use a squadron of 22 SF260 Genet. One helicopter squadron has 24 SA319 Alouette III, and six Mi-35, while a second helicopter squadron has seven Agusta-Bell 412 and two VIP AS532UL Cougar. A transport squadron has five BN-2A Islanders and nine CASA C212-200 Aviocars, of which one is used for VIP duties.

Glossary

AAM - air-to-air missile
AEW - airborne early warning
AEWC - airborne early warning and command/control
ALCM - air-launched cruise missile
ASM - air-to-surface missile
ASuV - anti-surface vessel
ASW - anti-submarine warfare
COIN - counter insurgency
CSAR - combat search and rescue
EEZ - exclusive economic zone
ELINT - electronic intelligence
FAC - forward air control
JSOW - joint stand-off weapon
MAP - Military Aid Programme
MCM - mine countermeasures
MLU - mid-life update
MR - maritime-reconnaissance
SAR - search and rescue
SIGINT - signals intelligence

NOTES

Sterling/US Dollar conversions are at £1 to US $1.43.
Manpower figures are from *THE MILITARY BALANCE*, OUP and The International Institute for Strategic Studies.

NATO designations for Soviet aircraft and missiles

The secrecy with which many aircraft were developed during the Cold War meant that there could be an interval between an aircraft being detected and its design bureau and designation becoming known to the West. To enable identification to be made in intelligence reports, aircraft were given designations beginning with 'F' for fighters; 'B' for bombers; 'C' for transports; 'H' for helicopters; 'M' for maritime-reconnaissance and, perhaps confusingly, 'M' for trainers. Air-to air (AAM) missiles were designated with names beginning with 'A', Air-to-surface (ASM) with 'K', and surface-to-air (SAM), with 'S'. A 'U' suffix indicated a training variant of a combat aircraft.

This gave designations, often with suffixes such as 'A', 'B', etc for uprated variants of an aircraft, or missile, with notable examples including:

An-2: 'Colt'
An-12: 'Cub'
An-14: 'Cold'
An-22: 'Cock'
An-24: 'Coke'
An-26: 'Curl'
An-32: 'Cline'
Be-6: 'Madge'
Be-10: 'Mallow'
Be-12: 'Mail'
Il-14: 'Crate'
Il-18: 'Coot'
Il-22: 'Coot'
Il-28: 'Beagle'
Il-28U: 'Mascot'.
Il-76: 'Candid'
Ka-25: 'Hormone'
Ka-27: 'Helix'
Li-2: 'Cab'
Mi-4: 'Hound'
Mi-6: 'Hook'
Mi-8/-17: 'Hip'
Mi-10: 'Harke'
Mi-24/-35: 'Hind '
Mi-26: 'Halo'
MiG-15: 'Fagot'
MiG-15UTI: 'Midget'
MiG-17: 'Fresco'
MiG-19: 'Farmer'
MiG-21: 'Fishbed'
MiG-23: 'Flogger'
MiG-25: 'Foxbat'
MiG-29: 'Fulcrum
MiG-31: 'Foxhound'
Mya-4: 'Bison'
Su-7: 'Fitter'
Su-9: 'Fishpot'
Su-11: 'Flagon'
Su-24: 'Fencer'
Su-25: 'Frogfoot'
Su-27: 'Flanker'
Tu-14: 'Bosun'
Tu-16: 'Badger
Tu-20: 'Bear'
Tu-22: 'Blinder
Tu-104: 'Camel'
Tu-114: 'Cleat'
Tu-134: 'Crusty'
Tu-154: 'Careless'
Yak-11: 'Moose'
Yak-40: 'Codling'

Missiles included:

AA-2: 'Atoll'
AA-6: 'Acrid'
AA-7: 'Apex'
AA-8: 'Aphid'
AA-10: 'Alamo'
AA-11: 'Archer'
AS-4: 'Kitchen'
AS-7: 'Kerry'
AS-10: 'Karen'
AS-11: 'Kilter'
AS-12: 'Kegler'
AS-13: 'Kingbolt'
AS-14: 'Kedge'
AS-15: 'Kent'
AS-17: 'Krypton'
AS-18: 'Kazoo'
SA-2: 'Guideline'
SA-4: 'Ganef'
SA-5: 'Gammon'
SA-6: 'Gainful'
SA-8: 'Gecko'
SA-9: 'Gaskin'
SA-11: 'Gadfly'
SA-12: 'Gladiator/Giant'
SA-13: 'Gopher'

Index

NOTE: Nations are listed alphabetically, and within each national entry, the order is: air force; naval air arm; army air corps. Aircraft manufacturers are listed in the index, but due to mergers or licence production, aircraft types may be credited to more than one manufacturer, sometimes depending on the date taken into service with any particular operator.

Index

Index

Index

Index